U.S. Army Intelligence
in the
Mexican Revolution, 1910–1915

Joe Lee Janssens

Volume 3

To
Those Who Debunk
Conspiracy Theories

Contents

Graphics vii

9. Armed Intervention 1

10. Civil War 34

11. The Bandit War 53

Conclusion 60

Documents Cited 65

Document Index 309

Index 313

Graphics

9. Armed Intervention

Photo 3.1 Marines marching to station - Veracruz, 1914.............................. 4

Photo 3.2 Blue jackets resting on sidewalk - Veracruz, 1914....................... 9

Documents Cited

Map 3.1 Hand-drawn Map of Bachimba.. 240

Map 3.2 Fight at Barroterán.. 257

Photo 3.3 Mexican Central Railroad Bridge..................................... 266

Photo 3.4 Mexico-Northwestern Railroad Bridge................................ 268

Photo 3.5 El Paso & Southwestern Railroad Bridge.............................. 271

Photo 3.6 Southern Pacific and E.P. & SW Railroad Bridges...................... 273

Photo 3.7 Southern Pacific Railroad Bridge..................................... 273

Photo 3.8 Stanton Street Wagon Bridge.. 276

Photo 3.9 Santa Fe Street Wagon Bridge.. 278

Photo 3.10 Suspension Bridge Northwest of El Paso.............................. 280

9. Armed Intervention

April

When the U.S. armed intervention finally came, with the taking of Veracruz, it did not occur as Burnside had advocated, on a massive scale with the military having "a sufficient free hand." On the contrary, the invasion of Veracruz came as the result of a "typical civilian order" to take control of the Customs House, "one building in the center of a fair-sized city, without taking the remainder of the town."[1]

"*All* [emphasis added] Navy plans" leading up to the invasion had contemplated taking the entirety of Veracruz and the surrounding area precisely because street fighting was "highly popular" among the Mexicans and they were "well known to do their best fighting when scattered and without uniforms and from housetops." "In one case some armed Mexicans removed boards from a floor, went under the floor of a house with a solid stone foundation, and fired up through the floor as the troops entered the door of the house." In another case, the Navy had to shell a house near the sea front with three- and five-inch guns in order to clear it of armed Mexicans, and a second time after they returned. Such a narrowly-targeted, even surgical, objective in the heart of the city proved "impracticable" and resulted in "the loss of 17 killed and 63 wounded." Major combat only concluded the day after the landing when the American military decided to expand its radius of action to capture the entirety of the city, the surrounding areas, and the water works to "clear the whole place of snipers." Had the order originally been given to occupy the larger area, "with

---

[1] RG 165, 5761-974, May 6, 1914.

1

the means available, it would probably have been a much simpler matter...and the loss of life would have been very much less; perhaps trifling." As the captain remarked, quite logically, *"Civilian military orders that contemplate neither war nor peace are difficult of execution* [emphasis added]." Consequently, he urged a study of the orders given by the White House to Admiral Fletcher, "for future guidance."[2]

The taking of Veracruz placed the Americans in Mexico City in serious jeopardy. About 150 Federal soldiers closely surrounded the American Embassy and then removed the "doves of peace" (two Maxim machine guns with associated accoutrement and approximately eight thousand .30 caliber rounds for same) that had begun to settle into the embassy only two weeks earlier. The Federals also confiscated 230 Model 1896 .30 caliber rifles with approximately 62,500 cartridges. Burnside later wrote that the "importation of these articles had been facilitated by Huerta's government for protection against rioters, etc., and the Huertistas had perfect knowledge of their existence," but they did not know about the "miscellaneous ammunition, cleaning rods, and magazines left over from the time of Madero's administration and these were not turned over to the Federal Army."[3] Burnside had never wanted the weaponry in Mexico City and instead had "advised that the arms be sent no further than Veracruz." The Federals also collected the arms of other foreign embassies, an event that received widespread coverage in the dailies with the undesirable effect being that "the native population immediately became 'muy bravo' and started the rock-throwing and window-breaking."[4]

The same day it confiscated the weapons, April 22, the Mexican Government gave the Embassy employees their passports to leave, which the American expatriates did the following day aboard a special train with an escort commanded by a General Corona and Colonel Alberto Braniff.[5] Burnside deposited all the confidential records of the Office of the Military Attaché in the safe of Mr. Francis J. Kearful's office in the Quirk Building, Avenue San Francisco, Mexico City,[6] and destroyed the War Department's telegraph and cable code No. 525 just prior to leaving Mexico City in order to avoid the code falling into "unauthorized hands" while he was in transit to Veracruz.[7] The State Department's Louis D'Antin assumed responsibility for watching over the American Embassy, and at the urging of American expatriates the British took charge of representing U.S. interests, but neither arrangement lasted. The Wilson

---

2 RG 165, 5761-947, May 1, 1914, from Burnside, Veracruz, Headquarters U.S. Naval Forces on Shore (quotes). Ibid.
3 RG 165, 6931-79, December 26, 1915, from Burnside, Kingsville, Texas, to Secretary War College (first quote). O'Shaughnessy, *A Diplomat's Wife*, 294, "doves of peace."
4 RG 165, 5761-974, May 6, 1914.
5 Ibid.
6 RG 165, 6931-75, December 7, 1914, from Burnside to Chief, War College Division Staff.
7 RG 165, 6931-67, May 6, 1914, from Burnside to Chief of Signal Corps via Chief, War College Division Staff.

administration replaced the British with the Brazilians and Mr. D'Antin ultimately had to seek asylum with the Brazilian Legation because of baseless charges leveled against him by the Mexican government that he had carried on communications with the Constitutionalists.[8]

As the Americans rolled out of Mexico City headed for Veracruz, the Wilson administration also renewed the arms embargo on April 23 against all factions in the revolution, including the Constitutionalists, undoing Lind's policy.[9] Therefore, none of the American actors in Mexico were satisfied, not Lind, who had wanted to facilitate a Constitutionalist victory, not Burnside, who had hoped for a full invasion, and certainly not the O'Shaughnessys, who had misplaced their hope in Huerta.

Other than leers and jeers from the locals along the way, the embassy staff arrived without incident at Tejería, where the Mexicans had destroyed the rails for two kilometers to retard a farther advance by the American military. From there, the Americans transferred over to trains operated by the U.S. military to reach Veracruz late in the afternoon of April 24.[10]

The American military still had not entirely secured the port of Veracruz, and the next day Rear Admiral F. F. Fletcher issued a proclamation to the people of Veracruz saying that "since people continue taking pot-shots at my soldiers from certain buildings, and desirous of re-establishing order and tranquility, I invite all those with arms and ammunition in their possession to turn them into the police headquarters, located in the basement of the town hall, as soon as possible and no later than noon of April 26," in exchange for a receipt. After that deadline, "all those found possessing them," as well as anyone who harassed the American troops would "be severely punished."[11]

Meanwhile, Burnside continued to express gloom and doom back to Washington, predicting the slaughter of "many isolated Americans and other foreigners" which could only be prevented by an "efficient intervention" to be made immediately, before it was too late.[12] Further reasons justifying an immediate expanded invasion were the fifteen thousand Zapatistas, about half of them mounted and many of them female, who had joined with the Huertistas and entered Mexico City on April 24; the continued training of poorly-armed men in Mexico City; the ongoing destruction of railroads by the Federals; and the weakness of the Federals, only about 5,000 men with little

---

[8] RG 165, 5761-957, May 15, 1914, Burnside No. 1118: Summary of Military Events from May 10 to May 15, 1914.

[9] Janssens, *A Revolution in Military Affairs*, 2: 171.

[10] RG 165, 6931-65, April 25, 1914, from Burnside to Chief, War College Division Staff. RG 165, 5761-947, May 1, 1914, from Burnside. RG 165, 5761-974, May 6, 1914.

[11] RG 165, 5761-947, April 25, 1914, Proclamation from Admiral F.F. Fletcher to people of Veracruz to surrender their arms and ammunition due to continued sniping at U.S. personnel.

[12] RG 165, 5761-942, April 25, 1914, Burnside's telegram from Veracruz to War College Staff regarding foreign intervention and safety of foreigners.

military instruction, guarding the Mexican Railroad; all of which combined to demand and invite a full-scale invasion. In concluding, Burnside pressed, "if the present crisis is passed without a complete armed conquest of Mexico, the future residence of all classes of foreigners in this country will be most difficult if not impossible."[13]

Photo 3.1 Marines marching to station - Mexican War. Llave Mexico Veracruz, ca. 1914. May 16. Photograph. https://www.loc.gov/item/2010651723/

The captain also mentioned another reality that indicated the need for a more complete U.S. intervention, the fact that a mere change in administration would not bring about a lasting peace and instead would only lead to more infighting of the type seen over the last three years if Mexico were to be "left entirely to domestic control."[14] On the one hand, the revolutionaries would "always remember the happy times when they roamed at will with stolen horses and carrying rifles," and would be hard pressed to give up "the free enjoyment of plunder, wine, women and song" in exchange for peace. "On the other hand, the majority of the present federal armed forces offered

---

[13] RG 165, 5761-941, April 27, 1914, Memo to Chief of Staff re Burnside's telegram from Veracruz, (quote). RG 165, 5761-948, May 8, 1914, Burnside's telegram to WCD recommended "that steps be taken now looking to an agreement for U.S. troops to occupy important cities for a period following any change of administration that may come about. Once successful, any of the present rebels will be as difficult to influence as was Madero and Huerta, and the same lack of trust and cooperation between the various factions will exist, and the revolution will continue." RG 165, 5761-957, May 15, 1914, Burnside No. 1118: Summary of Military Events from May 10 to May 15, 1914, Burnside again urges an expansion of the U.S. military effort.
[14] RG 165, 5761-953, May 9, 1914, Burnside No. 1114: Summary of Military Events.

little or no future promise; they cared little for the welfare of Mexico and their education had been largely confined to devising methods of graft; the present federal army could be viewed in no other light than as being suitable material for the scrap heap." Accordingly, Burnside recommended "having some foreign armed force temporarily occupy important places in the republic immediately following the next challenge of administration." Otherwise, "the successful revolutionary leaders will immediately start quarrelling among themselves about the division of spoils, and as a result the revolution will not cease."[15] In all this Burnside would be proved correct, it would take some twenty more years for the revolution to congeal. But he was incorrect in his belief that it was "probable that the next two generations [of Americans] will be fighting battles around Veracruz."[16]

The Federal response to the invasion of Veracruz had been to position groups of Mexican Army troops just beyond the American lines, although not in force, and to cut all four railroads departing from Veracruz to within a few kilometers of the port city.[17] In the isthmus, the Secretary of War had suspended all traffic on the Tehuantepec National Railway on April 21, 1914, and practically took over all the physical assets of the company, ordering that it concentrate all rolling stock at San Geronimo, where a General Gamboa had established his headquarters for all the Federals in the region. The Mexicans then destroyed one bridge and lifted the rails by sections over a distance of about eight kilometers between Salina Cruz and Tehuantepec, apparently to forestall any possible American invasion from the west; arrested all the American employees of the railroad company, although General Gamboa later freed them and expelled them from the area; and appointed Federal Colonel Francisco Correa, in charge of the forces at Rincon Antonio, as the new General Manager of the Tehuantepec National Railway Company.[18]

In Mexico City, "several Germans of high intelligence and a good knowledge of local conditions" noted "a very marked change...in the public sentiment there." In the immediate aftermath of the U.S. invasion of Veracruz, Huerta had enjoyed an increase in popularity. But then on Saturday, April 25, the people's support for the regime started to flag when it became known that he "had deliberately played for intervention knowing that if he were captured by Americans his life and property would be spared while if he fell in the hands of Villa he would be given short shrift." Moreover, it also became common knowledge that any Mexicans volunteering to fight the Americans were instead being sent north to fight Villa.[19] The abrupt drop in support for Huerta

---

[15] RG 165, 5761-974, May 6, 1914.
[16] RG 165, 5761-968, June 4, 1914, Burnside's letter to Major Crawford.
[17] RG 165, 5761-974, May 6, 1914.
[18] RG 165, 5761-955, n.d., Statement of I.C. Sánchez taken by Burnside regarding April 21 rail activity.
[19] RG 165, 5761-943, April 27, 1914, Burnside on conditions in Mexico City, from Veracruz.

was "fully confirmed" four days later. This coup in psychological warfare had been orchestrated by the revolutionists who circulated a "large number of hand bills giving information about the capture of Veracruz." The same flyers accused "Huerta of having played a trick on the people and advertised that all of the volunteers would be sent North to fight Villa." Additionally, the earlier story about Zapatistas entering Mexico City to join with the Huertistas in common cause had come into question. These may have been Federal irregulars made to give the impression that the Zapatistas were joining with the Huertistas against the Americans.[20]

To the North, the American invasion of Veracruz placed U.S. Army commands stationed along the border in a heightened state of alert. The day before the Veracruz operation, April 20, General Bliss reported on the disposition of the forces, conditions of the rails, and the state of telegraphic communications in northeastern Mexico along the Nuevo-Laredo-Monterrey, Monterrey-Saltillo, and Piedras Negras-Monclova-Saltillo lines. He also provided information on concentrations of Mexican soldiers along the border to the west at Nogales, Agua Prieta, Casas Grandes (in the Palomas sector), Ciudad Juárez—which could be reinforced quickly by ten or fifteen thousand Constitutionalists with a large number of field guns from Chihuahua and Torreón—and other garrisons in northern Sonora. Constitutionalist General Álvaro Obregón had his main body of about ten thousand men at Navojoa, south of Guaymas, and Bliss expected him to begin moving south to attack Tepic soon. To the informed observer, two aspects of Bliss's report stand out. First, Bliss made no distinction between Constitutionalists and Federals because, in his words, the:

> Attitude of Constitutionalists a few weeks ago indicated combining with Huerta in case of hostilities with this country [the United States] and Mexican Federal Government. This attitude seems to have changed recently. It appears now Constitutionalists will remain neutral so long as the territory controlled by them is not invaded or embargo on munition of war not restored.[21]

This was bad intelligence as there was no evidence that the Constitutionalists ever considered uniting with Huerta, and in fact, they did not join with Huerta post-U.S. intervention even after the embargo was resumed on April 23. Second, the figures for troops in the Northeast (along the Texas border) did not even include González's Constitutionalist Army Corps of the Northeast, another intelligence failure if the idea was to present all Mexican forces—Federal or revolutionary—that might oppose an American invasion.

In the days immediately following, the U.S. Army Southern Department relayed specific data on the Monclova Division of the National Mexican Railroad, including a

---

[20] RG 165, 5761-947, May 1, 1914, from Burnside.
[21] RG 94, AGOMI Box 7473, April 20, 1914, General Bliss to Adjutant General, Washington.

marked-up blue print that showed sections of the line that were "out," and narrative that gave information about bridges, rolling stock, fuel supply, roadbed (gauges and weights of rails and ballast), sidetracks and yards (for passing trains), repair shops, and, most importantly, water.[22] The focus on this particular stretch of railroad, rather than indicating an upcoming operation along that railway more likely seemed to present intelligence of opportunity. The Americans had easy access to that information.

In the meantime, the Federals evacuated Piedras Negras and Nuevo Laredo following the Constitutionalist concentration of forces in the vicinity of Monterrey and Saltillo. For Bliss, the intersection of these two facts gave "color to rumors that the two sides will join."[23] This was a complete misreading of events. In fact, the Constitutionalists had begun attacking Monterrey on April 20 and had positioned forces east of Saltillo to cut off any retreating Federals trying to escape to the capital of Coahuila State, while the Federals had pulled back from the border to initiate a scorched earth retreat against a possible invasion by the United States, which explained the burning of Nuevo Laredo (the commander of Piedras Negras had ignored his orders to torch the town).[24]

In a panic, Bliss also stated, quite erroneously, that "Villa's about to reinforce Juárez and can quickly bring ten thousand men there," and confessed that "I can get no reliable information as to what happens across the border and must rely on Washington for it." In sum, at the first sign that the Constitutionalists and Federals might have joined forces, Bliss urged Washington to "be prepared to immediately occupy Nogales, Juárez, Piedras Negras, Nuevo Laredo, and Matamoros" because there was "no other way to protect women and children in our towns of Nogales, El Paso, Eagle Pass, Laredo and Brownsville which are in pistol range of Mexican towns opposite." If the U.S. government allowed "unfriendly troops" to occupy Mexican border towns, then the Americans would have to evacuate all non-combatants from U.S. border towns.[25]

The Adjutant General of the Army replied to General Bliss, instructing him that as a state of war with Mexico did not exist, American soldiers were not to "invade" Mexico until such time as ordered to do so, and if attacked should "resist and defend to the limit" of their abilities. Bliss pointed out to the Adjutant General the very real possibility that circumstances on the ground might compel a detachment commander to respond in a manner inconsistent with the War Department's wishes. He reminded Washington that all, or almost all, legitimate commercial activity had ceased in the

---

[22] RG 165, 8529-1, April 21, 1914, from Captain G. W. Biegler, 14th Cavalry to Chief War College Division, Washington, D. C. through Chief of Staff Southern Department.
[23] RG 94, AGOMI Box 7473, April 23, 1914, General Bliss to General Wotherspoon, Chief of Staff.
[24] For Federal scorched-earth strategy, see Janssens, *A Revolution in Military Affairs*, 2: 155, 170.
[25] RG 94, AGOMI Box 7473, April 23, 1914, General Bliss to General Wotherspoon, Chief of Staff.

border towns on the Mexican side, that most of the women and children had left, and that the "houses on the edges of these towns" had been prepared for defense by infantry, and that "at various places, infantry trenches" had been constructed. Moreover, there were many places "in which scattered field guns" could be emplaced that could not be silenced "by fire from our side." Alternatively, the American border towns were "populous and rich" such that "infantry and artillery fire from the Mexico side could not miss its target, even in the darkness of night, since they would have the whole town of ours to fire at." In other words, the damage done would be lopsided, with the Americans coming out worse. He went on to explain that "in such a case as that suggested above, the art of war as it has been taught to our officers, requires them to immediately advance upon the enemy's position, drive them from it and capture their guns. Yet this is actually the thing which we are prohibited from doing under the instructions of yesterday." Ergo, Bliss urged that reactions to events "be left to the discretion of the trained officers on the spot."[26]

In this environment, Bliss hardly could have been comforted by the news from the Secretary of War that a shipment of 565,000 rounds of 7- mm. and .30-30 had arrived in Galveston and was to be shipped by rail to Laredo where it would probably cross the border.[27] Yet cooler heads prevailed, and the news that Constitutionalist General Francisco Murguía had defeated the Federals at Allende and entered Piedras Negras finally convinced Bliss days later that the Constitutionalists would not be making common cause with the Federals to fight the U.S. Army. Bliss then reported to the Adjutant General that the Constitutionalists along the border appeared to have taken the "conservative view" and he anticipated "no trouble from any present signs."[28]

Stasis

By May, both the Mexicans and Americans had settled into the new normal. In Veracruz, a "shot gun civil government" had been formed "with a squad of soldiers at the Governor's door...and all of the officers sworn in." This was replaced on May 1, by orders from Washington for a military government. The task of establishing a working

---

[26] RG 94, AGOMI Box 7473, April 25, 1914 (#1), General Bliss to Adjutant General, U.S. Army.
[27] RG 94, AGOMI Box 7473, April 25, 1914 (#2), Chief of Staff W. W. Wotherspoon to Adjutant General, U.S. Army.
[28] RG 94, AGOMI Box 7473, April 29, 1914, and AGOMI Box 7473, May 15, 1914, General Bliss to Adjutant General, Washington.

government was hampered by the reticence of Mexican civil servants and police to return to work, but they were "gradually coming around."[29]

Photo 3.2 Blue jackets resting on sidewalk - Mexican War. Llave Mexico Veracruz, ca. 1914. May 16. Photograph. https://www.loc.gov/item/2010651722/

After five days in Veracruz, Mr. O'Shaughnessy received a telegram from the State Department excoriating him for remaining in Mexico after the Huerta administration had given him a passport, even though he had promptly reported his arrival at the port to Washington and had received no specific instructions to return home.[30] O'Shaughnessy finally received a telegram, instructing him to leave "Mexican soil."[31] With the categorical severing of diplomatic relations between Washington and Mexico, the Diplomatic Pouch service was suspended indefinitely.[32]

Burnside took to using Veracruz as an intelligence-gathering center, especially with respect to information on "roads and general conditions." "Many intelligent men

---

[29] RG 165, 5761-974, May 6, 1914.
[30] RG 165, 5761-947, May 1, 1914, from Burnside.
[31] RG 165, 5761-974, May 6, 1914.
[32] RG 165, 6931-66, May 1, 1914, from State Department Chief, Division of Mails, to the Secretary of War.

from all over the republic" had assembled in the port city and Burnside had "all these willing workers on the go," especially locating "new trails not indicated on maps at the present time."[33] Still, he was not the intelligence officer for the American Expeditionary forces. That honor fell to Captain George E. Thorne, 7th Infantry, whom Brigadier General Frederick Funston appointed as the officer in charge of the Information Division.[34] But information proved difficult to collect. As Burnside noted:

> One thing…must always be remembered: people in Mexico have never been closely bound to facts, and at the present time there are some very racy stories going around; in many cases it is a difficult matter to pick the substantial from the fairy tales. The origin of all news items is on a breezy basis, and as a result, we who are here, as well as the world in general, have but a hazy notion of actual conditions and happenings.[35]

And yet, Veracruz was now finally quiet.[36]

Military intelligence indicated no buildup in forces by the Federal Army taking place and the only preparations appeared to be sporadic destruction of the rails on the initiative of individual commanders. In early May, a Federal major appeared at the water works under a flag of truce and ordered the American Marine major in charge of the line to either surrender or stack his arms and leave within ten minutes. The American major suggested that the Mexican major might want to use those ten minutes to get as far away as possible. Later three shots were exchanged, two from the Mexicans and one "accidental discharge" from the Americans.[37]

The troops opposite the American outposts were "mostly local home guards, not formidable," as demonstrated in the following story. An American squad going to the water works at Tejar took the wrong road and reached Mexican lines. After exchanging cigarettes and taking pictures, the Mexicans gave the American soldiers directions to the water works.[38] In light of this interaction, the reports from newspapers in Mexico City that the Federal Army had captured the water works at El Tejar, just outside of Veracruz, took on a comical air. Such reports were intended to indicate that the Federals were making progress against the Yankees because, as Burnside stated, "the destruction or holding of City Water Works is in Mexico frequently regarded as a highly efficient military measure."[39] Indeed, revolutionaries practiced the cutting of a city's water supply as a frequent tactic, as demonstrated in the battles for Casas Grandes

---

[33] RG 165, 5761-974, May 6, 1914.
[34] RG 141, MGV Entry 12, File 277, May 1, 1914, Special Order No. 1 of the U.S. Expeditionary Forces, Vera Cruz.
[35] RG 165, 5761-974, May 6, 1914.
[36] RG 165, 5761-947, May 1, 1914, from Burnside.
[37] Ibid. RG 165, 5761-974, May 6, 1914.
[38] RG 165, 5761-947, May 1, 1914 (quote). RG 165, 5761-974, May 6, 1914.
[39] RG 165, 5761-958, May 11 1914, Burnside memorandum on cutting of water at Veracruz.

(1911), Ciudad Juárez (1911), Cananea (1913), and Chihuahua City (1913), just to name a few.

Burnside went on to say that the newspapers customarily revealed upcoming operations being planned by the Federal Army, "as a rule, announced from two weeks to a month before actual execution." Some of these operations never did happen, however, and the captain discounted the possibility that the Federals would make a play for the water works. If such an operation did happen to take place, Burnside predicted that it would be directed from Tierra Blanca along the Veracruz al Istmo Railway since "railways are almost universally followed in all Mexican military operations, on account of the small mules used for transporting artillery" and the "use of artillery has recently become so general that scarcely any expedition is attempted without artillery of either 75 or 80 millimeter caliber." The Federals always positioned 80-mm. guns on railroad flatcars or emplaced them in defensive positions. They employed the 75-mm. guns with a bit more flexibility but because of their small mules could maneuver only short distances over roads. The 70-mm. mountain guns had "in the past been almost universally condemned by Mexican Army officers."[40]

Throughout May, Burnside continued to give reports on railroads, suspected deliveries of arms and ammunition, and most especially information about Mexico City.

The Huerta administration had taken over control of the railroads and fired all the American employees, whom Burnside recommended as scouts, guides, and interpreters in the event of a push into the interior. Also, if the Americans were going to advance into the country, they would need to ship locomotives to Veracruz and then assemble them after arrival as the port cranes could not lift assembled locomotives. It took about seven days to assemble an engine.[41]

Official estimates by U.S. Army officers of the arms and ammunition aboard the *Ypiranga* on April 21, according to the number of cases listed on the invoices that naval officers had inspected aboard, included a total of 14,750,000 rounds and probably 10,000 Winchester carbines. Rumors also circulated about possible receipts of ammunition by the Federal Army via Manzanillo, probably overstated in Burnside's opinion, and at 15-20 million rounds almost certainly a fiction in the opinion of this historian. Other information sources suggested that the Mexican government was "systematically" confiscating all materials that might be used in the manufacture of powder and cartridges, as well as alcohol, from the railroad shops and haciendas. Burnside assumed that the alcohol would be used as liquid courage for the troops, "a well-established custom in Mexico." The only documented case of drunkenness by Federal soldiers, however, occurred in the battles at San Pedro de las Colonias and

---

[40] Ibid.
[41] RG 165, 5761-947, May 1, 1914. RG 165, 5761-953, May 9, 1914. RG 165, 5761-957, May 15, 1914.

Zacatecas, and some ex-Federal officers argued that in the latter case no alcohol had been distributed. Meanwhile, the output from the National Cartridge Factory in Mexico City, continued to be paltry and of subpar quality.[42]

By the end of May, it would seem that the Americans still considered expanding the scope of the intervention to include an advance on Mexico City. General Funston ordered all the Officers of the Services under the command of the U.S. Expeditionary Forces, Veracruz, to forward to the Intelligence Officer "all matter of whatever nature in their possession that possesses any military value for operations between Vera Cruz and the City of Mexico."[43] During this time, a peace commission arrived from Mexico City that could not really be described as composed of "Huertistas or politicians," but rather prominent men who had been in an interview with Huerta for only five minutes. The president had given them no specific instructions and it seemed as though the commission's only concern was the inevitable occupation of Mexico City by the Constitutionalists. The moneyed classes apparently preferred American occupiers to the Norteños.[44]

By now the Americans had also deemed it "reasonably certain" that the Zapatistas would not make common cause with the Federals against the Americans. Those previous reports of Zapatistas coming into Mexico City made by an employee of the Tramways Company in Mexico City were now being discounted as a ruse carried out by the federal government, which had armed certain individuals to look like Zapatistas.[45] According to cablegrams intercepted in Veracruz, the Americans also had information about some 7,000 Federals being sent south from Mexico City to check the advance of the Zapatistas. Of course, the Mexican government knew that the Americans would intercept the messages and Burnside considered that this might have been yet another attempt at disinformation and calculated the true number at more like 2,000 or 3,000 sent southward for purely defensive purposes, not to undertake an offensive against the Zapatistas.[46] All this activity by the government with respect to the Zapatistas made the request from the U.S. Secretary of War for information about what, if any, defensive works—"entrenchments, entanglements,

[42] RG 165, 5761-953, May 9, 1914. See also, RG 165, 5761-968, June 4, 1914, Burnside's letter to Major Crawford, "The sporting rifles and ammunition...entering Mexico by way of Puerto Mexico" did not concern Burnside as much as the 22,000 high-powered Japanese rifles and ammunition rumored to have entered the country via the West Coast. In reality, the *Ypiranga's* load of arms and ammunition entered Mexico via Puerto Mexico on May 28, Janssens, *A Revolution in Military Affairs*, 2: 179.

[43] RG 141, MGV Entry 12, File 277, May 29, 1914, memorandum of the U.S. Expeditionary Forces, Vera Cruz.

[44] RG 165, 5761-952, May 13, 1914, Burnside on peace commissioners.

[45] RG 165, 5761-953, May 9, 1914.

[46] RG 165, 5761-957, May 15, 1914. RG 165, 5761-960, May 20, 1914, Burnside Update.

or other defenses"—had been prepared around Mexico City to defend the capital against an attack by the Zapatistas, appear quite timely.[47]

Prior to his leaving Mexico City, Burnside had not received any reports of defenses built up around the capital against a possible Zapatista offensive. The captain felt quite certain that if any such works had been prepared, then they would have been advertised through the government's official news organs. Still, nothing certain could be stated on this matter. Burnside could report, however, that ever since the fall of Cuernavaca the Zapatistas had ranged farther north, and the Federals had been compelled to raze several pueblos in the Ajusco Sierra to deny aid and comfort to the Zapatistas. Also, "wild stories" circulated about the Federals stockpiling food in the Citadel of Mexico City and rigging government buildings with dynamite to explode.[48] Burnside was inclined to believe the "wild stories" about the mining of government buildings for explosion, especially in order to intimidate or counter a possible coup by government employees,[49] but many of the residents of Mexico City disbelieved the reports.[50] It must be said that the preparation of the Citadel for a prolonged defense obeyed Federal Army tactical doctrine, and Federals were not above blowing up government buildings during the battle for a plaza. There had been similar rumors in April of the Federals wiring buildings to explode during the battle for Monterrey, and a couple of months later, June 1914, the Federals actually did dynamite the building containing their magazine—the statehouse no less—during the Battle of Zacatecas.[51]

In the meantime, barbed wire had been arriving at Veracruz that the Huerta administration intended to employ in creating "entanglements and defenses along railroads." The Americans had allowed some of this material to continue to Mexico City, even though it might be used against them in the advent of future operations in the interior, probably because Washington had only mandated that lethal merchandise be detained and kept from leaving customs in the port.[52]

Burnside estimated that Huerta could count on about 17,000 men to defend Mexico City. Fifteen hundred of these would be regulars with the rest being employees of the government, the national railroad company, and volunteers with "nothing but the most rudimentary military instruction." Reports indicated that 5,500 men recently recruited in Mexico City had been sent to the North, and that only about 1,500 garrisoned the rails between Veracruz and Mexico City on the Mexican Railway, with

---

[47] RG 165, 5761-949, May 12, 1914, W. W. Wotherspoon to Chief, War College Division.
[48] RG 165, 5761-951, May 12, 1914, Burnside's answer to request from Secretary of War on defensive works about Mexico City, "wild stories" quote.
[49] RG 165, 5761-957, May 15, 1914.
[50] RG 165, 5761-960, May 20, 1914, Burnside Update.
[51] Janssens, *A Revolution in Military Affairs*, 2: 148-49, 300, 429.
[52] RG 165, 5761-951, May 12, 1914, Burnside's answer to request from Secretary of War on defensive works about Mexico City.

an equal number along the Interoceanic, mostly near Perote.[53] Of all the volunteers being trained in Mexico City by the Federals, Burnside considered that the railroaders would be the best since they had an axe to grind with their former American bosses, and because they presented good physical specimens. Those volunteers from the middle class, however, would be intelligent and determined, but not physically imposing.[54]

General Ángel García Peña commanded the Mexican Railway line with about 1,500 Federals at Córdoba and only a few mountain guns and assorted troops around Atoyac. All Federals had been removed from between Veracruz and Tlalixcoyan. Allegedly García Peña had received the command in Veracruz to get him out of Mexico City and keep him from realizing his aspirations to become president. Other concentrations of Federals and volunteers could be found at Puebla, with troops and heavy artillery reportedly moving along the Mexican Railway toward Mexico City. It appeared that the Federals intended to defend Mexico City against the Constitutionalists and, if defeated there, would fall back on Puebla. Estimates of how many Huerta could muster for that final defense ran to as high as 35,000 men.[55]

Further reports on the Federal Army troops and equipment stationed along the rails from Veracruz to Mexico City, and in the latter city itself, came from the journalist Edwin Emerson, who filed his report with the U.S. Secretary of War on May 14. Emerson traveled to Mexico City at the beginning of May to interview President Huerta, and as to the woeful state and decreased numbers of Federals stationed along the road from Veracruz to Mexico City, his information generally agreed with what Burnside was saying. Emerson also gave a report on Pancho Villa's forces, which he had personally observed in the field during the recent campaign in La Laguna. The Americans were concerned with the Federal Army's potential to develop further operations and interested in the capabilities of Villa's Division of the North, considered the best army in Mexico at the time, lest the two join forces and drive the Americans into the sea.[56]

Along the border, the intelligence generated by the U.S. Army units mirrored those from Veracruz in subject matter, providing information on the arrival of arms and ammunition, the rails, disposition of forces, and updates on possible alliances.

In May, First Lieutenant W. N. Hensley, Jr., Intelligence Officer, 13th Cavalry, filed reports on the Mexican Central Railroad from Ciudad Juárez to Torreón, the branch from Jiménez through Parral to Santa Bárbara, and the Northwestern Railroad in the El Paso sector. The report covered the "complete official track charts, showing distances and directions of the railroad, with a profile showing grades, locations of

---

[53] RG 165, 5761-953, May 9, 1914.
[54] RG 165, 5761-957, May 15, 1914.
[55] Ibid. RG 165, 5761-960, May 20, 1914, Burnside Update.
[56] Edwin Emerson Jr.'s report of May 14, 1914, RG 165, 5761-1091/31.

sidings, location of bridges, railroad crossings, and water tanks." These rail charts extended beyond their obvious use to include topographical information in the vicinity of the rails and landmarks such as "wagon roads, ranch houses, corrals, rivers and mountains." The exile of all American railroad employees to El Paso and the entirely warranted suspicion that Mexicans harbored toward all Americans had made getting intelligence about railroads quite difficult. As soon as "Mr. Collins," who was "the only foreigner working on this line between Juárez and Chihuahua," arrived in El Paso, the intelligence officer promised to update the information.[57]

The U.S. Army intelligence reports from Captain John W. Wright, 17th Infantry Intelligence Officer, in Eagle Pass focused on the roads, rails, and watering holes in Coahuila that might have been of interest in planning an American invasion through the Northeast. The information included the extent of operations, general conditions, strategically significant notes, estimated time to put into complete working order should the United States take over the railroad, and basic information, to include: Gauge, Track, Grade, Curves, Rails, Ties, Platforms, Storehouses, Watering stations and supplies, fuel type, Work-shops, Telegraphs, Rolling Stock, Tunnels, Bridges, and Material available for reconstruction. The Federals had been able to repair bridges by "cribbing" or filling in underneath to support the track. They had not been as successful in repairing rails that had been twisted by rebels because "twisted rails cannot be straightened anywhere in Mexico."[58] Much of the information was provided or verified by Mr. Stich, Superintendent of the Monclova Division, and the civil engineer Bynum E. Nourse, who had worked on bridge construction and repair for the line from 1901 to 1907, including updates to earlier maps regarding stretches of track that had been abandoned or removed.[59] Captain Wright spoke Spanish and had spent

---

[57] RG 165, 8532-1, May 25, 1914, Report on Mexican Central Railroad by W. N. Hensley, Jr., First Lieutenant, 13th Cavalry Intelligence Officer, 8th Brigade, for Commanding General, Southern Department (quotes). RG 165, 8532-2, May 26, 1914, Report on Mexican North-Western Railroad by W. N. Hensley, Jr., First Lieutenant, 13th Cavalry Intelligence Officer, 8th Brigade, for Commanding General, Southern Department.

[58] RG 165, 8529-2, May 16, 1914, Report on the Monclova Division, Mexican Railway, John W. Wright.

[59] RG 165, 8529-3, May 30, 1914, Report on Railroad Bridges by Captain John W. Wright, 17th Infantry, Intelligence Officer. RG 165, 8529-4, May 30, 1914, Report on Railroads of Mexico by Captain John W. Wright, 17th Infantry, Intelligence Officer. See also RG 165, 8529-7, June 27, 1914, from J.C. Gravis to Captain H. H. Robert, and June 29, 1914, Road notes, Soto la Marina-Tamaulipas, Mexico, from John W. Wright relaying information on roads, rivers, crops, and terrain from the mouth of Rio Soto la Marina to Ciudad Victoria, Tamaulipas. Updates included: RG 165, 8529-10, August 13, 1914; RG 165, 8529-11, September 18, 1914, Water supply along routes in Mexico from Captain John W. Wright, updated September 10, 1914, by Captain B. Simmons, both Intelligence officers of the 17th Infantry Regiment.

several years in Cuba where he was "most successful in dealing with Latin-Americans."[60]

From the Piedras Negras sector came the news that the Rio Grande was still at high stage around fords, meaning that it would be difficult for Constitutionalists to smuggle ammunition across the border now that the embargo had been reimposed.[61] However, in the Matamoros sector, the lieutenant colonel of the 3rd U.S. Cavalry reported that the Constitutionalists claimed to be "expecting large shipments of arms and ammunition" to be delivered on German boats through Tampico, and it was believed that they had "350,000 rounds, small arms ammunition, at Monterrey" ready for immediate use.[62] Similarly alarming, in a message to the Adjutant General, General Bliss bemoaned the political boss of Starr County, Manuel Guerra, who ran roughshod over local officials and small ranchers; he was Carranza's fiscal agent in the area and smuggled arms and ammunition across the border virtually at will without the U.S. Army being able to make a dent in his operations.[63] This intelligence about arms trafficking in the Northeast made the following statement from Captain Burnside to his friend, Major C. Crawford, all the more curious:

> There is mighty little news here. Gen. Funston sometimes asks Wash. questions—which, of course, cannot be answered. The Navy has orders to prevent the Mex. gunboats from stopping the Tampico Rebel arms arriving on the *Atilla* (S.S.) – This step is liable to stir things. Anything U.S. may do will be interpreted by Mexs. as being unsatisfactory.[64]

Therefore, the Constitutionalist Army Corps of the Northeast continued to receive arms and ammunition in spite of Wilson's arms embargo and the best efforts of Bliss's soldiers to stop the flow of this contraband through Matamoros. Meanwhile, the U.S. Navy had orders to prevent the Federal Navy's gunboats from impeding the delivery of arms to the Constitutionalists through Tampico. The Americans' policy seemed to lack coherence, or possibly they wanted the Constitutionalists to continue to hold Tampico in order to prevent Huerta from receiving arms and ammunition through that port. In other words, Tampico may have been a tactical exception to the strategic rule to halt the flow of all arms and ammunition into Mexico.

In the El Paso sector the big news about arms smuggling was the arrest of Rodrigo Quevedo and a subordinate by El Paso police on May 29 for violations of neutrality

---

[60] RG 165, 5761-362, December 27, 1911, Captain McCoy to General Wotherspoon regarding Captain John W. Wright (quote). RG 165, 5761-363, December 29, 1911, General Wotherspoon to Captain McCoy regarding Captain John W. Wright.

[61] RG 94, AGOMI Box 7473, May 21, 1914, Colonel T. W. Griffith to General Bliss.

[62] RG 94, AGOMI Box 7473, May 19, 1914, General Bliss to Adjutant General, Washington.

[63] For more information on smuggling in the Northeast, see RG 94, AGOMI Box 7473, June 17, 1914, report on Manuel Guerra from General Tasker Bliss to The Adjutant General of the Army.

[64] RG 165, 5761-968, June 4, 1914, Burnside's letter to Major Crawford.

laws. The police planned to turn the two over to military authorities who would then send them to Fort Wingate in order to discourage "this and other bands" from trying to "launch a new revolution against the Constitutionalist Government." The absence of Villa's Constitutionalists in much of northern Chihuahua had "led to new accessions to these Huertista bands," although it seemed that the Constitutionalists had finally "started in to disperse these gangs." At the time, Villa only had about four hundred men in Ciudad Juárez.[65]

It should also be noted that as early as May 21, 1914, the commander of the U.S. 17th Infantry, Colonel T. W. Griffith, commented in a special report that the divide between Villa and Carranza was becoming more apparent and the Americans only expected it to grow.[66] Furthermore, Carranza and Pablo González flatly refused "any scheme for mediation," one presumes with the Huertistas.[67] Therefore, already in May the Americans were getting a clear sense that Mexico would remain fractured politically and militarily, with the Zapatistas and Constitutionalists refusing to join with the Huertistas, and a possible schism among the Constitutionalists.

Intelligence from the U.S. 3rd Cavalry that Villa was not going to attack Saltillo but rather would reach San Luis Potosí in the next three days and would attack that city proved incorrect.[68] After the victory at Paredón and a short occupation of Saltillo, Villa's division backed up to Torreón in anticipation of an attack on Zacatecas. Observers who had been in the field with General Villa commented favorably on his ability to move an estimated 18,000 men, including 15,000 irregular cavalry with all their horses, from Torreón to Paredón in just three days. Villa's men also gave favorable impressions with the conditions of their horses, armament and supply of ammunition, and uniforms.[69]

After the Battle of Paredón and Tampico in May, military activity slowed down. In the Northeast, González's army corps was occupied with importing arms and ammunition through Tampico in order to outfit Jesús Carranza's new 2nd Division of the Center. Villa's Division of the North remained in Torreón, recuperating, refitting, and recruiting. In the Northwest, General Álvaro Obregón kept busy trying to assemble a sufficient number of carts and livestock to make the overland trek into Jalisco State. General Pánfilo Natera's Constitutionalist 1st Division of the Center undertook the only operation of significance with its assault on the plaza of Zacatecas. When that effort failed, General Villa took his division to Zacatecas, against the wishes of First

---

[65] RG 94, AGOMI Box 7473, May 30, 1914, General John J. Pershing, Eighth Brigade, to General Bliss (quotes). For a discussion of counterinsurgency activities and response in Chihuahua at this time, see Janssens, *A Revolution in Military Affairs*, 2: 80–81, 485–86 (endnotes 19-21).
[66] RG 94, AGOMI Box 7473, May 21, 1914, Colonel T. W. Griffith to General Bliss.
[67] RG 94, AGOMI Box 7473, May 19, 1914, General Bliss to Adjutant General, Washington.
[68] Ibid.
[69] RG 94, AGOMI Box 7473, May 30, 1914, General John J. Pershing, Eighth Brigade, to General Bliss.

Chief Venustiano Carranza, and attacked and took the city in a battle that produced perhaps the worst carnage of the revolution.[70]

Villa's insubordination ultimately provoked a schism in the Constitutionalist Army. The Constitutionalist "Agent," a Mr. Grey, offered Colonel T. W. Griffith, 17th Infantry, a unique take on the break between Carranza and Villa, saying that Natera had ignored orders not to attack Zacatecas until reinforced by Villa. Natera's rash actions ended in a failed attack, and one thousand Federal prisoners deserted from Natera and returned to Federal service. This story about Federals deserting *from* the Constitutionalists *to* the Federals was novel, and highly unlikely. And, of course, Mr. Grey erred in his entire sequence of events leading up to the break between Villa and Carranza, since it was Villa who disobeyed orders by refusing to send reinforcements to Natera while the battle was in progress. Grey said that Villa had stopped taking orders from Carranza, who was trying to "sidetrack him," and decided to go it alone, attack Zacatecas and, having prevailed, continue on to Guadalajara and ultimately to Mexico City.[71]

The U.S. Army's anonymous report on the Natera phase of the Battle of Zacatecas, dated Torreón, June 15, 1914, was very uneven in its accuracy. The intelligence overstated the number of Federals, placing the estimate at 12,000 to 16,000 when in reality they numbered closer to 10,000, and 120 to 180 machine guns on hand can only be regarded as astronomical. The report's author also quoted a high-ranking Constitutionalist officer as saying that "the Federals had made up their mind to defend Zacatecas to the limit of their resources" given that Huerta surely "recognized the strategic importance of preventing the Constitutionalist Army from gaining a good foot-hold in central Mexico."[72] To wit:

> The agricultural and industrial development in that part of the republic is such that the invaders could cut loose from the railroad and by virtue of their organization, cavalry brigades, overrun the district and so force the entire Federal Army to concentrate at, possibly, no more than two or three points, one of them Mexico City. In this manner, thought the officer, most of central Mexico would fall into the hands of the Constitutionalists, and this General Huerta, whom he gave credit for being no mean tactician, would try to prevent it at all costs. Zacatecas, virtually, he said, was the last point offering every feature favorable to this plan. Aguascalientes,

---

[70] Janssens, *A Revolution in Military Affairs*, 2: 245-49, 331-333.

[71] RG 165, 8529-6, June 19, 1914, Special Report by Colonel Griffith, 17th Infantry.

[72] RG 94, AGOMI Box 7473, June 15, 1914, anonymous report (quotes). Miguel A. Sánchez Lamego, *Historia Militar de la Revolución Constitucionalista*, 5 vols. (México: Talleres Gráficas de la Nación, 1956-1960), 5: 252-53. Janssens, *A Revolution in Military Affairs*, 2: 249.

seventy-five miles further south and on the edge of the rich central region, being unsuited for defense.[73]

The report put Villa's forces for the operation against Zacatecas at "about 29,000 men, fifty-one pieces of field artillery and ninety machine guns." These numbers were overstated, but not in any meaningful way. Villa probably had just over 20,000 men and dozens of field pieces, although the bloated number of machine guns seems highly suspect.[74]

Washington Intelligence

In June and July, Washington started producing analyses based on the intelligence gathered from various sources. The first of these reports was the biographical sketch of Francisco Villa by Colonel John Biddle, Chief of War College Division, for the General Staff. It was difficult to get personal data about any revolutionary generals, because even the Mexicans knew so little about them. Biddle had developed the profile from "personal conversations with men of long residence in Mexico, reports, press dispatches and letters." It began with the retelling of the legend of Villa's bandit days, then enumerated his campaigns, his near-execution in 1912, incarceration in Mexico City, escape to Tucson, and rejoining the revolution in March 1913. In the summer of 1913, Villa had moved into the hills of Chihuahua south of Columbus, New Mexico, to organize his troops and U.S. Army officers "in that vicinity were struck with the efficient manner in which Villa organized, armed, and equipped his force of some 1,000 men."[75] American military officers always commented favorably on Villa's organizational skills.

As regards tactics, the report credited the incompetence of the Federals more than Villa's own efforts for his success. Villa's tactics against:

> large bodies of federals have consisted in breaking their lines of communication, so that they could not be re-supplied, then by piece meal attacks making them fire away all their ammunition until an evacuation took place. In no case has he captured or entirely destroyed a federal column of any size in the open field. The evacuation of Chihuahua and Torreón being examples. In both cases

---

[73] RG 94, AGOMI Box 7473, June 15, 1914, anonymous report.
[74] Ibid.
[75] RG 165, 5761-968, June 12, 1914, Memorandum to the Chief of Staff from Colonel John Biddle, Colonel, General Staff, Subject: Sketch of Francisco Villa.

the federals' retreating columns were entirely vulnerable but nothing was done to inaugurate a strong pursuit.

Biddle should have noted that in the case of Torreón, one hundred percent of Villa's troops had been engaged for days on end and they needed a rest, and two holding brigades were already in place to block the arrival of Federal reinforcements, AND he did send more brigades the day after Velasco evacuated in the direction of the Federals' general line of retreat. Moreover, Velasco was originally headed toward Zacatecas overland, which would have made pursuit difficult.

Biddle continued:

> At Torreón during the latter part of March and first part of April 1914, Villa really was whipped to a finish by Velasco. He had nothing left for defense or offense and had only a small reinforcement of ammunition and men been sent [to] Velasco, 'one hundred men and 1,000 rounds of ammunition' as one who knew the conditions said, Villa's army would have been scattered.

The colonel goes on to say that Villa only won at San Pedro de las Colonias because of the internal dissension among the Federal generals and the incompetence, in particular, of General Joaquín Maass. Villa did not continue southward after Maass toward San Luis Potosí because Zacatecas offered more in the way of booty and because of turmoil in the ranks of his own generals.[76] This is simply shoddy analysis. Velasco's men were down to fifty or sixty rounds on average, meaning twenty men alone would have had the requisite one thousand rounds, and it has been strongly suggested, probably correctly, that the large explosion on the last day of the battle was Velasco blowing up his magazine, which the Federals often did when a battle was lost.[77]

The section related to Villa's personality proved the most accurate and relevant parts of the report. The colonel attached several qualities to his style of leadership: bravery, daring in undertaking operations, excellent horsemanship, physical strength, generosity in sharing plunder with all, and his ability to win battles. But Villa was given to fits of rage and not good at taking advice, and even when he did it was only for a time, until the spirit moved him to "do whatever his fancy for the moment dictated." His "inability to enforce discipline or 'get along' with his companions both inferiors and superiors had always been true of Villa, from his bandit days when he killed his partners up to the present. Villa's ability to command a party of bandits of from 200 to 500 was unquestioned. His ability to command and keep together a force of upwards of 16,000 men for any length of time was a serious question." Additionally,

---

[76] Ibid.
[77] Janssens, *A Revolution in Military Affairs*, 2: 117-19.

Obregón and González would have nothing to do with Villa, so the odds that they would "ever really get together seemed quite impossible," which turned out to be true.[78]

The colonel then engaged in some Federal Army style calculus, figuring that Villa only commanded the loyalty of a group of fickle revolutionaries from Chihuahua and part of Coahuila who had "fought at various times under Orozco, Salazar, Rojas, Quevedo, Salas, Mercado, Huerta and others." Even if Villa commanded the loyalty of the entirety of those two states, which he did not, that would only give him sway over 767,357 of Mexico's 15,114,305 people, or about 1/19 of the population of Mexico. Villa's future following depended entirely on his ability to keep winning—something even Villa seemed never to learn: "With comparatively small reverses however it rapidly melted away as an organization, the men split into bands of from 50 to 200, took to the hills, and existed as bandits."[79]

In closing, the colonel let slip the first clue to the intelligence failure in this, and later Captain Mitchell's, report: "As things stand at present Villa has more of a 'punch' than any other individual military leader in Mexico. He is the embodiment of all the elements of the Northern revolutionist of Mexico."[80] At the same time, Captain Burnside wrote that none of the revolutionaries worried him except for Villa, who demonstrated "more military talent than is combined in the Federal Army." However, if Huerta could get the arms and take personal command of the Federal Army in a condensed area instead of "scattered over the enormous extent of northern Mexico" then Huerta would "give the northern rebels a surprise party one of these days."[81] The Americans focused almost solely on the capabilities of Villa, the Federals, and to a much lesser extent González's Northeasterners, completely ignoring General Álvaro Obregón's formidable Constitutionalist Army Corps of the Northwest.[82]

Toward the end of June, the War College Division requested permission for Captain Burnside to come to Washington D.C. and be debriefed. Captain Howard L. Laubach, who was in charge of a "monograph work on Mexico," believed that his project would "be very much facilitated if Captain Burnside, now at Vera Cruz, can be ordered to Washington for a few days." Laubach also thought that Burnside's work would "be very much improved by a personal conference with us at the War College." It was feared, however, that if the application for orders for Burnside came from the War College Division, General Staff, it might "interfere with his usefulness in his relations with the Mexican authorities who might subject him to greater, and perhaps annoying, surveillance." Therefore, the General Staff wanted the request for Burnside's return to come from the State Department "to have him ordered simply to Washington"

---

[78] RG 165, 5761-968, June 12, 1914, Biddle's memorandum.
[79] Ibid.
[80] Ibid.
[81] RG 165, 5761-968, June 4, 1914, Burnside's letter to Major Crawford.
[82] RG 165, 5761-968, June 12, 1914, Biddle's memorandum.

temporarily with the reason unspecified. The answer to this proposal from Army Chief of Staff Wotherspoon was curt: "Secretary does not approve at this time. W."[83] This request and denial made clear two points: first, and perhaps most significantly, claims to the contrary and absent official recognition, the U.S. government consistently treated the Huerta administration as though it were the established "Mexican authorities," and still considered the Constitutionalists as "rebels," even at this late date; second, the U.S. War Department's fears that the "Mexican authorities," meaning the Huertistas, might discover the true reason behind Burnside's return to Washington belies the possibility of any purported covert operations that the U.S. military might undertake to collude with the Constitutionalists. If the military attaché could not get away without being discovered by Huerta's many spies in Veracruz, how could the Americans ever have hoped to carry off a large sealift of war materiel to supply the Constitutionalists undetected?[84]

The U.S. Army's declassified file also contains an undated and anonymous Intervention Memorandum that was inexplicably included among correspondence relating to the Torreón Accords and the location of Ambrose Bierce, dated in September 1914. The gist of the memo included a warning not to invade across the border because it was desired to unite all the factions against the Americans at one point, in Mexico City, so that in the event of victory all the various factions could be party to negotiations. Multiple columns sent to invade across the border, in contrast, would force the various factions to disperse in order to face the various threats. Instead, invasion should come from Vera Cruz and consist of a smaller, rather than a larger, army—which would necessarily require use of the rails and more men to guard same—with a healthy number of cavalry, to make use of roads and cross country to increase mobility against the Mexicans who are often tied to the rails, an obvious reference to the Federals and not Constitutionalists or Zapatistas. The force should also be comprised solely of regulars, and with a healthy contingent of artillery.[85] An American expatriate probably crafted this useless memo, but it challenges the imagination that the U.S. Army should have preserved it.

The keystone invasion memorandum came in July 1914, in the form of the "Notes on Mexican Constitutionalists, Or The Northern Mexican Insurgents" by Captain William Mitchell, General Staff.[86] Captain Mitchell's report essentially consisted of two components, the first being methods and equipment employed by Mexican

---

[83] RG 165, 6931-68, June 26, 1914, memorandum for Colonel Hodges from the War Department Chief of Staff.

[84] This directly rebuts the conspiracy theory that the Americans were colluding with the Constitutionalists, see Hart *Revolutionary Mexico*, 297-98.

[85] RG 165, 5761-1091/8, n.d., Anonymous, Memorandum on Intervention and Invasion recommendations.

[86] RG 165, 5761-975, July 1914, Captain William Mitchell's "Notes on Mexican Constitutionalists Or The Northern Mexican Insurgents."

soldiers in which the infantry and artillery analysis applied almost wholly to the Federals, and the mounted infantry (which in Mexican military literature are also referred to as "dragoons" and "cavalry," interchangeably) to the Constitutionalists. The memo ended with a "Comments" section which recommended measures and countermeasures that might be employed against Mexican soldiers in the event of war, including expanded discussions of Mexican tactics and terrain.

It must be said that the report was singular in its perspicacity in some areas, egregious in its errors in other regards, and notorious for its inherent racism (although it closes with a rare encomium); moreover, it serves to memorialize a general intelligence failure. Because the United States did not have diplomatic relations with Victoriano Huerta's government, the U.S. Army could not send observers into the field with the Federals during the 1913-1914 time frame. Accordingly, virtually all its knowledge about the Federals came from American news correspondents such as "Colonel" Edwin Emerson and E. V. Stoddard, and most especially from Huerta's 1912 campaign against Pascual Orozco Jr. Similarly, essentially all the U.S. Army's intelligence about the Constitutionalists pertained to Villa's Division of the North, which had its own special State Department envoy, George C. Carothers, and other American civilian observers. Knowledge about Álvaro Obregón's Constitutionalist Army Corps of the Northwest and Pablo González's Army Corps of the Northeast were limited to what American officers observed from afar (i.e., not embedded) and on the periphery (Tampico, Nuevo Laredo, and Santa Rosa— probably not the May 1913 Battle of Santa Rosa, but rather a minor border affair). Therefore, Mitchell's report placed undue attention on the Federal Army, which ceased to exist one month after the captain filed it, and it was entirely silent on the capabilities of the Yaqui infantry that fought under General Obregón, among the finest light infantry in the world at the time.

The Federal Army Surrenders

In August 1914, General Bliss had become concerned that the Constitutionalists were "making strenuous efforts to secure large quantities of arms and ammunition" along the Chihuahua State-U.S. border. At no other point on the border were the Constitutionalists making such an effort. While the Constitutionalists elsewhere were importing arms and ammunition via Tampico, Villa's forces were "forced, for some reason, to rely on smuggled munitions from the United States," and the U.S. Army was unable to prevent it.[87] It would appear that Bliss did not know that after Villa's insubordination in attacking Zacatecas, Carranza had blocked Villa from using Tampico as a port of entry for his ammunition purchases.

Five days later, Bliss apparently had been brought up to speed and expressed concern that Villa was trying to stockpile large amounts of ammunition for a war against Carranza. Therefore, he urged a "restoration of embargo along entire frontier west from Brownsville." He did not think that Carranza would object because the Constitutionalists were importing all their arms and ammunition via Tampico.[88] This advice seems curious because the arms embargo had resumed on April 23, 1914, and the stated problem was an inability to physically halt the smuggling. Perhaps by "President's embargo," Bliss meant a complete closing of the border.

From Veracruz, General Funston expressed different fears. In a personal letter to General W. W. Wotherspoon, he confessed that he saw no hope for peace because he did not think that Carranza's generals would allow him to grant amnesty to the Huertistas. Ergo, he believed that the Federals might make a stand in Mexico City, where they were concentrating, continue to fight while retreating to Puebla, or simply flee into the American lines at Veracruz. Funston considered his intelligence in this matter quite reliable because all manner of foreigners, as well as Mexicans, fleeing Mexico City and other points in the interior brought information.[89]

Based on reports received, Funston knew that the forces of Jesús Carranza and Obregón were "making all possible haste to occupy the Capital" and had already advanced beyond Querétaro. If they continued pushing the Federals to Veracruz, Funston would have to intern and then "feed and guard thirty or forty thousand men with their 'soldaderas' and other camp followers." If the Constitutionalists showed up and demanded entry into the port or that the Americans surrender the Federal prisoners, then Funston would have no choice but to fight the Constitutionalists, a prospect that he did not relish:

> I suppose the Federals would sit down on their haunches and let us fight it out. I know that I would be afraid to trust any part of the line to them. Against four or five times our number of such troops as the rebels [Constitutionalists] have shown themselves to be we would have a red-hot time of it. All talk of our being assisted by the fire of the fleet is moonshine. Those people would have too much sense to operate against our flanks where they would be exposed to fire from the fleet, but would confine themselves to an attack on our center and on cutting off the water supply. The terrain is such that if our center were being attacked the fleet would not dare fire as there

---

[87] RG 94, AGOMI Box 7473, August 4, 1914, report from General Bliss.
[88] RG 94, AGOMI Box 7473, August 10, 1914, General Bliss to Adjutant General, Washington.
[89] RG 165, 5761-1091/49, August 1, 1914, Frederick Funston to Wotherspoon.

would be a confused struggle in the sand hills back of the port, where friend and foe could not be distinguished. Where the fleet would be of great help would be in the men and guns that it could land to assist us, and with this help I would have no fear as to the result. I have reason to believe that it is the intention of the Federal troops if an amnesty is not arranged to retire on Vera Cruz, as the only place in the Republic where they can seek shelter. Of course, we are bound by international law to receive them.

Still, Funston thought that if the Constitutionalists gave guarantees then the Federals would surrender.[90]

Under the present condition, with Interim President Francisco S. Carbajal in charge and Huerta gone (Huerta had fled the country in July 1914), Funston believed that war with the Federals was now "quite unlikely" and if the Americans were to "get in a mix-up" it would "be with the present rebels" and under the conditions just mentioned. In the event of such an occurrence, Funston said, "this little [U.S.] force will be thrown quickly on the defensive owing to the great numbers and morale of the rebels," and given the "several thousands of the men who have been winning the recent victories in the north and center of Mexico it would behoove us to be pretty careful about getting too far from Vera Cruz with the troops that we have."[91]

As it happened, Carranza did grant amnesty to all who applied for it, except those officers who had participated in the Citadel uprising, and the Federal Army surrendered on August 13, 1914. Less than a month later, the Americans lifted the embargo, permitting all manner of arms and ammunition to flow into Mexico into the hands of all factions,[92] and cancelled the "daily reports of sales of arms and ammunition" required of the Winchester Repeating Arms Company.[93]

The Americans also repatriated the former members of the Federal Army who had fled after various battles across the border, especially from Ojinaga, and were still incarcerated in the United States. Villa wanted the ex-Federal prisoners at Fort Wingate released into Chihuahua since, he claimed, "we have much work and can easily place all of them," but as Carranza already had requested that those soldiers come back through Piedras Negras, Villa said he would not press the issue.[94] The U.S. War Department did authorize Bliss to tell the ex-Federals that Villa had given them

---

[90] Ibid.

[91] Ibid.

[92] RG 94, AGOMI Box 7473, September 8, 1914 #1, from Adjutant General to Secretary of the Treasury. RG 94, AGOMI Box 7473, September 8, 1914 #2, from U.S. Secretary of the Treasury to Secretary of War.

[93] RG 94, AGOMI Box 7473, September 14, 1914, U.S. Army Chief of Staff to Winchester Repeating Arms Co.

[94] RG 165, 5761-1091/32, September 15, 1914, Villa to General Hugh Scott regarding Federal Prisoners at Wingate.

full guarantees in territory under his control, and some of them asked to return via Ciudad Juárez, which Bliss was inclined to approve since that would "save much expense." But the Adjutant General's Office responded the next day with a rush telegram that "Internes [sic] cannot be sent Mexico via Juárez,"[95] probably because Carranza was recognized as the First Chief and his wishes were that the ex-Federals should return via Piedras Negras.[96] The Americans made an exception for Colonel Kosterlitzky's command interned at Fort Rosecrans after the 1913 Battle of Nogales to return to Mexico via Nogales because it had been confirmed that those men were actually customs guards and other civilians pressed into Federal service who had ties to Nogales.[97]

Villa certainly could have put the ex-Federals to work. By September 26, 1914, he and the Governor of Sonora, José María Maytorena Jr., had both rebelled against Carranza. In Douglas, Arizona, Colonel Hatfield reported:

> General [Benjamín] Hill with about three thousand Constitutionalist troops have arrived at Naco Sonora and evidently intend to make his stand there. Period. I am ordering Col. [John] Guilfoyle with four of his troops and machine gun platoon to proceed at once by marching to control situation there. Period. General Hill's requests to send about twenty of his wounded by train to Douglas. Period. I have refused his request but ask instructions as to entry of his wounded. Period. End of Quote.[98]

The Mexican consul in Arizona later requested permission to bring two or more carloads of wounded Constitutionalists into U.S. territory. General Bliss said that the State Department had to make that decision, but he urged the War Department not to support the request because, otherwise, "we will soon have another large force of prisoners on our hands."[99] The remaining ex-Federal prisoners left Fort Wingate on September 27, 1914, at 4:20 p.m.[100]

---

[95] RG 92, OQG September 18, 1914, Exchange between General Bliss and Adjutant General's Office.
[96] Manuel W. González, *Contra Villa: Relato de la Campaña 1914-1915* (México: Ediciones Botas, 1935), 153-55, the First Chief had been trying to repatriate General Mercado's Federal Division of the North, which had crossed into the United States during the Battle of Ojinaga in January 1914, through Piedras Negras. Most of these ex-Federals had been drafted according to "la leva" and "almost all" accepted an invitation to join Constitutionalist General Antonio I. Villarreal's Division of the Bravo, and the rest received free passage home.
[97] RG 92, OQG September 28, 1914, #237, General Bliss to Adjutant General, Washington, regarding Kosterlitzky. RG 92, OQG September 29, 1914, #237, Adjutant General, Washington, to General Bliss.
[98] RG 92, OQG, September 26, 1914, #232, Exchange between General Bliss and Adjutant General, Washington.
[99] RG 92, OQG September 26, 1914, #233, Exchange between General Bliss and Adjutant General, Washington.
[100] RG 92, OQG September 27, 1914, and September 28, 1914, #234, exchanges between General Bliss and General Staff, Washington.

The point to be made here was that the United States did not support the Constitutionalists, but rather pursued its own interests. The Americans had agreed to repatriate the ex-Federals to Piedras Negras according to the wishes of the First Chief of the Revolution. In the midst of the process, on September 23, Villa broke with Carranza and the Maytorenistas attacked the Constitutionalists in Naco. The Americans completed the repatriation of ex-Federals and then refused to admit Constitutionalist wounded, without prejudice, to enter the United States. Thus, on the one hand the Americans' decision favored the Constitutionalists, and on the other it went against them.

Leaving Veracruz

After the resignation of President Victoriano Huerta and the defeat of the Federal Army, the Americans no longer had a reason to remain in Veracruz. On September 17, 1914, General Funston started the process of evacuation by cabling to the Adjutant General the number of men and tonnage and dimensions of freight to be transported back to the United States. He recommended selling the thirty mules that the Marines had purchased in Mexico instead of bringing them to the United States since they were "very small," confirming what Burnside had earlier reported about Mexican mules being too small for Federal artillery and therefore confining that arm to the rails. Funston also suggested having the Constitutionalists designate "new officers who will take over various departments" to work under the Americans in order to "have sufficient time to familiarize with their duties." Funston offered an October 10 evacuation date, saying that it was possible to leave earlier, but he needed the extra time to train any new officials.[101]

The Secretary of War personally responded to Funston's proposed transition to Constitutionalist leadership, saying that he considered "it inadvisable to adopt suggestion in your Number One Fifty with respect to having new officials working in the respective offices under or with our officials." "The procedure to be followed in the turning over of the various governmental departments" was yet to be "determined upon." In the meantime, the secretary urged Funston to observe "extreme caution...in any interviews you may have with Mexican authorities."[102]

The "large number of Mexicans who feel that they have to leave the country" for fear of reprisals by the Constitutionalists posed the biggest sticking point for the

---

[101] RG 94, AGOMI Box 7478, September 17, 1914, General Funston to Adjutant General.
[102] RG 94, AGOMI Box 7478, September 22, 1914, Secretary of War to General Funston.

American withdrawal. According to Funston, a "great deal of alarm among the inhabitants of Vera Cruz" had been incited by "irresponsible persons" who claimed that once the "Constitutionalists got in control they would punish them severely because they submitted quietly to our rule, and in many cases fraternized with us and assisted us." Although Funston considered these fears unwarranted, he did not want to leave until First Chief Venustiano Carranza had made a statement guaranteeing the safety of these officials.[103] He must not have known that all Mexicans who had "served the Americans in any official capacity whatsoever" were considered "traitors according to Mexican law" and would be dealt with accordingly unless "specific provisions" could be made to assure their safety before the American evacuation. Additionally, very few of these Mexicans had the economic resources to go into exile.[104]

This sense of loyalty toward the porteños of Veracruz with whom they had toiled under very trying circumstances became the greatest factor in delaying the departure of the Americans. How to handle customs duties paid to the port authority during the American occupation created another hurdle. Importers feared that they would have to pay twice—the second time to the incoming Constitutionalists. Indeed, the Constitutionalists would have the authority to nullify all official business during the American occupation, even civil marriages, which would have visited "untold loss and confusion and embarrassment" upon everyday citizens after the Americans left.[105]

Carranza refused to negotiate with the Americans over these issues, and reached an understanding directly with the Veracruz Chamber of Commerce, essentially excising the Americans from the process of reconciliation. Satisfied with this outcome, the Americans would eventually leave the country—but they would take all the customs duties collected under their administration as a guarantee that Carranza would keep his word, and also for safekeeping during the abandonment of the city that might be as chaotic as the invasion had been.

In October, another consideration for the departing Americans was the large number of political refugees and formerly wealthy men who now had become impoverished. These would most likely be executed if they could not acquire the means to escape. The total number of persons affected by the withdrawal of U.S. troops from Vera Cruz was 649 (not including family members, which made the number higher), with 242 refugees and the remainder being government employees, mostly police, postal, and health workers.[106]

The American Consul, William W. Canada, had submitted an earlier memo, dated September 21, 1914, that included two lists, one called the "Partial List" and the other

---

[103] RG 94, AGOMI Box 7478, September 17, 1914, General Funston to Adjutant General.
[104] RG 165, 5761-1091/28, October 5, 1914, Memorandum for the Chief of Staff: Persons affected by the withdrawal of U.S. troops from Vera Cruz.
[105] Ibid.
[106] Ibid.

the "Additional List." Most notable among the first list were: Deputy, Senator and Writer, Francisco Bulnes; Secretary Foreign Relations Federico Gamboa; Federal General Gaudencio G. de la Llave; Military Governor, State of Veracruz, Eduardo Cauz; Federal General Juvencio Robles, Federal General J. Rasgado; author of *La División del Norte*, José Juan Tablada; and Governor of the Federal District, Carlos Rincón Gallardo. Included in the second list were: Federal and Colorado Generals Francisco del Toro, José Alessio Robles, Enrique Goroztieta, Ignacio Bravo, Gustavo Salas, Félix Terrazas, Marcelo Caraveo, and the former Governor of Chihuahua, Enrique Creel Terrazas. Canada did not list the numerous Catholic priests and sisters, at least 220, who risked persecution if they fell into the clutches of the Jacobin Constitutionalists.[107]

Additionally, there were an estimated 1,500 ex-Federal soldiers and officers, some of the latter "in danger of execution unless they can escape," and the soldiers apparently of no use since it was said that they had been "originally sent to Veracruz...by the Carrancistas in order to be gotten rid of." Time and money were of the essence because it took time to book passage on steamers, and if the U.S. Army exited under a quick schedule, many of the refugees would be caught unprepared.[108]

As time dragged on and the Americans still had not left Veracruz, the Constitutionalists grew increasingly impatient. The situation with the Convention of Aguascalientes—a conference between the major revolutionary factions to agree on a new government—continued to deteriorate and rumors about an imminent attack on Veracruz reached the U.S. Secretary of War. General Funston reassured the secretary that the "unsubstantiated rumors" about a pending Constitutionalist attack on Veracruz had persisted for some time but he maintained "cordial relations with General [Cándido] Aguilar." Nevertheless, Aguilar's subordinate, General Agustín Millán, who commanded the Constitutionalist troops opposite the Americans' right flank had "shown bad disposition" and seemed "inclined to make trouble." As a precaution, Funston had issued "appropriate orders" such that any fighting that did start would have to be initiated by the Mexicans and finished by the Americans: "At first we would have a hard time, but would finally win." There were no other signs of concern to report except that Jesús Carranza had gone to "Tehuantepec to collect and bring North troops he left there a few weeks ago." His Constitutionalists were probably going north to fight Villa but it was possible that they intended to fight the Americans. Accordingly, Funston had made arrangements to get more information about Jesús Carranza's activities.[109]

On November 16, satisfied that the Constitutionalists would respect the rights of local officials and official business transacted under the military government, and after making arrangements for other refugees, the Americans finally gave the First

---

[107] Ibid.
[108] Ibid.
[109] RG 94, AGOMI Box 7478, November 7, 1914, General Funston to Secretary of War, Washington.

Chief a definitive withdrawal date of November 23, 1914.[110] There would be no training of the Constitutionalist officials who would take over local administration, nor would there be any official handover ceremony. In the interest of neutrality, the Americans simply intended to evacuate and leave the plaza to whichever faction might move in and occupy it.

The planned pullout infuriated Venustiano Carranza, who sent a memo typed in all capitals and dated November 22, 1914, Córdoba. The First Chief charged that "by simply leaving the city without making proper delivery of the port" General Funston's "action would amount to leaving that city [Veracruz] in the hands of enemies of the Constitutionalist government." Carranza maintained that the local police force that presently controlled Veracruz had been organized by the Americans during the occupation and consisted mostly of political refugees from the Huerta government, "such as agents Villavicencio, Bolanos, Velez, etc." Moreover, under these Huertista policemen, the remainder of Huerta's army that had revolted in Puebla on August 21 had been permitted to take up residence in the port under the command of General Higinio Aguilar. Carranza claimed to be "at a loss to know how it is possible that these troops [were as] well provided with arms and machine guns as they really are," insinuating that either the Americans had armed these ex-Federals, or permitted them to be re-armed using the war materiel known to exist inside the city's warehouses. Since many Americans openly favored Villa over Carranza, the First Chief feared that the Americans were literally allowing a new alliance to flower between the Villistas and ex-Huertistas, which was presently occurring in other places of the Republic. To put it bluntly, Carranza wrote:

> The evacuation that in such conditions as Gen. Funston intends to carry out, amounts to delivering the port to the remainder of the Federal Army which was under Gen. Huerta. It may be said, that Gen. Funston in refusing to make a formal delivery of the port to any political party, simply fulfills his duty, but taking in consideration that by abandoning the port they would practically deliver it to the Huerta element, who to-day pretend to call themselves Villa followers, it will be clearly seen that this procedure would be as objectionable as any other.

Carranza juxtaposed the planned course of action for evacuation with possibly handing over the port to General Cándido Aguilar, who was the only one with the title of civil governor and military commandant of Veracruz, and the only one with de facto control of the state. Handing over control of the port to him would "avoid further troubles,"

---

[110] Janssens, *The Military Agricultural Complex*, 1:47.

which if escalated might be seen as a pretext for the Americans to remain in Veracruz.[111]

Carranza recommended that Funston have a military changeover "in the same manner as he has made delivery of the Civil Departments."[112] However, the civilian transition did not work out so well. The Americans used the customs duties and other taxes to pay the civilian employees through the end of the pay period, with the understanding that they would work with whatever government might take over. But on November 23, as the Americans loaded into their boats the Mexican officials fled their posts in terror over possible reprisals from the Constitutionalists, such that the port authority complained that the service had become "practically abandoned" before the Americans had completed the pullout.[113]

At a minimum, Carranza requested that Funston "ascertain for himself that there were no armed groups or buildings in readiness to resist Gen. Cándido Aguilar's army,"[114] yet the Americans did not even provide that minor courtesy. And now it becomes clear why the Carrancistas were so upset about the U.S. troops simply abandoning the city—there were Huertistas inside the city, and tons of arms and ammunition contained inside warehouses and the customs house with no one to watch over them until the Constitutionalists could arrive to establish order. This was not merely a perceived slight in protocol.

Fortunately, cooler heads prevailed. Although the Americans had given refuge to Huertistas and Porfiristas, they had not armed them. As the Americans pulled in their outposts and filtered through the city to the docks to board launches and return to their ships, the Constitutionalists marched in on their heels. There was no loss of life, and the people of Veracruz, four times heroic, greeted Carranza's soldiers with joy.

The documents—both unclassified from the Military Intervention file and declassified from the Military Intelligence Division—related to the last three months of the American occupation of Veracruz demonstrate, definitively and categorically, that the Americans were not working with the Constitutionalists inside Veracruz, in direct contradiction to the thesis postulated by John Mason Hart. Moreover, not only did the Americans not orchestrate a massive sealift to arm the Constitutionalists against Zapata and Villa, as Professor Hart theorized, but in fact the Constitutionalists feared the opposite: that the Americans had armed a large contingent of Huertistas

---

[111] RG 165, 5761-1091/19, November 22, 1914, Evacuation of Veracruz, from Carranza in Córdoba, upset over transfer of power, or lack thereof.
[112] Ibid.
[113] From Administrator of Customs to Military Governor, Civil Affairs, RG 141, MGV Entry 12, File 236, November 23, 1914, regarding of abandonment of the service. See memorandum dated November 17 in NARA-MGV, Entry 12, File 277, for arrangements to hand over civil government.
[114] RG 165, 5761-1091/19, November 22, 1914.

and left Veracruz in their hands.[115] In fact, all logic and documentation point to the official story being the factual one: the Americans simply abandoned Veracruz to whichever faction could take it, which just so happened to be the Constitutionalists.

* * *

In the aftermath of the American intervention in Veracruz, U.S. Army intelligence generated from the border and Veracruz focused on railways and roads, the distribution of Mexican forces, the importation of arms and ammunition into Mexico, and the possible uniting of the various factions against the Americans. By May, it had become clear that the Zapatistas, Huertistas, and Constitutionalists would not be making common cause, and instead there might even be a schism among the latter group.

Without a military attaché in Mexico City, a greater reliance had to be placed on intelligence developed by U.S. Army units based on what they could observe from the periphery, in the battles for border towns and port cities, especially with regard to Pablo González's Constitutionalist Army Corps of the Northeast. The most influential source of intelligence, however, would come from American agents who observed Pancho Villa's Division of the North in the field. The might of Villa's division combined with the eminent accessibility that Villa afforded to Americans meant that the U.S. Army compiled a detailed knowledge of that Constitutionalist's division. Grounded in these sources, the Mitchell report skewed toward Villa's cavalry, and the Federal Army's infantry and artillery, but it missed the most formidable ground force, Obregón's Army Corps of the Northwest and its Yaqui infantry. This myopic view of Villa's army as the greatest ground force in the Mexican Revolution was the greatest intelligence failure of the revolution which, luckily for the U.S. Army, it never had to discover empirically since it withdrew from Mexico in November 1914 without expanding the military effort.

The evacuation of Veracruz demonstrated the United States government's intention to remain neutral without choosing sides. After the fall of the Huerta government the U.S. Army gave protection to Huertistas and Porfiristas in Veracruz and then abandoned the port city without an official handover, both to the disgust of Venustiano Carranza, the constitutionally elected governor of Coahuila. Yet the U.S. government respected Carranza's request to return ex-Federal prisoners to Mexico through Piedras Negras, Coahuila, over the wishes of Pancho Villa, who rebelled

---

[115] See Hart, *Revolutionary Mexico*, 290-4, 300-1 for this wrongheaded thesis, based on the misreading of documents and incomplete research. For a thorough debunking of Hart's conspiracy theories based on numerous additional sources and methods, see Janssens, *The Military-Agricultural Complex*, 1: 54-61; 2: 135-38.

against Carranza in the midst of the repatriation process. The only consistent bias that the Wilson administration ever exhibited was against the personage of Victoriano Huerta, whom the revolutionaries finally succeeded in forcing from office.

Burnside had been correct that the overthrow of the Huerta administration would not bring peace to Mexico, but he was wrong in assuming that the Americans would be in Mexico for generations: his vision of an expanded military incursion never did come to fruition. Instead, the Americans pulled out and watched as Mexico devolved into an internecine civil war, as per Lind's policy proposal. In order for Lind's prescription to play out the Americans would have to be neutral and allow all sides to fight it out on a level playing field. That required the lifting of the arms embargo, which happened in September, and for the Americans to leave Mexico, which they did in November.

Accordingly, Professor John Mason Hart's claim that the Americans provided the Constitutionalists with a massive sealift of arms and ammunition to use against the Zapatistas and Villistas would have impaired President Wilson's aim of Mexican self-determination. Moreover, as the small-scale "doves of peace" operation—and even the proposed recall of Burnside to Washington that the U.S. Secretary of War feared might be discovered—exemplified, it would have been virtually impossible to keep a massive buildup of arms from the attentions of locals, spies, and the ever-present U.S. press corps in Veracruz. As Burnside had once remarked, "Everything that takes place will eventually be known as there is no such a thing as a secret in Mexico."[116] Finally, as if circumstantial evidence were not enough, we have recourse to copious unclassified and declassified material in the archives to disprove Hart's wrong-headed conspiracy theory. Americans were not conniving with Constitutionalists in Veracruz in August, or even November, but rather feared that they would have to fight them. Meanwhile, the Constitutionalists believed that the Americans had armed the Huertistas-turned-Villistas and left Veracruz in their possession, which turned out not to be the case.

---

[116] RG 165, 5761-861, October 1, 1913, Burnside No. 962: Summary of Military Events from Sept 25 to Oct 1, 1913.

10. Civil War

One month before the Americans left Veracruz, on October 23, Captain Burnside was relieved "from further operation of instructions detailing" him "to obtain military information from abroad, and from duty as Military Attaché at the American Embassy, Mexico City, Mexico."[1] It would be 1917 before the United States would have another military attaché to Mexico. Until that time, limited military intelligence would come from Captain Walter F. Martin, the military attaché to Guatemala. Most of Martin's information was based on his State Department connections and related to Mexico's southern border, across which arms and ammunition flowed northward, and the Huertistas later mustered out at Salina Cruz.[2]

---

[1] RG 165, 6931-71, September 22, 1914, Memorandum for the Chief of Staff, from Brigadier M. M. Macomb, Burnside's future as military attaché was decided on September 22, 1914. Since the captain decamped from Mexico City on April 23 after receiving a passport from the Huerta government, he was directed to proceed straightaway to return to Mexico City to collect his "personal baggage" and "for the purpose of closing up the affairs of the office of military attaché." Thereafter he was to proceed to Washington to report to the Chief of Staff for a period of fifteen days. His detail as military attaché would conclude on November 13, 1914, and the next military attaché to Mexico would be appointed "when diplomatic relations are again established with...that country." The date of November 13 was later cut short, see RG 165, 6931-72, October 23, 1914. See other correspondence related to Burnside's service: RG 165, 6931-69, July 17, 1914, from Burnside to the Adjutant General through the Chief, War College Division Staff; RG 165, 6931-70, August 1, 1914, from the Adjutant General of the Army to Burnside, in Veracruz.

[2] RG 165, 5761-915, February 5, 1914, Burnside No. 1036: Summary of Military Events from Jan 29 to Feb 5, 1914, concerning Federal Army efforts to stop the flow of arms and ammunition coming up from Guatemala. RG 165, 5761-996, October 17, 1914, Federal Evacuation from Mazatlán as forwarded by Captain Walther F. Martin, military attaché in Guatemala.

November-December

In November, while the Americans were busy preparing for and executing the evacuation of Veracruz, the Constitutionalists were doing the same in Mexico City. The forces of the Convention—Villistas, Zapatistas, and followers of Governor José María Maytorena, among others—had begun pressing southward into the Bajío and northward and eastward from the South toward Mexico City and Puebla, throwing the Constitutionalists back on their heels and sending them fleeing toward Veracruz.

The Conventionist forces occupied Mexico City on November 24, and forced Pablo González out of Pachuca five days later, on November 29, scattering the Constitutionalists to the periphery of the republic. It was a low point for the Constitutionalists, and utterly free of scrutiny from the agents of the U.S. Army, who were absent from the scene. By mid-December the Conventionists were on the offensive, with the Zapatistas taking Puebla City from the cowardly socialist, Constitutionalist General Alvarado Salvador, on December 14, General Villa capturing Guadalajara in Jalisco State, and General Ángeles driving from Torreón against Saltillo, Coahuila.[3]

In a special report dated December 16, 1914, Colonel Henry Clay Hodges Jr., 17th Infantry, informed the Commanding General, Southern Department, that an American who was in Mexico City when General Villa made his entrance reported on the professionalism, modern equipment, and abundant supplies of the Division of the North. This unnamed American also said that the people in Mexico City were receptive to Villa and that General Ángeles was busy recruiting ex-Federal artillery officers. According to articles from the previous week's editions of *El Demócrata*, the current distribution of Constitutionalist forces was as follows: General Obregón had his Division of the Northwest in Puebla with 10,000 men; General Cándido Aguilar's Division of the East was in Veracruz State with 12,000 men; and General Pablo González's Division of the Northeast was in the State of Chihuahua with 20,000 men.[4] González had that many in Coahuila, Nuevo León, Tamaulipas, and San Luis Potosí, combined, but none in Chihuahua.

In Coahuila, Lieutenant Colonel José [Fortunato] Zuazua was reportedly recruiting a battalion of sharpshooters (Zapadores) ["*Zapadores*" are actually "Sappers," combat engineers, not "sharpshooters"] in Piedras Negras. The American Consul in Piedras Negras reported that Lieutenant Colonel Sebastian Carranza, had left Piedras Negras with some of his troops for Monclova, where the Constitutionalists were supposed to

---

[3] Janssens, *The Military-Agricultural Complex*, 1: 37-57, 76-81, 88-92.
[4] RG 165, 5761-999, December 16, 1914, Report on Villa from Camp Eagle Pass based on word from "an intelligent American who recently arrived from the City of Mexico" and reports from *El Demócrata*.

be concentrating from points south along the Piedras Negras–Coahuila Railroad to oppose a body of Villistas coming from Ojinaga.[5] Ojinaga was not an invasion point for northern Coahuila; this was poor intelligence probably created by someone who took a quick glance at the map without knowledge of the terrain.

January

In the first week of January, General Ángeles smashed General Antonio I. Villarreal's Constitutionalist Division of the Bravo at Ramos Arizpe, achieving a great victory for the Conventionist Army. But in the South, Constitutionalist General Álvaro Obregón recaptured Puebla City, and in Mexico City the President of the Convention, General Eulalio Gutiérrez, rebelled against his own army, effectively splitting the Convention in two, North and South. In the ensuing mayhem, caused on the one hand by Gutiérrez's defection, and on the other by the U.S. Army's demand that Pancho Villa come north to negotiate an end to the Maytorenista attacks on Naco, Sonora, the Constitutionalists under Generals Manuel M. Diéguez and Francisco Murguía recaptured Guadalajara. By the end of the month, Obregón had returned to Mexico City.[6]

Villa learned about the defection of Gutiérrez and the loss of Mexico City on his way back from the border after negotiating the Naco Accord. He stopped in Aguascalientes to attend to administrative duties at the beginning of February and then went back into Jalisco State. Once again Villa forced the Constitutionalists out of Guadalajara and then thrashed the combined forces of Generals Diéguez and Murguía at Cuesta Sayula. Villa intended to keep driving toward the coast and destroy or capture the Constitutionalist Division of the West, or at the very least scatter its soldiers into the hills of Guerrero State and recover all its rolling stock. Instead, General Ángeles urged him to return to the Northeast and, in a change of strategy, conquer the entirety of the North, as the Constitutionalists had done in 1914, and then methodically drive southward and retake Mexico City.[7]

---

[5] Ibid.
[6] Janssens, *The Military-Agricultural Complex*, 1: 104–160.
[7] Ibid., 1: 178–192.

March

Reluctantly, Villa accepted Ángeles' proposal and returned to Monterrey, throwing back the Constitutionalists who had been besieging the capital of the State of Nuevo León, and then initiated a multi-pronged offensive to drive the Constitutionalists out of the Northeast. The Villistas completed the conquest of the State of Coahuila and then launched offensives against Nuevo Laredo, Matamoros, and Tampico, the first two on the U.S. border and the third to the south. The Villistas never made it to Nuevo Laredo or Tampico, but under the leadership of General José Rodríguez they attacked Matamoros on March 27. The assault failed miserably, and the Villistas went into siege mode.[8]

At that point, the gunboat diplomacy warrior, General Frederick Funston, showed up on the scene, taking charge of the American reaction to the assault on Matamoros and the Villista threat against Nuevo Laredo, making observations, and managing the military situation from his post at San Antonio, Texas.[9] The Constitutionalist consul, José Z. Garza, invited Funston to meet the commander of the Constitutionalists defending Matamoros, General Emiliano P. Nafarrate, on the International bridge, but General Funston declined saying that "he had no instructions authorizing such a meeting, and the matter was dropped."[10] This seems curious, because later he would go to the international line to meet Pancho Villa firsthand and without reservation.

With this renewed activity in the Northeast, on March 25, 1915, Captain Henry H. Robert of the Corps of Engineers released a report based on information provided by Mr. A. J. Ruckman, Manager, Mexican Coal and Coke Company. Throughout 1915, Captain Robert became a clearinghouse for intelligence, most likely because so much of the information pertained to infrastructure, such as rivers and bridges, roads, coal mines, railroads, and telegraphs, details that would interest an engineer. Ruckman informed Robert that the Villistas were "threatening to confiscate the mines of the Mexican Coal & Coke Company at Las Esperanzas and the Coahuila Coal Company at Palau," the only ones that were in a condition to operate after General Luis Gutiérrez's Constitutionalists had dynamited others in order to deny them to the invading Villista forces under General Rosalío Hernández. Ruckman refused to concede ownership to Villa's manager of railroads, Mr. Zamora, and instead was negotiating for the delivery of coal at prices favorable to the Villistas that would most likely result in a loss for his

---

[8] Ibid., 1: 192-201.
[9] For news accounts of Funston, see "Brownsville Faces Shells," *The Galveston Daily News*, March 29, 1915, 2; "Mexican Aeroplane Hopelessly Divided," *The Galveston Daily News*, April 2, 1915, 1; "Carrancistas Claim Victory at Celaya," *The Galveston Daily News*, April 9, 1915, 2; "Armies Mark Time in Matamoros Lines," *The Galveston Daily News*, April 15, 1915, 1.
[10] "Villaistas delay Matamoros Attack," *The Galveston Daily News*, April 1, 1915, 1.

company. Still, he deemed this preferable to handing over operations to the Villistas, who being ignorant of how to operate a mine would damage it. A Mr. Walter K. Adams, former bridge engineer, National Railways of Mexico, was present during the debriefing of Mr. Ruckman and offered comments and clarifications to Ruckman's testimony.[11] American businessmen, and especially engineers, who presently or formerly operated inside Mexico or worked with Mexican enterprises became key sources of intelligence for the U.S. Army.

April

At the beginning of April, Villa received news that General Obregón had penetrated the Bajío, a conservative region eminently Conventionist in sympathies. Aside from the obvious political implications, vis-à-vis the United States government, that the loss of the heartland would entail, the abandonment of Villistas to their own devices compelled Pancho Villa to return to the Center. But his forces were worn out from the trip back south, and in confronting Obregón he had to rely overwhelmingly on regional forces that were not on par with his Norteños. Accordingly, he lost signature battles at Celaya that, combined with the defeat of the Villistas trying to take Nuevo Laredo and Matamoros, turned the tide in the Northeast.[12]

May

By the last week of May, the Constitutionalists had reconquered all the capitals of the Northeast—Saltillo, Monterrey, and Ciudad Victoria—and the months-long Battle of El Ébano was coming to a close. General Jacinto B. Treviño's Constitutionalist Army Corps of the Northeast had successfully defended its possession of the strategic petroleum facilities in and around Tampico against the Villista campaign to capture them.[13] This change in the military situation produced a round of military intelligence, mostly related to logistical conditions in Coahuila.[14] Doctor W. E. Quinn, a former

---

[11] RG 165, 8529-12, March 25, 1915, Henry H. Robert to Chief, War College Division, General Staff.
[12] Janssens, *The Military-Agricultural Complex*, 1: 248-259.
[13] Ibid., 2: 7-27.
[14] RG 165, 8529-13, May 3, 1915, John W. Wright to Chief, War College Division, General Staff. RG 165, 8529-15, May 14, 1915, John W. Wright to Chief, War College Division, General Staff.

resident of Monclova, reported that the Constitutionalists were concentrated in Las Vacas in order to begin a campaign to roll up the various Villista commands along the railroad from Piedras Negras to Monterrey, beginning with Allende. Captain Robert stressed that this information was not definitive and needed to be taken in context with that from other sources.[15] In fact, Colonels Fernando Peraldí and Ramón Múzquiz commanded the Constitutionalist forces in question, but they were not in any position to challenge the Villistas for control of the rails in Coahuila.[16]

June-July

Going into June, the forces of Villa and Obregón had been faced off for weeks in combat around Trinidad Station, Guanajuato. In the midst of this pivotal battle, President Wilson made a public announcement to all the revolutionaries on June 2, 1915, "serving notice that unless they themselves compose the situation, some other means will be found by the United States to bring about the establishment of a stable government in the republic." The Americans had grown concerned as it was becoming known "that the food situation was serious, that crops had failed, and industries were exhausted."[17]

Around this time, Brigadier General C. A. Devol, U.S. Army and general manager of the Red Cross Society, directed the efforts by the American Red Cross to move food and other relief into Mexico. The Constitutionalist consul in Eagle Pass, a Sr. Sequin, was very much opposed to any relief aid being supplied by the Red Cross, because any such aid from the Americans usually prepared the way for an armed intervention, Cuba being the best example. General Devol countered these suspicions by saying that "our efforts are prompted solely by motives of good will and humanity." Yet if the Red Cross did import food, then it would also have to send armed escorts to make sure that the food was not confiscated by soldiers and sold back into the United States in exchange for arms and ammunition. Also, there was plenty of beans and corn inside the country and Mexico still continued to export food, all of which demonstrated that only those

---

[15] RG 165, 8529-14, May 19, 1915, Henry H. Robert to Chief, War College Division, General Staff. Later, for unknown reasons, Dr. Quinn provided Captain Wright with a map of Monclova and confirmed infrastructure locations such as the street car barn, electric light plant, continuation of the Tramvia, a wagon road, an irrigation ditch, the railroad station, and La Bartola Hill, "an important feature...about one hundred and forty feet above the wagon road running at its base" that had "always been fortified," functioned as "the citadel of the town," and therefore would "be intimately connected with any defense or attack of the town"; RG 165, 8529-21, October 4, 1915, Report to Captain H. H. Robert from John J. Hainsworth.
[16] Janssens, *The Military-Agricultural Complex*, 1: 289.
[17] "President Adopts New Mexican Policy," *The Galveston Daily News*, June 2, 1915, 1.

interested in precipitating an armed intervention were spreading the rumors of starvation and the need for relief. Captain Robert thought that this cynicism about foreign aid was hogwash, that people truly were starving inside the country, and that Mexico continued to export food so that its armies could take advantage of higher prices on the U.S. side.[18]

General Villa finally did allow limited amounts of relief to enter the country, but in July, General Devol closed the Red Cross offices in San Antonio, disgusted by the task of "forcing food down Mexico's throat," and left for Washington to report to the Red Cross and, significantly, to the U.S. Army Chief of Staff, not the State Department as he had earlier threatened.[19]

General Devol wrote in the subject line of his memorandum "Report upon conditions in Mexico as they exist today for information of the Secretary of War," which included a summary of the efforts made to export food to Mexico. During his activities on the border, Devol met with many Mexican commanding generals and he made it perfectly clear that he "represented entirely the Red Cross" and that his "visit had absolutely no military significance." And yet, included in Devol's now-declassified report was information on: the railroads; food supplies throughout the country, with the most acute shortages occurring in those "sections of the country that are constantly being occupied and reoccupied by contending forces," most especially in Mexico City; estimates of food exports from Mexico to the United States and Cuba, with rough amounts of value in export taxes to the Villista military machine; military control of the food chain; and the estimate of a fifty percent decrease in plantings that augured continued misery into the future, with Americans incapable of making much of an impact on the situation.[20] Constitutionalist General Pablo González, who recaptured Mexico City in July, had been one of those who often resisted efforts by the American Red Cross in his country, considering them a prelude to armed intervention.[21] Had he been privy to Devol's report he would have been more convinced than ever.

---

[18] RG 165, 8529-17, June 10, 1915, Henry H. Robert to Chief, War College Division, General Staff. "Villa Sends Reply by Trusted Advisors," *The Galveston Daily News*, June 10, 1915, 1. "Fall of Mexico City Is Said To Be Near," *The Galveston Daily News*, June 14, 1915, 1 (quote).

[19] "Villa Orders Lift Embargo on Relief," *The Galveston Daily News*, June 15, 1915, 1; "Villa–Zapata Agree on National Leader," *The Galveston Daily News*, June 16, 1915, 2; "Carranza's Claim Accepted as True," *The Galveston Daily News*, June 25, 1915, 2; "Villa Plans Campaign in Two States," *The San Antonio Light*, July 17, 1915, 2 (quote).

[20] RG 165, 5761-1010, July 21, 1915, Report of Brigadier C. A. Devol, Red Cross, regarding posting to border and report of Mexico.

[21] Pablo González, *Informe que el General de División Pablo González Rinde al C. Venustiano Carranza, Primer Jefe del Ejército Constitucionalista y Encargado del Poder Ejecutivo de la Nación sobre su gestión en la parte administrativa, como General en Jefe del Cuerpo de Ejército de Oriente y con motivo de la recuperación y ocupación de la Ciudad de México y poblaciones cercanas* (México: Imprenta J. Chávez y Hno., 1915), 5-6: "The elevated class, exploiter of its own [people] through a tradition of centuries and enemy in almost its totality of the revolution, had taken advantage...to make it appear to

Devol's mission coincided with the beginning of a new battle in the Northeast that rivaled the Battle of El Ébano in duration, if not scale, although the roles were reversed. In the Battle of Icamole, which ran from May 23 to September 4, 1915, General Jacinto Treviño's Army Corps of the Northeast was on the offensive, trying to invade Coahuila to take possession of its coal fields in the north and cotton bounty in the southern part of the state from the Villistas commanded by Generals Rosalío Hernández and Raúl Madero.[22] On July 5, General Treviño executed an assault on the Conventionist forces at Icamole in combination with a strategic invasion of Coahuila to the north carried out by General Fortunato Zuazua that provided the U.S. Army with some "idea of Mexican tactics."[23]

In his July 12, 1915, "Report of Barroterán Engagement," Captain John W. Wright, 17th Infantry Intelligence Officer, enclosed a "rough sketch" made by a civilian to show the specifics of an engagement fought on July 8 near the railroad station at Barroterán. About 1,200 Constitutionalists were moving northward and had burned some bridges farther south when they stopped at the heights to the south of Barroterán station. A similar number of Villistas in Barroterán became aware of the presence of the Constitutionalists and decided to attack straightaway. Sixty Villista cavalry deployed in a line of foragers advanced in the center. At a point about six hundred yards from the Constitutionalists they opened fire with their rifles and carbines as they advanced at a trot. At the same time, another group of about 250 Villistas, also deployed in a line of foragers with a five-foot interval, attempted an enveloping movement around the Constitutionalist left. The Villistas held no reserves and were only supported by about two hundred infantry left behind at the station to protect the trains. The frontal attack force had just made it to about halfway up the high ground when it halted and fled in disorder. The flanking force reached its objective on the heights and fought for maybe five minutes before likewise dispersing to the northwest, possibly because they had run out of ammunition. The Constitutionalists, who were almost all mounted and supported by two machine guns, then advanced on the town of Barroterán, occupying a part of it. Then, more Villistas arrived from Piedras Negras by rail and drove the Constitutionalists out.[24]

The key takeaway for the intelligence officer in this case was that the Villistas attacked on a single line of foragers against the center in combination with a flanking movement without support or reserves. They moved forward at a trot and then opened

---

foreigners that the situation of the country—not that of Mexico City, dominated by Zapatismo and Villismo—made us similar to the inferior people over whom international charity intervenes as a preliminary means to moral and armed intervention by the stronger nations."
[22] Janssens, *The Military-Agricultural Complex*, 2: 188-197, 206-214.
[23] RG 165, 8529-18, July 12, 1915, John W. Wright to Henry H. Robert, Engineer Officer, Southern Department.
[24] Ibid.

fire about eight hundred yards out, continuing to advance and shoot.[25] Although Mexican tactical doctrine required infantry to maintain a minimum 1.5 meter interval (essentially the five feet observed) and called for a second and third echelon (called the "support" and "reserve," respectively), cavalry regulations had no such prescriptions, especially when deployed in the open order of foragers.

Treviño's July offensive did not succeed because his men failed to capture Icamole, and thereafter the contending forces entrenched and, with the exception of rare assaults and maneuvers, waited for the revolution's combat center of gravity to continue moving north.[26] By this time, General Obregón had definitively defeated the forces of Pancho Villa, first at Trinidad Station, and then at Aguascalientes, while General Pablo González had pushed the Zapatistas out of Mexico City. Thereafter, the Villistas were in full retreat back toward their base in Chihuahua and La Laguna, closer to the border. There is virtually no information in the U.S. Army's declassified Mexico file about what became known as the Battles of the Bajío, and it seems likely that the War College Division kept abreast of developments just like every other American, by reading the newspapers.

August

General Villa spent most of the month of August attending to matters of administration in Chihuahua. The last of his forces arrived at Torreón from the Bajío on August 18, and General Obregón followed after them, repositioning his Constitutionalists to be able to carry out the final stages of the First Chief's strategy.[27] The main effort would come from the east, with Obregón attacking Saltillo from the south in tandem with a push by Treviño at Icamole. Meanwhile, farther to the west, General Francisco Murguía's cavalry division drove north from Zacatecas toward Torreón, with the ultimate objective being the conquest of Durango State, where he hoped to link up with the Constitutionalists commanded by the Arrieta brothers in the western part of the state. At the same time, General Diéguez commanded the Constitutionalist effort to drive up the West Coast, conquering Tepic and Sinaloa and, ultimately, Maytorena's Sonora.[28]

---

[25] Ibid.
[26] Janssens, *The Military-Agricultural Complex*, 2: 196-97.
[27] Ibid., 2: 95-96, 184, 462
[28] Ibid., 2: 194-96, 202-209.

All these advances came off without a hitch, with Obregón attacking Saltillo in cooperation with the forces of Constitutionalist General Luis Gutiérrez on September 4 while General Treviño took Icamole and then Paredón.

With Villa and Zapata once more reduced to mere regional status, the possibility of granting recognition to the Constitutionalists as the de facto government in Mexico was gaining momentum. It was probably within this context that the U.S. Army's southern command drafted a spreadsheet showing all the ammunition that the Villistas and Constitutionalists had imported into the country from March to September. According to that document, titled "Report of Exportations of Arms, Ammunition and Munitions of War through Border Ports into Mexico, March 15, 1915, to September 4, 1915," 20,057,237 Cartridges had passed through El Paso to Villa's men in Ciudad Juárez with another 1,255,150 cartridges crossing the border in Coahuila for the forces of Villista General Rosalío Hernández. Villa's ally, Governor Maytorena, had received 1,909,200 cartridges through Nogales while his foe, General Plutarco Elías Calles in Agua Prieta had imported only 1,048,000 rounds. In Tamaulipas, the Constitutionalist Army Corps of the Northeast had crossed 5,644,822 cartridges over the border during the time frame in question.[29]

Villa's Sonora Campaign

Villa knew that his diplomatic situation was precarious and that he had to make a bold move to reverse his fading fortunes. Instead of making a stand at Torreón, as he had originally intended, Villa adopted the plan to invade Sonora and Sinaloa, link up with Governor Maytorena's forces, and then march down the West Coast with the ultimate goal of capturing Mexico City.[30] From El Paso, the intelligence officer of the U.S. 6th Infantry, Major F. A. Wilcox, was in a prime position to observe the movement of Villa's forces as they passed through Ciudad Juárez on the way to the staging area at Casas Grandes. The major gave information on the condition, morale, and numbers of troops by combat arm, their weaponry, equipment, aircraft, and communications and confessed that while the exact amount of ammunition was not known, it appeared to be "moderate. During the past six months *much ammunition is reported to have passed over the border* [emphasis added] but here such ammunition has been mostly shipped on further south."[31] These comments about ammunition, as well as the report

---

[29] Ibid., 1: 348.
[30] Ibid., 2: 262.
[31] RG 165, 8532-4, September 17, 1915, Information on Mexican Northwestern and Central Railroads and Villista troops in Ciudad Juárez by Major F. A. Wilcox, 6th Infantry, Intelligence Officer to

mentioned in the previous paragraph, directly contradict Villista lies about an alleged U.S. arms embargo selectively applied against Villa's forces.

Normally, the movement of such large numbers of Villa's men and materiel toward the border portended battles that might endanger American interests on the U.S. side of the border, but Major Wilcox wrote on September 17 that:

> Just at present there are no operations in progress in this vicinity. From authentic and confidential sources, it appears Villa, with his remaining forces, is planning to withdraw from his present field of operations. He is trying to arrange for the Mexican Central rail-road to bring him north to Juárez, thence taking the Mexico Northwestern R.R. to the south-west with an apparent view of getting into Sonora Province, where he appears to have his greatest hold and where he will be in more friendly territory. The indications are that should Villa not make this move before October 1st, conditions in Juárez may have changed so that he would find it very difficult to pass through Juárez.[32]

If Obregón's memoirs are to be believed, he did not confirm Villa's intentions to attack Sonora until around September 23, which would mean that the Americans had better intelligence about Villa's plans than the commanding general of the Constitutionalist Army of Operations.[33]

Major Wilcox also provided information on the Northwestern Railroad from Ciudad Juárez to Chihuahua City and the Mexican Central Railroad from Ciudad Juárez to Torreón, the condition of the Rio Grande River, its river bottom, fords, and bridges as well as the water supply of El Paso, Texas, and Ciudad Juárez, Chihuahua, and general facts regarding the construction and layout of the two cities.[34]

As Villa executed his Sonora campaign, two veterans of the age of high imperialism sprang into action: General Frederick Funston and Captain MacKinlay, who had left the army and was now employed by the pro-Carrancista *San Antonio Light*. In his

Department Engineer, Southern Department, Fort Sam Houston, Texas (quote). RG 165, 8532-5, September 25, 1915, Information on Mexican troops in or about Ciudad Juárez by Major F. A. Wilcox, 6th Infantry, Intelligence Officer to Department Engineer, Southern Department, Fort Sam Houston, Texas.

[32] RG 165, 8532-4, September 17, 1915.

[33] Álvaro Obregón, *Ocho Mil Kilómetros en Campaña*, (México: Fondo de Cultura Económica, 1959), 442-43.

[34] RG 165, 8532-3, September 4, 1915, Information on Rio Grande River at Ciudad Juárez and El Paso by Major F. A. Wilcox, 6th Infantry, Intelligence Officer to Department Engineer, Southern Department, Fort Sam Houston, Texas. RG 165, 8532-4, September 17, 1915. RG 165, 8532-6, September 28, 1915, Information on the cities of El Paso and Ciudad Juárez by Major F. A. Wilcox, 6th Infantry, Intelligence Officer for Department Engineer, Southern Department, Fort Sam Houston, Texas.

capacity as a journalist, MacKinlay was now funneling intelligence to the U.S. Army under the *nomme de plume* of W. L. Gibson.[35]

On September 25, 1915, Major General Funston ordered General John J. Pershing to have his intelligence officer, Major F. A. Wilcox, verify "translations of certain extracts from local Mexican papers furnished by" Mr. W. L. Gibson. One key piece of information provided by MacKinlay was that on September 23, the division commanded by General José Rodríguez had passed through Ciudad Juárez headed for Casas Grandes with more than 2,500 men, and as of September 25 fifteen "train loads of troops, arms, ammunition, horses, live-stock, supplies and camp followers" had already passed through Juárez, totaling between eight and ten thousand men bound for Sonora, all well-armed and munitioned. More were on the way from Chihuahua City, which would bring the grand total to nineteen trains by September 29. Per Wilcox, MacKinlay's estimate of 8,000 to 10,000 men seemed about right.[36]

Some inside Mexico suspected that Villa's purpose in sending his troops through Ciudad Juárez on the way to Casas Grandes was to prove to the Americans that his army still had more than 25,000 men. Accordingly, the trains only traveled as far south from the border as Mesa Station on the Northwestern before stopping so that Villa's troops could detrain, march overland to the Central Railroad, and then return to Ciudad Juárez. According to Wilcox, no one "conversant with the facts" believed the story of this ruse as it was "too absurd to be considered," especially since much of the U.S. Army's data in this matter came from confidential sources within the Northwestern Railroad. Additionally, Villa's men were being refitted and resupplied in Ciudad Juárez, and the superior condition of the Central Railroad accounted for the path taken to Casas Grandes.[37] Wilcox's judgment was subsequently confirmed by the fact that even after the disastrous Sonora campaign more than 25,000 Villistas mustered out of service in December.[38]

There did seem to be a certain breakdown in discipline among the Villistas, since according to reputation General Villa always had required abstinence of his men. Yet Wilcox refuted reports that Ciudad Juárez had gone dry upon the arrival of Villa's men saying that "much liquor has been drunk in Juárez since the arrival of the troop trains and drunken men have been much in evidence."[39]

---

[35] For W. L. Gibson's undercover identity see RG 165, 5761-1029, October 2, 1915, and RG 165, 5761-1048, November 8, 1915.
[36] RG 165, 8532-7, September 29, 1915, Translation of articles by W.L. Gibson and verification of facts by Major F.A. Wilcox.
[37] Ibid.
[38] Janssens, *The Military-Agricultural Complex*, 2: 339.
[39] RG 165, 8532-7, September 29, 1915, (quote). See translations of reports from Spanish language papers in Texas about Villa's Sonora campaign and the capture of Torreón at: RG 165, 5761-1023, September 28, 1915, Report to Captain H. H. Robert from W.L. Gibson; RG 165, 5761-1027, September 30, 1915, Report to Captain H. H. Robert from W.L. Gibson; RG 165, 5761-1025,

U.S. Army Intelligence in the Mexican Revolution

Observers believed that Villa's campaign ultimately would end in failure because his men did not know the territory and Constitutionalists were already flooding into Sonora and would wipe out the Maytorenistas before the Villistas could arrive. A Sonora campaign, at any rate, would force the Villistas to abandon Torreón—which they finally completed on September 28—and Chihuahua City. Wilcox accurately commented that Villa needed to quit Chihuahua because after five years of warfare it was depleted while Sonora provided resources, and he hoped to count on the support of the Yaquis, who weren't going anywhere any time soon.[40] Not so with Maytorena.

On October 1, 1915, General Funston telegraphed Washington that according to news from a Colonel Frier in Nogales, Governor Maytorena had quit the revolution and crossed over into the United States bringing his staff and all the funds from the state treasury except for three hundred and fifty odd dollars. Some things never change; this was a rehash of Maytorena's cowardly behavior in February 1913. Maytorena claimed that he was on his way to Washington and the Pan American conference when arrested by U.S. authorities. In contrast, the former state treasurer and now acting governor of Sonora, Carlos Randall, planned to wait and see how badly the loss of the staff had hurt the war effort before deciding whether to join them in exile. General Francisco Leyva, lionized for his resistance to the U.S. Navy's threats to land troops in order to protect American interests in the Yaqui River Valley during the summer of 1915, also went into exile accompanied by several of his officers, which further indicated a "disintegration" of the Conventionist Army. General Francisco Urbalejo took over as the commander of Villista forces in Sonora.[41]

Meanwhile, Villista authorities in Ciudad Juárez confirmed that all the Conventionist forces had evacuated La Laguna, with the last to leave being Generals Juan N. Medina and General Manuel Madinaveytia with 2,000 infantry and cavalry, most likely headed for Chihuahua. As of September 29, Villa was still in Casas Grandes, "directing the sending of his troops toward the state of Sonora," but he would return to Chihuahua City as soon as he had finished giving orders for the Sonora campaign. Other sources said that he would go directly to Sonora and set up his headquarters in Nogales. Before abandoning Chihuahua, Villa allegedly intended to destroy all fifty locomotives and 2,200 passenger and freight cars in his control in order to prevent them from falling into the hands of the Constitutionalists. Previously, Villa had appointed Conventionist General Juan Banderas to the post of Governor and Military Commander of the State of Sinaloa and Banderas had left Chihuahua City on September 26 (an alternative departure date of October 6 has been given by other sources) headed down the Northwestern Railroad with a column supposedly containing five thousand

---

September 30, 1915, Report to Captain H. H. Robert from W.L. Gibson; RG 165, 5761-1039, October 12, 1915, Report to Captain H. H. Robert from W.L. Gibson.
[40] RG 165, 8532-7, September 29, 1915.
[41] RG 94, AGOVR Box 7646, October 1, 1915, Funston to Adjutant General U.S Army.

men. Banderas intended to abandon the rails to the southwest and threaten the rearguard of Constitutionalist General Manuel Diéguez, who was marching north from Mazatlán into Sonora, and cut him off from Mazatlán.[42]

MacKinlay also reported that on October 3, Villa went to Santa Rosalía Camargo to confer with General Rosalío ("Chalío") Hernández, who had "been negotiating with the enemy for some time," and the next day a fight broke out in which many Villistas deserted and several were killed or wounded. After that, Villa decided to leave for the Sonora campaign, with plans to be in Ciudad Juárez on October 7 or 8. MacKinlay followed this news in his memorandum with the location and suspected strength of other key Villista generals. The bulk of Villa forces, consisting of 12,000 to 15,000 men and eighteen field pieces, had been reported at Casas Grandes, where General José Rodríguez was said to be in command. Some sources believed that Juan Banderas had joined Rodríguez there, although Banderas could just as easily have been at La Junta or Ciudad Guerrero. MacKinlay considered the stated total of Banderas's command at 5,000 men to be "very doubtful." Another report from New Mexico put Villa's forces at Carretas, west of Casas Grandes, on the Sonora line. Meanwhile, the only Constitutionalists believed to have penetrated the state of Chihuahua were those of General Luis Herrera, who may have occupied Parral on October 4; a small band of fifty to two hundred men at the Gavilondo ranch on the New Mexico line, near Laings Ranch, New Mexico; and the men of General Rosalío Hernández, who was "probably now a Carrancista."[43] The distribution of forces and dates associated with these key events are probably as accurate as any.

In Sonora, Villista Governor Carlos Randall reported that the Southern Pacific Railroad from Nogales to Guaymas had been put in good order, and that after consultation with General Francisco Urbalejo, he had agreed to send more than one thousand Villista soldiers, four pieces of artillery, and some machine guns to reinforce Guaymas in order to defend the port in the event that Constitutionalist General Manuel M. Diéguez should attempt to land there to continue his advance into Sonora. The gunboat *Guerrero* arrived offshore Guaymas the night of October 11, carrying troops under General Diéguez, who demanded the surrender of the port. His flotilla consisted of five armed transports, including the *Korrigan II*, and the gunboats *Guerrero* and *Pacific*. The commanding officer of the U.S.S. cruiser *Chattanooga* "informed General Diéguez that sufficient notice should be given the towns of Guaymas and Empalme before commencing bombardment to permit non-combatants to reach a place of safety." Diéguez did not think that a bombardment would be needed, but said that if it came to that, he would designate a neutral fire zone for the non-combatants to take

---

[42] RG 165, 5761-1028, October 1, 1915, Report to Captain H. H. Robert from W.L. Gibson.
[43] RG 165, 5761-1033, October 7, 1915, Report to Captain H. H. Robert from W.L. Gibson.

refuge. The next day, General Diéguez reported to Constitutionalist General Plutarco Elías Calles in Agua Prieta that the attack upon Guaymas was "imminent."[44]

Then came the stunning news, on October 19, 1915, that the Wilson administration was granting recognition to the Constitutionalist Army as the de facto government in Mexico. When Villa found out, he flew into a rage and ordered the railroad stations, round house, rail yards, fourteen locomotives, and all the rolling stock in Chihuahua City put to the torch. His men also destroyed rail bridges to the north and south of the state capital in order to impede the advance of the Constitutionalists advancing from Torreón. Passengers soon began arriving at El Paso in coaches, carts, and on horseback after "suffering all kinds of privations and molestation." MacKinlay wrote that Villa had departed from Casas Grandes on October 21, leaving the Casas Grandes District virtually devoid of troops after his men had slaughtered more than two thousand head of cattle belonging to an American cattle company for the trip. Villista General Manuel Madinaveytia remained behind at Ciudad Juárez and once joined by General Eduardo Ocaranza with the troops from Chihuahua City both would proceed to Casas Grandes and then Sonora.[45]

The Constitutionalists immediately took advantage of their new diplomatic recognition and applied to Washington for permission to move men through United States territory to reinforce Agua Prieta, where the Villistas would strike first. In response, Villistas threatened to dynamite the railroad in the vicinity of Columbus, New Mexico, to block any Constitutionalists trying to reach Agua Prieta via U.S. territory.[46] As the locus of fighting moved westward, Major General Funston left for Douglas, Arizona, on the border with Agua Prieta, traveling through Alpine and Marfa, Texas. He actually met with Villa at the international border and reported that Villa was very disgusted at being informed that the Americans had allowed Obregón to move his men through U.S. territory. Once the last of the Constitutionalists had crossed back into Mexico, the U.S. Army pulled its guards that had been stationed along the rails.[47]

Villa's attack on Agua Prieta failed in short order, and U.S. Army officers who witnessed the attempt thought that he merely had "nibbled" at the border town before giving up the fight and deciding to move on to Hermosillo and Magdalena.[48] From Agua Prieta, Villa drove west and south. After failing to capture one single town, including Hermosillo, the capital of Sonora, on November 25, Villa blew up his excess arms and ammunition and returned to Chihuahua.[49]

---

[44] RG 165, 5761-1040, October 12, 1915, Report to Captain H. H. Robert from W.L. Gibson.
[45] RG 165, 5761-1043, October 21, 1915, Report to Captain H. H. Robert from W.L. Gibson.
[46] RG 59, USDS 812.2311/195, October 30, 1915, Report on Villista planned sabotage as reported by State.
[47] "Continuous Fall of Bullets Will Draw Americans' Fire," *The Galveston Daily News*, November 2, 1915, 1. "Intention of Villa Puzzle to Officers," *The Galveston Daily News*, November 4, 1915, 1.
[48] "Intention of Villa Puzzle to Officers," *The Galveston Daily News*, November 4, 1915, 1.
[49] Janssens, *The Military-Agricultural Complex*, 2: 296-97.

On November 26, 1915, MacKinlay provided Captain Robert with a translation of General Álvaro Obregón's report to First Chief Venustiano Carranza released by the Constitutionalists and dated Torreón, November, 17, 1915. But the report had already appeared in English in the November 24 edition of the *San Antonio Light*, and it is probable that MacKinlay had also translated Obregón's report for the newspaper.[50]

Demise of the Convention

Concomitant to Villa's Sonora campaign, what remained of the Convention in Toluca definitively split into its North and South parts, effectively destroying any legitimate claims to national sovereignty by the organization. General Benjamín Argumedo led the Norteños in their trek back north to La Laguna, leaving Toluca on October 10 and hoping to join Conventionist Generals Margarito Salinas, Canuto Reyes, and Rodolfo Fierro operating in the region. According to information gleaned from passengers arriving at El Paso from the south, Villa had sent the three to Zacatecas on or about September 21, indicating that he was not yet disposed to yielding Torreón to the Constitutionalists.[51] In actuality, Fierro was most likely already bound for Casas Grandes, and Salinas and Reyes were from Zacatecas, in effect conducting guerrilla warfare on home territory, although the reports by refugees made it appear that Villa's forces were still operating over a larger expanse of territory.

MacKinlay gave the following account of Argumedo's odyssey based on information from an "absolutely reliable source." After his column reached Tamaulipas, it linked up with the command of General Alberto Carrera Torres. Then, from November 15 to 17, the force of "Conventionist and Carrerista" troops, estimated at from 3,000 to 6,000, attacked Tula, where "only the scarcity of ammunition and the bad condition of the men and horses prevented the defeat of General Manuel [Miguel] Zapata of the Carranza forces, who defended the place." After that failed attempt, Argumedo's Conventionists fled to the northwest, crossed the state line into Nuevo León, and attacked the town of Doctor Arroyo [unconfirmed] but were repulsed. Next, they turned to the southwest and tried to take Matehuala [unconfirmed], and once again failed.[52]

---

[50] RG 165, 5761-1052, November 26, 1915, Translation of Obregón's report to Captain Robert from W.L. Gibson. Note: this report in its entirety appeared in the November 24 edition of the *San Antonio Light*.
[51] RG 165, 5761-1028, October 1, 1915. See Janssens, *The Military-Agricultural Complex*, 2: 325-29, for a history of Argumedo's column.
[52] RG 165, 5761-1054, December 11, 1915, Report from W.L. Gibson and follow-up note dated December 13 regarding Carrera forces.

Finally, the Constitutionalists dealt Benjamín Argumedo a tremendous defeat at the Hacienda de la Gruñidora, situated southwest of Mazapil, Zacatecas (Lat. 24° 30' N, Long 102°). Argumedo lost about eight hundred among killed, wounded, and prisoners, and booty reported to include six machine guns, one hundred saddled horses, and almost all the Conventionists' ammunition and impedimenta. MacKinlay disclosed a list of prisoners captured, with the note that the enlisted would be tried at Parras, Coahuila, by military authorities, and the civilians and officers at Saltillo by a court-martial and sentenced according to their rank and station.[53]

For some reason, MacKinlay thought that it might be worthwhile to reveal the opinion of a cleric in Mexico City, who feared that Argumedo's defeat at Hacienda Gruñidora only meant the breakup of his command into guerrilla-size elements and that, if allowed to reach La Laguna, it would be very difficult for the Constitutionalists to subdue.[54] This amounted to the worst of alarmism. It would appear that those few Conventionists who escaped Gruñidora took a course for the northwest and crossed into Coahuila and occupied Parras before General Luis Gutiérrez and others drove them out. At last report, the main body had dispersed with the remaining "generals" and their small escort and wounded reportedly headed for the United States. A group of men to the north of Aldamas, Nuevo León were said by Ricaut's information office to be from this group.[55]

Meanwhile, in the early days of December, Villa returned to Chihuahua. Constitutionalist General Jacinto B. Treviño directed the invasion of the state from the south, but his advance had been delayed by the condition of the rails. Upset with the pace of progress, Treviño left his infantry behind and rode into Chihuahua State.[56] A key setback to Villa ensued when Conventionist General "Máximo García surrendered to Treviño...at Corralitos, between Rellano and Jiménez," and joined his brigade "to the 'happy family.'" Some ex-Federals still rumored to be with Villa included Eduardo Ocaranza, whom Villa considered to be among the best, José Delgado, whom Villa personally killed, and "our 'friend of Tampico,'" Ignacio Morelos Zaragoza. The final ex-Federal mentioned in MacKinlay's report, General Arnoldo Casso López, had given up fighting; "he was a pretty good fellow in Morelos even if he did get 'good and licked.' I saw the entertainment 'some of it I was and all of it I saw.' [probably lyrics to a song]"[57]

By mid-December, MacKinlay figured that Treviño had about 8,000 to 10,000 men along the line between Escalón and Santa Rosalía, Chihuahua, and that Villa would try

---

[53] RG 165, 5761-1058, December 20, 1915, Memorandum to Captain H. H. Robert from W.L. Gibson.
[54] Ibid.
[55] RG 165, 5761-1054, December 11, 1915.
[56] See Janssens, *The Military-Agricultural Complex*, 2: 330-39, for the Constitutionalist invasion of Chihuahua.
[57] RG 165, 5761-1059, December 20, 1915, letter from Mr. Gibson to Captain Robert relayed by W. K. Adams.

to block his passage with 4,000 to 5,000 men at Bachimba Pass, *a la* Orozco. MacKinlay had been there during the battle of July 3, 1912, as was Captain W. A. Burnside, 26th Infantry, now at Kingsville, Texas.[58] MacKinlay also had news of a column of Constitutionalists "going from Piedras Negras via Eagle Pass to Marfa, thence *overland* to Ojinaga and 'on to Chihuahua.' Another column had or was going in at Palomas across the border from Columbus, New Mexico."[59] None of these movements by the Constitutionalists made sense, or seem to have actually happened, especially the last stand at Bachimba, which definitely did not take place. Instead, Villa retreated to Hacienda Bustillos and disbanded his remaining men, telling them that when the time was right, he would recall them to duty. The civil war of 1915 had concluded.

* * *

In general, the U.S. Army produced little information during Mexico's civil war of 1915, because campaigns ranged over the entire territory of the nation and the fighting mainly took place in the heart of the country where the U.S. Army had virtually no intelligence sources, especially after the recall of Captain Burnside, the military attaché, to America. Once diplomatic ties with Mexico had been broken, there could be no replacement military attaché, of course. Therefore, most "military intelligence" in 1915 continued to come from the State Department and reports in the newspapers that cited as sources: State Department dispatches (especially from George C. Carothers); Naval reports; information from international relief agencies; the President's special commissioner, Duval West; the official Villista and Constitutionalist news agencies and their consuls, representatives, and agents; American aviator "scouts" who had served with the Constitutionalist and Villista army, as well as other U.S. citizens who had crossed the southern border returning from Mexico;[60] and Mexicans who had served the various factions and had fled to the United States. The U.S. Army's military attaché to Guatemala—where Mexico had another international border exploited by revolutionaries to smuggle arms and ammunition—also provided limited intelligence. And yet, there seems to have been less urgency for the development of military intelligence because the possibility of another intervention decreased by an order of magnitude as Governor Lind's policy of managing responses to the Mexican Revolution through the lens of the U.S. Civil War came to fruition in 1915.

---

[58] RG 165, 5761-1057, December 15, 1915, letter from W.L. Gibson.
[59] RG 165, 5761-1059, December 20, 1915.
[60] RG 165, 6269-7, November 6, 1914, Brigadier R. K. Evans, commanding general Second Brigade, to Commanding General, Southern Department, Fort Sam Houston, Texas. The same report can be found at RG 59, USDS 812.24/31, November 6, 1914, Reports on Carranza's aviation assets according to Aviator C. F. Niles, the First Chief's Chief of Aviation.

U.S. Army Intelligence in the Mexican Revolution

The Wilson administration decided to let the Mexicans slug it out until one side had clearly come to control a dominant portion of the national territory before granting recognition to that faction, the same standard that the United States had demanded of foreign powers during its own civil war with the Confederate States of America.[61] The faction that gained that de facto recognition—which meant exclusive rights to import arms and ammunition from the United States and permission to move troops through American territory—came in October 1915. However, the Wilson administration showed signs of possibly breaking with Lind's policy in June 1915, perhaps considering military intervention, once the extent of misery taking place in Mexico became known. In that time frame, military intelligence on economic and military conditions came from the American Red Cross, which was supposed to be a neutral disaster aid organization without political allegiances, and certainly not a source of military intelligence for the United States. General Devol's indiscretion in this regard justifies Constitutionalist General Pablo González's suspicion of the American Red Cross and his disagreeability to foreign aid agencies attending to Mexico City's suffering population.

Finally, General Funston, once again reared his head, this time along the northern border, first at Matamoros, and then at Agua Prieta, to gain firsthand knowledge. The gunboat diplomacy warrior would be dead within two years, Huerta would die in Texas in January 1916, and two months later the U.S. Army would invade Mexico's northern border in pursuit of Pancho Villa in what would become the futile Punitive Expedition.

---

[61] Janssens, *Maneuver and Battle*, 1: xix.

11. The Bandit War

By the second half of 1915, U.S. Army intelligence had become concerned with a plot—called the Plan de San Diego, after the city of its origination in San Diego, Texas—to spread the Mexican Revolution into the southwestern part of the United States. The idea, apparently, was to incite a rebellion within the United States' border states that might revert that territory to Mexican control. Some contemporaries believed that the movement might have been initiated by the contending forces of the Mexican Revolution, with the intention of forcing the United States to choose sides. Constitutionalist General Nafarrate and certain brigades within the Constitutionalist 5th Division were implicated in the plan, causing friction with the American authorities in Brownsville, but this distraction, which demanded the attentions of Constitutionalist Generals Obregón and Treviño exactly when they were driving on Torreón for a presumed showdown with Villa, was related in Obregón's memoirs as an utmost irritant.[1] Ergo, for this, and other reasons culminating in the pronouncement of the Carranza Doctrine in Matamoros, it appears highly unlikely that the Plan de San Diego had any authorization from high up in the Constitutionalist Army's hierarchy.

W. E. W. MacKinlay had his own ideas about the Plan de San Diego. He had been in McAllen, Texas, since August 18, employed by the *San Antonio Light* and working with "our friend," Major W. H. Hay. The *Light* was owned and edited by Mr. Charles S. Diehl, previously a lieutenant colonel of the 1st Illinois Infantry, and Mr. Harrison L. Beach formerly a captain in the 2nd Illinois. The two had worked for the Associated Press for many years in Chicago and both had been in Cuba in '98 where Diehl had been in charge of the AP boat while Beach was ashore with the army.[2] MacKinlay had concluded that the "so called 'bandit problem'...originated in Los Angeles, California, with the Mexican branch of the I.W.W. known as Magonistas." He then proceeded to

---

[1] Janssens, *The Military-Agricultural Complex*, 2: 327.
[2] RG 165, 5761-1048, November 8, 1915, Report to Major Crawford from W. E. W. McKinlay.

give a history of the "'Plan de San Diego, a famous document on the anarchist style, drawn up by a group of Magonistas, at San Diego, Texas." Then came the attack on Matamoros "and all the Mexicans and a lot of Americans on the Texas side did all they could to help [General José] Rodríguez, the Villista commander, take Matamoros and defeat Gen. Emiliano P. Nafarrate, the Carrancista commander there. This left Nafarrate bitter towards the Americans and ready to wink at injuries to them." Thereafter, a former deputy sheriff of Cameron County, Luis de la Rosa, formed a bandit gang on the Mexican side from which to raid Texas with the connivance of Nafarrate. That led Constitutionalist Consul José Z. Garza of Brownsville to recommend to Carranza's Minister of Foreign Affairs that Nafarrate be relieved as the jefe de las armas of Matamoros. Carranza agreed, and on September 22, Nafarrate received a promotion to general of brigade and reassignment to command the 5th (Tamaulipas) Division of the Army Corps of the Northeast. Nafarrate's politically-motivated promotion to the rank of general of brigade just prior to his being relieved infuriated his corps commander, General Jacinto B. Treviño, who appointed Brigadier Eugenio López de Lara, "formerly a milkman at Matamoros and now only 24 years old," to replace Nafarrate.[3]

The truth, most probably, was much simpler. Nafarrate hated Americans for typical reasons unrelated to the Battle of Matamoros, supported the PLM, like many Constitutionalists, and therefore backed the Plan de San Diego and its military campaign—known as the Bandit War—against innocent Americans. The Americans knew this, and demanded that he be replaced.[4]

Another source of intelligence for the U.S. Army in this period, aside from MacKinlay, was John J. Hainsworth, Chief Engineer of the Port Brownsville Sugar Lands Company. Hainsworth's greatest claim to fame was that he happened to live next door to the son of ex-Federal General Rómulo Cuéllar, former commander of the Division of the Center. During the month of June (surely Hainsworth meant July), 1914, the Constitutionalists had put so much pressure on Cuéllar's Federals that he had remained in the mountains and in the saddle for six days during the retreat to Mexico City. In fact, Obregón's cavalry smashed Cuéllar's division on July 31 at the Hacienda de Temascatío, in the last large-scale engagement of Huerta's war. Subsequently,

---

[3] Ibid., (quotes). RG 165, 5761-1032, October 7, 1915, Report to Captain H. H. Robert from W.L. Gibson, Treviño upset over Nafarrate's promotion.
[4] RG 165, 5761-1030, October 5, 1915, Report to Captain H. H. Robert from W.L. Gibson, López de Lara was sent to relieve Emiliano P. Nafarrate, as the Constitutionalist commander of Matamoros because, according to El Presidente (said to be the Clerical journal in south Texas), "the American authorities and public sentiment in general accuse him of being the real author of the disorders that have taken place along the banks of the Rio Grande," referring to the Plan de San Diego.

Cuéllar escaped Mexico with President Carbajal to arrive in Brownsville on September 6, 1914, after having amassed quite a fortune in the State of Tamaulipas.[5]

According to Hainsworth, Cuéllar was of the opinion that neither Carranza nor Villa would be capable of bringing peace to Mexico, and he predicted that the United States would intervene within ninety days—that is, before the end of 1914. While it later proved true that the process of national reconciliation did not begin until the defeat of Villa and death of Carranza, the U.S. Army did not invade Mexico again until 1916, after the ultimate winners of the revolution had been determined. Hainsworth's intelligence proved not only dubious in many cases, but also not very valuable. For example, Hainsworth predicted that Carranza would be forced out and replaced by Ernesto Madero, the former Minister of Finance, who would ascend to the presidency, and he alleged that Carranza had been in open revolt against Francisco Madero before Federal Army officers ousted him from office and then murdered him.[6] While a few Maderistas accused Carranza of having been in rebellion against President Madero in 1912, this story amounted to little more than an exaggeration and misinterpretation of the facts, in what actually had been nothing more than a typical state-versus-federal disagreement over the control of forces raised by the State of Coahuila and paid for by the Federal government.[7]

Hainsworth's intelligence did not improve. In early October, 1915, the engineer reported to intelligence officer Robert that a member of Carranza's secret service had informed him that the bandits along the border mainly operated out of Cerralvo. Hainsworth blamed Spanish mine owners for causing "them to start out," in the hopes of provoking an American intervention. Additionally, Americans around Matamoros thought that without recognition for Carranza soon, it might get "ugly" for Americans in the area.[8] Ergo, Hainsworth located the actors—Spaniards—responsible for the Plan de San Diego banditry and placed the subversive effort within the context of diplomatic recognition and intervention while, again, the true actors clearly were radical Mexicans pushing a nationalistic agenda.

Meanwhile, in his capacity as a representative of the *San Antonio Light*, W. L. Gibson (MacKinlay) interviewed Brownsville's Constitutionalist Consul, Garza, who said that Major Cavazos of General Nafarrate's staff had located numerous bandits at Las Flores Ranch, on the Mexican side of the river, and that General Nafarrate had telegraphed General Jacinto B. Treviño requesting instructions. MacKinlay placed the location of the ranch about one-half mile southeast of Las Cruces Ranch, in the vicinity of Progreso, on the road from Matamoros to Monterrey. Major Cavazos belonged to an

---

[5] RG 165, 5761-1020, August 31, 1914 and September 7, 1914, Report to Brigadier Kingman from John J. Hainsworth, Sugarlands Company. Janssens, *A Revolution in Military Affairs*, 2: 348-49.
[6] RG 165, 5761-1020, August 31, 1914 and September 7, 1914.
[7] Janssens, *Rise of the Praetorians*, 508-510.
[8] RG 165, 5761-1036, October 3, 1915, Report to Captain H. H. Robert from John J. Hainsworth.

established family that owned the Cavazos Ranch, the namesake for the border crossing south of Mission, Texas. MacKinlay could not understand why Nafarrate, who held the rank of general of brigade and title of "Chief of the line of the frontier of the Bravo" (i.e., Rio Grande), needed permission to attack the bandits. He could only reason that many of Nafarrate's own men came from the area and might have had conflicts of interest in such an operation and therefore Nafarrate needed troops from Treviño, who was in Torreón, that had been recruited elsewhere.[9]

A couple of days later, Hainsworth reported that the change of command had begun when about four hundred Constitutionalists arrived at Matamoros from Torreón under the command of General Eugenio López de Lara. Hainsworth estimated that about half of López's men were Huasteca Indians, but several among the new command had worked around San Benito and elsewhere in Texas in the past, meaning that this new command suffered from the same defective loyalties as the Constitutionalists already inside Matamoros, and probably explains why López later disarmed many of his men who had previously lived in Texas.[10] The Carrancista organ in San Antonio, Texas, *La Raza* corrected Hainsworth's information, stating that López brought to Matamoros close to three hundred men, including about one hundred "Tehuantepec [not Huasteca as Hainsworth had stated] Indians, considered to be warriors of a first class quality, especially for brush fighting and broken ground," and therefore had been "sent at once to patrol the river for the purpose of stopping the Mexican-Texan rebels from taking refuge on Mexican soil."[11]

The change of command between López and Nafarrate started on October 4 and finished the next day. After celebrating at a banquet held in his honor, Nafarrate, who had "given the press so much to talk about on account of his attitude in the Texas [Plan de San Diego] conflict," was to take his own brigade—"as with them he can succeed in his task"—to Ciudad Victoria for the campaign against Conventionist General Alberto Carrera Torres. The "Carreristas" currently had their headquarters in Peotillas, San Luis Potosí, and had been defeating small Constitutionalist commands and capturing arms and ammunition. Once in Ciudad Victoria, Governor Luis Caballero, gave additional troops to Nafarrate, who had never directed a campaign but supposedly had been a member of the Rural Police (there is no evidence that Nafarrate was a Rural Police guard during the Porfiriato) assigned to the region during the administration of President Díaz. Therefore, it was presumed that he knew the terrain well.[12] General Miguel Zapata, who successfully defended Tula against the attack by

---

[9] RG 165, 5761-1029, October 2, 1915, Report to Captain H. H. Robert from W.L. Gibson.

[10] RG 165, 5761-1035, October 4, 1915, Report to Captain H. H. Robert from John J. Hainsworth. RG 165, 5761-1046, October 25, 1915, Report to Captain H. H. Robert from John J. Hainsworth, disarming of certain of López's men.

[11] RG 165, 5761-1030, October 5, 1915.

[12] RG 165, 5761-1034, October 7, 1915, Report to Captain H. H. Robert from W.L. Gibson.

the combined forces of General Carrera Torres and Benjamín Argumedo in November, 1915, belonged to Nafarrate's command.

In the end, the murderous General Nafarrate finally got his just desserts. He was assassinated in 1918 in a restaurant, or according to his enemies in a brothel, possibly on orders from Carranza.[13] It must be said, however, that Nafarrate had made many enemies after participating in the Plan de San Diego, needlessly executing the well-connected General Eugenio Aguirre Benavides in June, 1915, and ruthlessly gunning down the young volunteers who had surrendered after defending Matamoros against the Constitutionalist attack in June, 1913.[14] The list of those who wished Nafarrate dead was long indeed.

After removing Nafarrate, "Carranza sent out orders for the local garrisons to get busy after 'bandits'" and "the love feast" ensued, with various Constitutionalist officers crossing over to confer with their U.S. Army counterparts on joint patrols.[15]

On October 24, Hainsworth had breakfast with the chief of General López's secret service, who reported that "La Rosa" had been in Reynosa in disguise a few days earlier. López had about 800 men between Matamoros and Ciudad Camargo of which 600 were in civilian clothes with orders to kill bandits on sight. Therefore, bandits had started to avoid coming into towns.[16]

Then, in November, MacKinlay accompanied the First Chief's entourage as it traveled from Reynosa to Matamoros, and relayed the information to the Southern Department that he had previously telegraphed to the *Light*, such as the trains, equipment, escorts, and staff accompanying Carranza. Surprisingly, MacKinlay had nothing to say of the Carranza Doctrine, pronounced by the First Chief in Matamoros on November 28, 1915, which makes more sense in context with the Plan de San Diego. The Carranza Doctrine protested the meddling of any nation in the affairs of another, and by going to Matamoros—ground zero for the Bandit War—Carranza essentially disavowed any rumors of complicity by the Constitutionalist high command in the Plan de San Diego. Conversely, Carranza subtly affirmed that the Constitutionalist government was not a puppet, client, or Cat's Paw of the United States. In this same vein, MacKinlay's note that "there are no Americans with Carranza as officers to my

---

[13] Ignacio Muñoz, *Verdad y Mito de la Revolución Mexicana (relatada por un protagonista).* 4 vols. (México: Ediciones Populares, S.A., 1962), 3: 12-14.

[14] See Janssens, *A Revolution in Military Affairs*, 1: 149; and Janssens, *The Military-Agricultural Complex*, 2: 188-89.

[15] RG 165, 5761-1048, November 8, 1915, Report to Major Crawford from W. E. W. McKinlay.

[16] RG 165, 5761-1045, October 24, 1915, Report to Captain H. H. Robert from John J. Hainsworth. RG 165, 5761-1048, November 8, 1915, Report to Major Crawford from W. E. W. McKinlay, MacKinlay supplied a more-detailed breakdown in terms of numbers, locations, and jefes of López's command than Hainsworth's.

knowledge," should be acknowledged by conspiracy theorists alleging manipulation of Carranza by the United States.[17]

MacKinlay also talked with Joseph De Courcy of the *New York Times*, who was traveling with Carranza and commented that under Carranza things were improving with the "leading" Constitutionalists. He specifically mentioned Cándido Aguilar; Brigadier Alberto J. Machuca; Aguiles Juárez; Luis Caballero; General Daniel Gómez La Madrid, who was chief of military police and impressed MacKinlay as "very good, a man of 40 yrs old, tall, black moustache, white, formerly commissioner of 6th Police Dist. Mex. City"; and Major Almarante, a "bandit chaser" on General Alfredo Ricaut's staff. The last two claimed to know who was behind the Plan de San Diego disturbances on the border and "seemed to think Nafarrate had not tired himself out doing anything and Lopez not much better." That probably explained why the Constitutionalists had already announced that the brigade of General Ricaut, Carranza's nephew, would soon relieve Eugenio López's 2nd Brigade in Matamoros.[18]

MacKinlay also relayed information about the many conspiracies gaining in currency, including a telegram from Monterrey regarding a plot in the headquarters there against Carranza that implicated about six hundred men on both sides of the border. MacKinlay did not believe it, saying "The Villa and Huerta crowds here on the border are naturally very 'hot' over the recognition of Carranza and will do all they can to keep rumors alive of possible trouble." Americans, too, liked to keep alive reports of possible disturbances because "practically everyone down here wants all the troops (with the pay day attachment and detachment of said pay) down here indefinitely." Alternatively, "a lot of gentlemen who enjoyed themselves in Aug + Sept during the 'open season' for Mexican speaking residents are now wondering what might happen if too many guards left," indicating that those racists feared retribution. Still others discounted the protection that the U.S. Army supposedly afforded to U.S. citizens, saying that "'all the soldiers are good for is to shoot *our* quail' but they still like the pay day of the 'quail killers.'"[19]

Nevertheless, it would appear that by year end U.S.-Mexico relations were at an all-time high, at least according to MacKinlay. Luis de la Rosa was reported "at or near Jiménez, Tamaulipas, between San Fernando and the R.R. from Victoria to Tampico," and although MacKinlay put no hope in Nafarrate or López catching him, he thought that Ricaut, who "'hits the booze' fairly often," might succeed "by some

---

[17] RG 165, 5761-1053, December 1, 1915, Report to Southern Dept engineer W. K. Adams from W.L. Gibson. See Janssens, *The Military-Agricultural Complex*, 2: 344, for more on the Carranza Doctrine.
[18] RG 165, 5761-1053, December 1, 1915. RG 165, 5761-1057, December 15, 1915, letter from W. L. Gibson, containing quote about La Madrid. See Janssens, *The Military-Agricultural Complex*, 2: 344, for more on the Carranza Doctrine.
[19] RG 165, 5761-1057, December 15, 1915.

stratagem." MacKinlay satisfied himself that "Ricaut is doing all he can to help."[20] In concluding, MacKinlay said that "Everything is 'mucho amigo' now except the rangers are pretty sullen over not being allowed on or near the river, ditto deputy sheriffs, Nafarrate ditto, everyone else pleased."[21]

* * *

Going into the end of 1915, the main concern for the U.S. Army, other than possible attacks on border towns, seemed to be the Bandit War. Therefore, one sees U.S. Army handlers in Texas running civilian agents who delivered intelligence based on their travels and business connections in northern Mexico dating from this period. The Plan de San Diego and the Bandit War also caught the attention of First Chief Carranza, who traveled to Matamoros and pronounced the Carranza Doctrine, which continues in force today, and explains why Mexico consistently refuses to join U.S.-led coalitions, no matter how just the cause, with the sole exception of the World War II crusade against fascism. It is only within the context of the Bandit War that one can understand the substance of the doctrine and Matamoros as the location for its release.

---

[20] RG 165, 5761-1059, December 20, 1915 (quote about Ricaut's drinking). RG 165, 5761-1057, December 15, 1915 (all other quotes).
[21] RG 165, 5761-1059, December 20, 1915.

Conclusion

History

The U.S. Army established a general staff and permanent intelligence function at the turn of the twentieth century when the armies of the world were going through a period of professionalization and the United States was poised to project power on the world stage. Congress worked with the War Department as it experimented with various configurations of command and control for the U.S. Army. Paradoxically, by 1915 the U.S. Army's intelligence function had gone into a state of atrophy precisely at a time when the military's leadership most needed it, when the intentions of Japan in the Far East were in question, as the European continent had descended into bloody world war, and the combat phase of the Mexican Revolution reached its pinnacle.

Nature of Intelligence

Most of the intelligence in the U.S. Army's declassified Mexico file (5761) for the 1910 to 1915 time frame, the conventional warfare phase of the Revolution, came from the U.S. State Department, especially with respect to events taking place in the provinces, and to a lesser extent from what it developed internally, with a smattering from the U.S. Navy as regards revolutionary activity on the coasts.

Along the border, the U.S. Army produced its only homegrown intelligence, but that intelligence tended to be tactical, that is, of questionable use for larger planning, with the best information obtained as a result of serendipity. The sheer volume of this "border intelligence," as it might be called, spiked in the winter of 1914 and late 1915, precisely when the military campaigns of the Constitutionalists moved toward the North. In part, this intelligence of limited use reflected the fact that, as opposed to the Navy, the Army tended to be more tactical and operational than strategic in purpose.

The lion's share of the strategic intelligence developed by the Army came from what the military attachés produced based on publicly available information, largely from the capitalino press—with all that entailed in bias, disinformation, and ignorance about events taking place far from Mexico City. In fact, attachés functioned more as intelligence analysts than intelligence officers, being enjoined from covert activities. Additionally, the Army simply did not have a network of contacts comparable to that of the State Department, nor close relationships with their Mexican counterparts. The lack of close ties with Mexican officers, probably because they could not hide their overt racism and contempt for the abilities of the Federal Army, most likely explained why U.S. Army military attachés did not receive invitations to events or to the facilities of the Federal Army. In fact, relations were so tenuous that Burnside, for one, suggested bribing low-level officers for information.

Cultures

The most enlightening revelation from the research that went into this book was that from the very beginning U.S. Army officers focused solely on intervention, that is, how to defeat the forces inside Mexico and then coopt elements that might serve the U.S. Army in the process of pacification. The U.S. Army was still in Cuba mode, its experiences in that country affecting the attitudes, behaviors, and even the very language that its officers used, such as use of the word "insurrectos" to describe rebels and revolutionaries, a word rarely used by Mexicans.

A more practical reason for the U.S. Army's perhaps fatalistic belief in an impending intervention may have been the compelling need to protect Americans—as dictated by the imperatives of domestic politics—and third country nationals living in Mexico, possibly because the United States had assumed a legal liability for the protection of foreigners and their property. Accordingly, all intelligence was geared toward the purpose of intervention.

Yet while U.S. Army war planners remained preoccupied with the possibility of occupying and pacifying Mexico, never once did the military attachés provide a report on Federal Army tactical doctrine or explain Federal Army strategy, and they described the Zapatista campaign in only the most general terms. Both Burnside and Sturtevant blamed Madero for causing and even fomenting Zapata's rebellion and harbored an abiding resentment toward the president—no doubt influenced by the surrounding anti-provincial chilangos—for causing the country's deteriorating military and economic conditions. Moreover, they probably infected the American ambassador, Henry Lane Wilson, with their prejudice and Burnside, for one, most likely provided colored military counsel to the ambassador during the Decena Trágica that resulted in Wilson's unseemly behavior that made Huerta's palace coup a *fait accompli*.

After Huerta took over as president of the republic, Burnside held out hope that the Federal Army would soon be able to restore peace, but that optimism faded quickly, especially since President Woodrow Wilson took a hardline stance against Huerta. In July 1913, President Wilson recalled the American ambassador and replaced him with his confidential agent, Governor John Lind, who arrived in Mexico with a list of demands that can only be termed a "political intervention." The U.S. government was not interested in cultivating a Cold-War style "proxy," and did not show any bias for or against any one particular faction in the revolution with the exception of trying to force Huerta to hold elections and turn over the instruments of government.

From this point on, Burnside assumed a more aggressive posture, no longer just collecting information to understand the Federal Army, but developing specific intelligence to be used in an armed intervention and then advocating the same—and on a grand scale. Governor Lind, however, had different ideas in mind. He intended to permit the free flow of arms and ammunition into the country and let the Norteños bring the revolution to its conclusion, even if that meant a fratricidal war on par with the U.S. Civil War.

The American diplomatic mission (Charge d'affaires O'Shaughnessy in particular) disagreed with Lind's solution. The career diplomats had become compromised and corrupted in their worldview with a capitalino bias for the South and against the North, even though the chilangos, as they often are even today, were in the wrong. It took an accidental diplomat (Lind) totally ignorant of Mexico (ridiculed by Mrs. O'Shaughnessy but respected by Captain Burnside) to make the correct decision to empower the Norteños to defeat Mexico City. Burnside finally came around to Lind's way of thinking, sort of, but he still advocated and remained oriented toward a policy of interventionism. In the end, granting access to the U.S. arms industry and thus bringing Huerta's illegal government to a swift termination proved to be the correct policy prescription.

Yet, Burnside and the U.S. Army also got its wish, to a limited degree, with the invasion of Veracruz.

Intelligence Failures

In April 1914 the Americans occupied Veracruz to stop the regime of Victoriano Huerta from receiving a large shipment of arms and ammunition. Immediately, the U.S. Army thought that the Wilson administration might have miscalculated and feared that the Federal and Constitutionalist armies might unite to repel the Americans and possibly invade selected spots along the border.

In this milieu the U.S. Army General Staff created an assessment of the military capabilities of the Federal Army and Constitutionalist ("Northern") forces as of summer 1914. This document, however, proved woefully flawed in its scope, because it used Pancho Villa's Division of the North as a proxy for the Constitutionalist Army and gave consideration to the Federal Army. As the nation's professional army and heir to the European tradition, the Federal Army had always captured an outsized portion of the U.S. Army's attentions even though the Federals were far from the most capable when it came to waging war and their army would be defunct within a month of the report.

Conversely, the report overlooked the other major military bodies, such as Zapatistas in the South and even more importantly Obregón's Army Corps of the Northwest, and only obliquely referenced irregulars from La Laguna (the area surrounding the twin plazas of Torreón and Gómez Palacio), mainly because the U.S. Army lacked effective intelligence inside Mexico. In other words, the U.S. Army's report really only took into consideration three of the six major Mexican ways of war based on regional defense models and traditions, while Edwin Emerson's report, filed in May 1914, obliquely made reference to a fourth. Consequently, the U.S. Army did not have a full view of all the Mexican ways of war and experienced an intelligence failure in scope since it missed two of the biggest traditions: the Sonoran and Suriano.

Fortunately, the U.S. military did not have to pay for this intelligence failure because it successfully withdrew from Veracruz on November 23, 1914, without having to engage in any further fighting. However, with the fall of the Huerta administration, the closing down of the embassy and recall of the American charge d'affaires, and the reassignment of the military attaché, the U.S. Army would have to rely solely on its border intelligence for a period of time that would include the peak combat of the revolution.

Going into 1915, the U.S. government continued Governor Lind's policy to permit all factions to import arms and ammunition and thus empowered the revolutionaries to control the direction of the revolution, and with it the fate of the republic. The campaigns of 1915 ranged across the entirety of the republic, with the biggest battles taking place in the Bajío, far from the border in central Mexico. Because of this distance and since the U.S. Army lacked intelligence sources in the heart of Mexico, its file 5761 was thin when it came to the first eight or nine months of the civil war. Accordingly, 1915 signaled rock bottom in a gradual decline in the quality of intelligence produced by the U.S. Army in both its military and diplomatic appreciations, myopically advocating intervention consonant with the age of gunboat diplomacy, and producing faulty intelligence for what might ensue militarily in the event of such an intervention.

Not until the fighting moved back toward the North, in August and September, 1915, did the Americans start receiving more intelligence, most significantly from the former covert U.S. Army intelligence officer W. E. W. MacKinlay, now working for the *San Antonio Light* and going by the pen name of W. L. Gibson. MacKinlay was on hand to report on the Bandit War and even accompanied Venustiano Carranza's train to Matamoros where the First Chief pronounced the Carranza Doctrine, which informs Mexican foreign policy to this day and in its own time heralded non-intervention by Mexico and the United States in each other's affairs.

* * *

In conclusion, when we look at the two main questions that provoked this study, as mentioned in the Preface, we can state categorically that the U.S. military did not undertake any covert operations to privilege any one faction over the others and, secondly, that the U.S. Army's declassified Mexico file is not an adequate substitute for the archives of the AHSDN when writing military histories. The best substitute for the AHSDN archives is the series of books written by General Sánchez Lamego that are based almost solely on documents from the AHSDN.

Documents Cited

9. Armed Intervention

5761-362

WAR DEPARTMENT.
OFFICE OF THE CHIEF OF STAFF.
WASHINGTON.

Dec. 27, 1911.

MEMORANDUM FOR GENERAL WOTHERSPOON:

General:

I enclose a personal letter from Captain John W. Wright, 17th Infantry, Fort McPherson, Ga. who desires to be considered available for military information work. From my personal experience with Captain Wright in Cuba for some years, I would vouch for him for that sort of work, especially in Spanish countries as he is most successful in dealing with Latin-Americans and speaks Spanish fluently. He also has the confidence of the Chief of Staff.

    McCoy
    Captain, General Staff.

5761-363

Dec. 29, 1911.

My dear McCoy:

I return you with this Captain Wright's personal letter to you and thank you for letting me see it. At the present I see no necessity for sending anybody to Mexico but I have always had Wright in mind as a very capable officer, particularly suitable for work of the character he refers to. Should occasion arise for sending additional officers to Mexico, I will be glad to let you know and consult Captain Wright's wishes.

Very sincerely yours,

W. W. Wotherspoon

Captain F. R. McCoy, General Staff,
Office of the Chief of Staff,
War Department.

5761-941

War Department

Office of the Chief of Staff

Washington

April 27, 1914.

Memorandum for the Chief of Staff:

The following is a copy of telegram just received from the American Military Attaché, City of Mexico, viz:

"Vera Cruz, Mexico,
April 27, 1914.

"War College Division,

General Staff,

War Department.

"The situation is such that the result will probably be the sacrifice of a great many Americans in Mexico City as well as in other parts of the country, Huerta having been practically defeated by the revolutionists and smarting under the treatment of the United States is taking advantage of the circumstances to try to unite his people in order to arouse enthusiasm. He is taking advantage of the natural hatred of his ignorant people for Americans and other foreigners. Army officers and agitators are inciting the people to murder Americans. This has the approval of the Government in spite of the constant repetition of the promises that foreigners are to be protected. Americans will suffer first, and afterwards foreigners of all nationalities. The Indian is being given to understand that he is to have an opportunity to satisfy his cruel nature. The disarming of all foreigners in Mexico City, and the public announcement in the press that this has been done, will embolden the masses through the belief that they now have the foreigners at their mercy. The movements of foreigners are now restricted, and that they are actually suffering will not be known until too late in Mexico City. I estimate between ten and fifteen hundred Americans, many from the interior, have attempted to reach Mexico City but have been detained en route, and in some cases it is known that they have been placed in jail by the authorities on the pretext of protecting them. Fifteen thousand Zapatistas from near Mexico City united with Huerta and entered the city on the 24th instant. About half of them were mounted; numerous armed women were among the ranks. Great numbers of men are receiving military instruction in Mexico City but they are poorly armed. There are very few troops along the Mexican Railway; estimate 5,000 poorly instructed. Very little military movement along the other route. Work is being done looking toward the destruction of these roads, but the dynamite being used is only 40%. If the present crisis is passed without a complete armed conquest of Mexico, the future residence of all classes of foreigners in this country will be most difficult if not impossible. BURNSIDE."

5761-942

Received at House Office Building, New Jersey Ave. & B St., S
86W BT 72 COLLECT 2EXTRA VIA GALVESTON
Vera Cruz APR 25TH 14
War Col Staff
(Mr. Keith 214 8th St SE) Washington DC

Five forty. There is little doubt but that temporizing with the situation in Mexico will eventually result in the slaughter of many isolated Americans and other foreigners and finally force efficient intervention this country must be put in order by foreign interference meanwhile the situation is highly uncertain and by lack of communication prevents definite information as to what is happening to foreigners if undertaken immediately the military task will be simplified. Burnside."

5761-943

Veracruz
April 27, 1914.

Warcolstaff,
Washington, D.C.

"A reliable observer leaving Mexico City yesterday reports my impression obtained through conversations with several Germans of high intelligence and a good knowledge of local conditions in Mexico City, is, that on Saturday April 25 a very marked change took place in the public sentiment there. During the early part of the week intense anti American feeling had prevailed and had caused a transitory enthusiasm for Huerta merely as head of the Republic. On Saturday however the idea began to spread that Huerta had deliberately played for intervention knowing that if he were captured by Americans his life and property would be spared while if he fell in the hands of Villa he would be given short shrift. It was further reported that many of the volunteers who had enlisted to fight the Americans were being shipped North to hold back Villa. The gross exaggeration of conditions in Veracruz as published early in the week by the leading newspapers were believed to have been inspired by the authorities as was the destruction of the Washington monument. Both of these moves were made for the purpose of fanning a race hatred. All this has combined to bring on a great revulsion of feeling and the sentiment now expressed regarding Huerta indicates profound contempt and hatred. Serious disturbances in the capital are apprehended by many.

> Burnside.
> Captain Infantry,
> Military Attache.

(Collect.)

5761-947

Proclamation from Fletcher to people of Veracruz about sniping and surrendering arms and ammunition:

"PROCLAMA AL Pueblo de Veracruz Como han continuado las agresiones a los soldados de mi mando, haciendo disparos aisladamente desde algunos edificios y deseando que el orden y tranquilidad queden absolutamente restablecidos, invito a todos los que tengan en su poder armas y parque a que los entreguen en la Inspeccion de la Policía, bajos del Palacio Municipal, a la mayor brevedad, entendiendo que de no hacerlo antes de las 12 del dia 26 del actual, serán castigados con toda severidad aquellos a quienes les sean encontrados, así como los que continúen hostilizando a las fuerzas de mi mando.

Al hacer entrega de las armas y parque les sera otorgado el correspondiente recibo. Veracruz, Abril 25 de 1914 EL CONTRA-ALMIRANTE F.F. FLETCHER"

> Headquarters
> U.S. Naval Forces on Shore.
> Vera Cruz, Mexico.
> From Capt. W.A. Burnside, Inf., Military Attache, Mexico City, Mexico.
> May 1st, 1914.

The enclosed statements of experiences of Americans who have recently made their way from the interior to Veracruz are forwarded for the purpose of giving an idea of the conditions that have prevailed during the past eight days.

No retained copies are being kept on account of lack of clerical force and stationery.

A new series of numbers will be started and this Memo will be given 1100. The old series from Mexico City was left behind and the last number cannot be recalled.

All attaches of the American Embassy in Mexico City were sent to Veracruz on April 23rd and reached this place late on the afternoon of the 24. The Embassy was furnished pass ports and transportation by the Mexican Government.

4. Communications from Veracruz have been as follows;-

Cablegram	April 25, 1914.
"	"	26	"
"	"	27	"

Copies of two of these Cablegrams are forwarded herewith.

5. Referring to Cablegram of the 26th- the information as to the Zapatistas came from the Tramways Company of Mexico City. The traffic Manager who handled the electric trains that transported the Zapatistas from Xochimilco, San Angel and from the Toluca direction gave me the statement. He said that he saw them with his own eyes and that there was no question about the armed me[n] coming into Mexico City. They may not have been the closely related followers of Zapata but at least they came from his territory. They may have been bunches of men collected and armed by the Government and brought into town for the purpose of creating the impression that there was about to be a United Mexico. The number was probably overestimated as is usual in such cases but my informant told me that he handled thirty trains hauling about three hundred each and that he estimated at least half of the total number as having been mounted and arriving by way of the roads leading into the City.

5. [misnumbering in original] The information contained in the Cablegram of April 27 is now fully confirmed. This change of sentiment in Mexico City appears to have been brought about partly by the circulation by the revolutionists of a large number of hand bills giving information about the capture of Veracruz (of a reliable nature and with at least the intention of stating facts) and stating that the Northern Revolutionists would continue to oppose Huerta. The same bills charged Huerta of having played a trick on the people and advertised that all of the volunteers would be sent North to fight Villa. etc etc. People began to look around and there was some improvement. But sentiment swings quickly from one side to another in Mexico. The important thing to keep in mind is that the lid is off- so anything may happen. Conditions are still very bad in spite of the partial change on sentiment among the Mexicans.

6. Veracruz is now quiet and with the exception of communication with the surrounding country everything goes on as usual. THE PEOPLE ARE SMILINGLY SULLEN but have not the enthusiasm that some of the numerous newspaper representatives are representing. Foreign reporters are referred to.

The local press is now in a quarrel with the press of Mex. Cy. The latter is heaping all sorts of insult on the people of Veracruz for not defending their town and calling the Cruzanos cowards and traitors. The V.C. press that formerly published stories of how "the women of V.C. will bare their tender breasts to American balls (the English translation of the Spanish must be pardoned)" is now reasonably favorable so far as anti-American expression is concerned.

7. Huerta has just consented to a four days suspension of hostilities with Carranza- or Villa. This from the Press Censor- Lieut. Greenwood, U.S. Navy.

8. One American came to Veracruz from Mexico City via Mex. Ry leaving the City on the evening of April 29. He is friendly with the Minister of Foreign Affairs, a man in whom I have confidence, and the gentleman who arranged the interview between Gov. John Lind and Portilla y Rojas (Minister of Foreign Affairs- Mexico) about six weeks to two months ago. Only American on the train. Accompanied by a prominent Felicista- the same one that was interested in arranging the Portillo- Lind interview. He reports 1500 to 2000 Americans still in Mexico City. Number cannot be known as people are remaining under cover. Notice had been posted in British Embassy that the Puerto Mexico refugee train was positively the last that would leave City (left 28 April arrived Puerto Mex. Apr. 30) All Americans in Mex. Cy Greatly confused and did not know what to do - bewildered due to lack of positive information and leadership - which under the circumstances of communication is not surprising. This American expressed the opinion that the greater part of the Americans would leave Mex. City were it possible to do so. In Mex. City it is reported that about 30 kilometers of Mexican Railway is out. This is not true. Only about land 1/2 kil. is out and this near Tejeria- ten miles from Veracruz. Some precautions were taken to prevent this man from seeing what was going on along the railroad. Did not see any display of military activity. Believes that there are less than three thousand troops between Mex. Cy and Soledad. Was put in jail at Soledad for safe keeping and to prevent observation. Soldiers outside of the jail all night with rifles at a ready believing that the Hydro Planes (Navy) that have been circling over the territory near V.C. were going to make a night attack. Soldiers greatly confused and alarmed. Such is the Mexican soldier. Trip made without any humiliating incidents. Probably due to the presence of the Mexican friend. No other American on the train. American reports that a great change of sentiment has taken place in connection with supporting Huerta and believes that he stands an excellent chance of meeting his end in Mexico City before the present incident is ended. Also that it appears to be fairly well established in Mexico City that Carranza and Huerta are not going to form any fast and firm friendship.

9. With the proclamation enclosed herewith a complete file of all issued to date have now been forwarded. In the Proclamation marked A (part lined) it will be seen that the part lined gives an excuse for the occupation. Cablegrams coming from Washington at the present time are calling for the incidents with the greatest insistence for absolute veracity. The future effect of the paragraph lined should be watched carefully. The haste displayed in creating a Civil Government where the Governor will be guarded by a squad of troops is another matter that may be observed with interest.

10. A study should be made of the instructions given to Admiral Fletcher from Washington. All Navy plans had been for the taking of Veracruz and contemplated an

efficient occupation of the whole town and surrounding country. It is believed that the instructions actually given were to "take possession of the Customs House." This is in the center of V.C. The practical result of this was that there was fighting for two days with the loss of 17 killed and 63 wounded. With the means available it would probably have been a much simpler matter to have occupied the whole town – and the loss of life would have been very much less; perhaps trifling. Civilian military orders that contemplate neither war nor peace are difficult of execution. This is not intended to be critical but is offered as a subject of study for future guidance. The good intentions in connection with the measures employed here probably outweigh all other considerations. Mexicans are well known to do their best fighting when scattered and without uniforms and from housetops. Street fighting is highly popular and it is generally the cities that are selected as battle grounds. The occupation of a single series of buildings in the center of the town is impracticable. It is to be observed that on the second day of the fighting - April 23- it was found necessary to occupy the whole town and clear the whole place of snipers. It is a remarkable fact that the Mexicans that were killed consisted almost entirely of those who were doing the firing. Only one woman and one child have been reported as killed. It would be a difficult matter to imagine better conduct in the way of sobriety than that of the sailors and marines during the past week on shore. I have observed no cases of drunkenness and have heard of but two. In one case some armed Mexicans removed boards from a floor went under the floor of a house with a solid stone foundation and fired up through the floor as the troops entered the door of the house. In one house near the sea front which had been shelled by three and five inch guns of the Navy and cleared of armed Mexicans firing was again discovered by armed Mexicans who had returned to the building.

11. Mr. O'Shaughnessy after having been in Veracruz for five days received a telegram from the State Department to the effect that he was showing very bad taste in remaining on Mexican Soil after having been given his passports by the Mexican Government. Although he had promptly reported his arrival here he had received no previous instructions.

12. There are but six narrow gauge switch engines in Veracruz at the present time. One crippled standard gauge engine. The Alvarado Engines (4) are not here. It is believed that they are in Alvarado; they are not of great value. Engines shipped from the states must be assembled after arrival as the available crains [cranes?] will not carry assembled locomotives. About seven days is required for assembling an engine after arrival at this port.

13—Information indicates that no unusual preparation of a military nature is being made in Mexico. The most notable step is the effort being made by the Government to create a patriotic spirit. Such enthusiasm as has been created will pass away within a short time. Some sporadic destruction of railroads is taking place but this is largely dependent on the fancy and state of mind of some local commander. The troops near V.C. are mostly local home guards and are not formidable.

[Attached to this are two statements by Americans about being arrested, etc. the first backs Ms. O'Shaughnessy's story about the British Captain Tweedy who got the Americans to Veracruz, and the other about certain Americans fleeing Pachuca for Veracruz. Neither is reproduced herein.]

<u>5761-948</u>

May 8, 1914.

Memorandum for the Chief of Staff:

The following is copy of telegram just received from the American Military Attaché to Mexico, viz:

"Vera Cruz, May 7, 1914."

"War College Division,
    General Staff,
    War Department.

"In the event of Rebel success and a change of administration can not see any promise of permanent peace through entirely local control. Permanent peace in Mexico must depend on some administration having time for making sweeping changes, and time for creating a dependable police force of dependable nature. Rebels will not be easily reformed, and offer but little promise of being satisfactory peace agents after coming into control. The majority of present Federal Army officers offer no present or future promise if armed intervention is to be avoided and Americans are to continue to reside in Mexico. Recommend that steps be taken now looking to an agreement for U.S. troops to occupy important cities for a period following any change of administration that may come about. Once successful any of the present rebels will be as difficult to influence as was Madero and as is Huerta, and the same lack of trust and cooperation between the various factions will exist, and the revolution will continue. BURNSIDE."

<u>5761-949</u>

War Department
Office of the Chief of Staff
Washington

May 12, 1914.

Memorandum for the Chief, War College Division:

The Secretary of War desires that you send a despatch, in substance as follows, to Captain Burnside, Vera Cruz:

"Report fully your knowledge, or best information, as to any, attempts having been made, either prior to your departure from City of Mexico or since, to provide entrenchments, entanglements, or other defenses for the City of Mexico against an approach by the Zapatistas. Period. Be accurate, and if you base your statements on rumor or hearsay, state sources, and the fact that they are rumors or hearsay. Period. Department is anxious to know what provision, if any, has been made, or is being made, in the immediate vicinity of the City of Mexico to defend it against a possible attack by the Zapatistas. Period. Show this to General Funston, and reply through him."

W. W. Wotherspoon,
Major General,
Chief of Staff.

5761-951

May 13, 1914.

Memorandum for the Chief of Staff:

The following is copy of telegram just received from the American Military Attaché to Mexico, viz:

"Vera Cruz, May 12, 1914."

"War College Division,
General Staff,
War Department.

"Through General Funston prior to departure from Mexico City no suggestion of Federal defenses to the south had reached me. Had they existed it would have been advertised, as was the locating of an imaginary division at Tlalpam. Nothing positive is known with reference to recent defensive preparations to south, but since the fall of Cuernavaca conditions are somewhat changed, and the main force of Zapatistas are now more threatening for Mexico City. Some months ago villages on Ajusco mountain were destroyed by Federals to drive out Zapatista sympathizers. Barbed wire has been arriving at this port for Government entanglements and defenses along railroads; some of this is now in Vera Cruz, but part has been forwarded to Mexico City. With recent successes of Zapatistas the providing of defenses is probable, but in no way confirmed. Will investigate further upon arrival of other foreigners from Mexico City, but this is now a two day journey by trains moving at night. Rumor now states that a large quantity of food is being stored at Ciudadela [the Citadel] in Mexico City, and other wild stories indicate preparation of Government buildings for dynamiting. BURNSIDE."

5761-952
War Department
Office of the Chief of Staff
Washington

U.S. Army Intelligence in the Mexican Revolution

May 13, 1914.

Memorandum for the Chief of Staff:

The following is copy of telegram received from the American Military Attaché to Mexico, viz:

"Vera Cruz, May 12, 1914."

"War College Division,
   General Staff,
   War Department.

"Peace commissioners are not Huertistas or politicians, but men having the confidence of educated and wealthy classes, selected on account of favorable local effect on these classes. President interviewed and instructed them for five minutes only before their departure declining to give specific instructions. Commissioners worried over prospect of occupation of city by rebels; would actually prefer U.S. troops immediately. Customary delays may thus be avoided. See W. F. Buckley with commission, confidential information. BURNSIDE."

5761-953

E.O. 1114 <u>Summary of Military Events</u>.　　　　May 9th, 1914.

1. The officers of the Navy state that the arms on the SS "Ypiranga," which arrived at this port on April 21st, are invoiced as follows: ammunition 10000 cases; 4,000 cases; 250 cases; 500 cases; 1000 cases carbines; 20 cases machine guns; 717 cases shrapnel; 1 lunette de bateria; 78 cases munitions of war; 1333 cases rifle ammunition. The exact contents of all of the packages is not clear; it is assumed that the 10000 cases which lead the list are cases of Winchester ammunition; that the 4,000 cases are of 7 millimeter Mauser ammunition; that the 250 and 500 cases are miscellaneous, this would give a total of ammunition of about 14,750,000 rounds. The last item on the list, 1333 cases of rifle ammunition, cannot be satisfactorily explained with the information at hand, it probably consists of miscellaneous rifle ammunition. In Mexico carbines are generally packed in cases of ten (10) each and the 1000 cases probably contain 10000 Winchester carbines.

2. The Mexico City government has now taken over all railroads in territory within its control. All American railroad employees have been discharged; vacant positions have been filled by Mexicans. All Mexican railway employees have been mustered into the service of the federal army, and the control of railroads is strictly military. The Tehuantepec National Railway was taken over by the Local Federal Commander on April 24th; former employees of this road inform me that there is plenty of repair material on hand, such as rails and ties, along the road at the present time, with the principal supply at Rincon Antonio.

Documents Cited

3. On April 24th nine (9) kilometers of track, two (2) culverts and one (1) bridge to the north of Salina Cruz, were destroyed, the point at which this destruction began is not known, and this information is not positive.

4. About April 24th, the Boca del Rio bridge, on the Veracruz al Istmo Railway, 13 kilometers south of Veracruz, was dynamited, and about one-half of the bridge destroyed; on April 29th the federals returned to this bridge and destroyed the remainder of the bridge bed by burning the ties and timbers; 7 cars standing (freight cars) near the bridge were also burned. The important Rio Blanco bridge, 56 kilometers south of Veracruz, is reported as having been prepared for destruction through drilling and charging with dynamite. The El Hule bridge, 156 kilometers south of Veracruz, is reported to have been similarly prepared for destruction. The destruction of the Boca del Rio bridge is certain; the information as to the Rio Blanco and El Hule bridges is believed to be correct, but is not absolutely certain.

5. Along the Mexican Railway similar reports have been received in connection with the preparation of important bridges for destruction. It is also reported that the steep rock ledges along this road, east of Atoyac, have been drilled and charged with dynamite for the purpose of obstructing the track, should it be necessary. Similar preparations have also been reported along the Interoceanic Railway. However, it must be remembered that reports at the present time are particularly unreliable; one eye witness will report incidents, while another will make absolutely contrary statements as to a certain condition.

6. A well informed and educated man, close to the Mexican government, reports that on May 2nd, 15 carloads of ammunition arrived in Mexico City from Manzanillo, and that this ammunition was stored in the Ciudadela; estimated number of rounds of ammunition 15 to 20 million. As to the same incident, a retired Mexican Army officer reports that this is positively a mistake, and that this shipment of ammunition consisted of 400 cases of 1000 rounds per case. This retired officer also reports that ammunition is badly needed in Mexico City.

7. [the bottom of the pages is torn with parts missing, as indicated by the elipses] The capacity of the National Cartridge Factory at Mexico City is now about...,000 rounds per day; some machinery had recently been installed, and the out...of this factory has been increased. The statement has repeatedly bee...the ammunition from this factory is of such an irregular character...be used only in rifles, and that it is not sui...

8. The government has systematically taken possession of all sulphuric acid, brass, copper, bronze and alcohol in the railroad shops and on Haciendas. It is believed that material of this class is being collected for use in connection with the manufacture of powder and cartridge cases. The question of the supply of sulphuric acid has in the past been frequently discussed by the newspapers. It has been stated that this acid is an important material for use in connection with the manufacture of powder; the

supply on hand in Mexico is very limited. Alcohol is probably being collected for the purpose of issuing it to troops – a well-established custom in Mexico.

9. It is definitely known that General Rubio Navarrete has been in charge of the force engaged in preparing the railroads out of Veracruz for destruction.

10. It now appears reasonably certain that Zapata will not under any circumstances join forces with General Huerta. The report as to some Zapatistas having come into Mexico City now appears to have been a mistake. The writer's statement in connection with this was based on information given by a reliable employee of the Tramways Company in Mexico City, who claimed that his information came from actual personal observation; he now insists that the armed men that he moved into Mexico City were men equipped by Huerta for the purpose of giving the appearance of being Zapatistas.

11. It is estimated that an armed force of about 17000 men can be raised in the City of Mexico for the support of Huerta; these will be composed of employees of the various government departments, regular troops, recruits, students and the employees of the street Railway Company. Outside of about 1500 regular troops, this force would be composed of men with nothing but the most rudimentary military instruction. Of the 100,000 men in Mexico City, capable of bearing arms, it is estimated that Huerta might obtain as supporters about 20 thousand; 80 thousand would be positively hostile to him, no matter what the cause for which he was fighting. It is reported 5500 of the troops recently recruited in Mexico City have been sent north. It is estimated that there are approximately 1500 troops now located along the Mexican Railway, between Mexico City and Veracruz. It is believed that an equal number of troops are now distributed along the Interoceanic Railway, with the greater part near Perote.

12. In connection with the present incidents in Mexico, it is important to keep in mind the establishment of a permanent peace, or steps that will lead to such a peace; a change of administration will not bring this about, if Mexico is left entirely to domestic control. A temporary peace, followed by revolutions and disturbances, similar to those we have had for the past three years, is not worth considering.

5761-955
Statement, I. C. Sanchez,
From Capt. W. A. Burnside, Inf., American Military Attaché to Mexico.

On the 21st of April, 1914, traffic was suspended on the Tehuantepec National Railway by order of the Secretaría de Guerra, and practically possession was taken of the entire rolling stock of the company, issuing orders to have all rolling stock concentrated at Rincon Antonio, by the federal forces on the Isthmus, General Gamboa at San Geronimo in command. The track was taken up in sections for a distance of about eight kilometers from Salina Cruz to Tehuantepec, and one bridge was destroyed. All Americans, employees of the Tehuantepec National Railway Company were arrested by order of the Jefe Politico at Juchitán; afterwards they were released by order of General Gamboa, but the Mexican element actually forced all Americans to leave the Isthmus, stating to them that their lives would be in danger if they remained; in fact, they were told that if they did not leave they would be murdered. The suspension of the traffic, the confiscation by the Mexican Government of the Tehuantepec National Railway and all its property; the arrest of all American employees at Rincon Antonio, caused a panic, and everything and everybody was in a state of indescribable confusion. We were cut off from Communication with Mr. Ryan, Vice-President and General, Manager, who was at Coatzacoalcos on the 21st of April, trying to persuade the Captains of the American Hawaiian vessels that were in port at the time, loading and unloading cargo, to continue the work of loading and unloading; and we were informed that Mr. Ryan actually did persuade the Captain of these AHSS Company vessels to resume operations after they had withdrawn from the port of Coatzacoalcos. I think that the last day we communicated with Mr. Ryan at Coatzacoalcos was on the 23d of April, when we were informed that all of AHSS Company vessels had left the port of Coatzacoalcos. Colonel Francisco Correa, commanding the federal forces at Rincon Antonio, issued a circular declaring the office of the Vice-President and General Manager vacant, alleging that the said Vice-President and General Manager had left; we insisted that Mr. Ryan had not left, and that at any rate, told Colonel Correa, that if Mr. Ryan was absent from headquarters, Rincon Antonio, his office was left in charge of his chief clerk; Colonel Correa named himself as superior officer in charge of the affairs of the Tehuantepec National Railway Company.

W. A. BURNSIDE

5761-957

Summary of Military Events – May 10 to 15, 1914.

E.O. No. 1118   May 15, 1914.

1. The editor in chief of the periodical called "El Imparcial" published in Mexico City is responsible for the violent anti American editorials that have until recently appeared in that newspaper.

In one editorial between April 23 and 26 the killing of all Americans before another Sun set on them was advocated. About the same time the story of the sinking of the Carlos V, a Spanish man of war that was anchored in the harbor at Veracruz on April 21, was published for the purpose of creating anti American feeling among the Spaniards and to incite them to demonstrations against Americans in Mexico City. Mexican demonstrations were at this time taking place. Had the demonstrations taken a violent form there would have been a joint responsibility and the Mexicans could have taken advantage of the circumstance to charge all crimes to the hands of the Spaniards.

Due to the protest of the British Minister the Government Organ- "El Imparcial" has now ceased publishing these violent editorials and now apologizes for not continuing to express similar ideas. This paper is absolutely controlled by the Government.

2. Louis D'Antin, the State Dept. employee left in charge of the American Embassy in Mexico City is reported as having taken asylum in the Brazilian Legation. The charge against him appears to have been communication with Carrancistas or Constitutionalists. The Legation protecting him was watched by Government (Mexican) detectives who on one occasion made an effort to search the building. A protest from the British and German Ministers against such a violation of extra territorial privileges and a statement from them as to the effect on Europe of such a violation appears to have caused the Mexican Government to make no further effort in this line.

3. After pass ports had been furnished to the members of the American Embassy by the Mexican Government on April 22, 1914, the question arose as to Leaving our Embassy in charge of some foreign diplomatic representative in Mexico City. The sentiment among Americans was universal for being left in charge of the British Minister. Many Americans made a point of coming to our Embassy and make this request. The Embassy was actually put in charge of the British Minister. This was later changed and on instructions from Washington the Brazilian Minister assumed charge of the interests of Americans. The change did not take place till after the departure of all of attaches of the American Embassy.

4. Great credit is due the British and German Ministers, Mr. Hoehler, 1st Secretary of the British Legation, Captain Tweedy (?) British Navy and Mr. Carl Heynen for the escape of Americans from uncomfortable positions in the interior of Mexico. Many other British and German subjects have also made most unselfish efforts and in many cases exposed themselves to danger in order to assist Americans in their unfortunate position.

5. The question of the temporary employment as scouts, guides, and interpreters of men who have been driven from the interior should be kept in view. In case of actual military operations in Mexico these men will be of great value on account of their personal knowledge of different sections of the country. Many suitable men for this class of employment still remain in Veracruz but many have gone to the United States. In case men of this class are needed in future it is suggested that widely circulated newspaper notices be inserted in papers in the United States. Such a call will bring forward a large number of men whose services will be of great value. The same applies to railroad engineers and conductors- all of whom should be Americans in the event of hostilities.

Documents Cited

6. Captain Franz von Papen, German Military Attaché, Mexico City, paid a visit to Veracruz for two days and returned to Mexico City on May 12.

7. The Archbishop of Mexico, Mora del Rio arrived in Veracruz on May 12. The departure of the head of the Catholic Church in Mexico from Mexico City at the present time is somewhat unusual and significant. The Constitutionalists are well known to be anti-clerical and at the present time much concern is felt over the possible fate of the Churches in case of the arrival of the Revolutionists in Mexico City. No on appears to know just how much financial support General Huerta is exacting from the Church at the present time. This source of revenue being always available it has probably been reserved for an emergency.

8. Foreigners leaving Mexico City, especially Americans, are now searched for Arms, Cameras and newspapers. Some few Mexico City papers do arrive at this port but the regular circulation has been effectually stopped.

9. Referring to your cablegram of May 12th and reply from this office intercepted cablegrams from Mexico City now indicate that 7,000 Federal troops have been sent from Mexico City to the South and along the railroad leading to Cuernavaca for the purpose of checking the advance of the Zapatistas on Mexico City. The interception of cable messages is not being advertised though the Mexican Government knows, of course, of the relay of their wire through Veracruz. If the Mexican Government advertises 7,000 troops the number actually sent can be estimated at from 2,000 to 3,000. The writer's [Burnside's] guess on the report as to preparing the public buildings for dynamiting is that it is being done for the purpose of further intimidating the people of Mexico City and to check the indications of a revolutionary outbreak in the City. Some of these buildings might be connected with a plot for a revolutionary outbreak and if they are charged with dynamite some discouragement would be offered. The numerous sources from which this report comes indicates there is a possibility of its having some foundation of fact.

10. In Mexico City activity continues in connection with giving instruction of a military nature to volunteers. The force that will develop the highest degree of efficiency will be "The Railway Volunteers" and with the limited operations of railroads the number will be rather great. These men have a well-developed grievance against Americans since they received their railroad education under the tutelage of American conductors and engineers and in a very rough school. These Volunteers will be strong physically and are the result of many years of careful selection. The middle class Mexico City volunteers will be without much physical development but will have some intelligence and determination.

11. Referring to my cablegram of May 12 re[garding] the Mexican Peace Commissioner. W.F. Buckley is from Texas, closely associated with Army Officers at Galveston and Texas City, has had much experience with Mexican law in Tampico and Mexico City and has been closely associated with Agustin Rabasa; the latter asked Mr.

Buckley to accompany the peace commissioners to the U.S. During the, Navy occupation of Veracruz, under the direction of Admiral Fletcher, the writer [Burnside] was associated with Mr. Kerr and Mr. Buckley in connection with the preparation of numerous proclamations and was impressed with the apparently substantial ideas of Mr. Buckley. The opinions of some of the other members of this board were much less substantial. By the time Mr. Buckley arrives in the U.S. he ought to have a fairly clear idea of what is in the mind of the commissioners (Mexican).

Mr. Buckley is not looking for any position or employment and is independent. He intends to return to Veracruz and in case an advance is made on Mexico City he especially desires to accompany the expedition. His services could be utilized to great advantage in connection with questions associated with Mexican Law.

12. General Garcia Peña has been placed in command of the troops at Cordoba and along the railroad (Mexican) towards Veracruz. General Peña was the last Secretary of War and Marine for the Madero Administration and since that time, Feb. 1913, has not till recently been on the active list. He is well known in Jalapa where for about fifteen years he had the direction of "El Comision" or what corresponds to our Coast and Geodetic Survey. Most of the map sheets of Mexico in common use were prepared under his direction and he probably has more knowledge of the topography of Mexico than any other Mexican. Since General Huerta has been President General Peña has spent most of his time in Mexico City. This is said to have been for the purpose of keeping his friends in Jalapa out of jail. Every time General Peña went to Jalapa rumors of revolutionary plots at Jalapa would arise and the General would return to Mexico City. This man is regarded as one of the ablest of the Mexican Generals. Present circumstances have compelled the Huerta Administration to cancel their suspicions of supposed Maderista tendencies on the part of Peña.

13. Arrivals in Veracruz on the morning of May 15 report:

A few 70mm guns at Atoyac.

Very few troops to the East of Atoyac along the Mexican Ry.

Approximately 1500 troops at or near Cordoba

Movement of troops and heavy artillery from along the Mexican Ry. towards Mexico City.

Fortification of Mexico City and Puebla.

Estimate that Huerta can muster about 35000 troops in Mexico City and Puebla.

Concentration of Federal troops and volunteers in Mexico City and Puebla.

A belief that Huerta intends to make a stand against the revolutionists 1st.- In Mexico City. 2nd- Failing in Mexico City a withdrawal to Puebla.

This is reported as a rumor from a fairly accurate observer.

There are no troops (Mexican) between Veracruz and Tierra Blanca.

        "   "   "   "   Tlalixcoyan.

On Veracruz al Isthmo Ry. bridge destroyed at Boca del Rio. A short section of track destroyed at Paso del Toro. The Rio Grande bridge mined and prepared for dynamiting.

One daily train running from Tierra Blanca to Guyabo, one to Santa Lucrecia and one to Orizaba. These trains do not carry military escorts.

At Tierra there are 18 standard gauge locomotives. Four (4) of these are of the "Consolidated" type – weight 134,000 pounds- will carry 500 tons on a 2% grade- are large enough to serve for use on the Mexican railway.

A limited supply of fuel oil at Tierra Blanca.

This comes from railroad sources. Some of it second hand and through Mexican railroad employees of al Isthmo.

14. A very practical wireless station for use on board military trains is being used by our troops at the present time and is worthy of further development. Two poor negatives of the essential parts are enclosed. This has been improvised with the materials at hand. Such an apparatus could be made far more compact and made to occupy a space about equal to that required for two passengers. A copy of the report of the Navy officer who installed this apparatus will be forwarded as soon as it can be obtained.

15. For the troubles of Mexico there is no local solution in sight and any military preparations made by the U.S. at the present time will be found of great value later. Hap hazard and forlorn hope expeditions should be discouraged and everything considered from the view of a high degree of efficiency. Troops of a moderate degree of training will prove efficient but the numbers should be great enough [to] cover much territory at one and the same time.

5761-958
Memorandum.

May 11th, 1914.

About ten (10) days the Mexico City papers commenced publishing reports to the effect that the Water Works at El Tejar had been recaptured by the Mexican troops. The destruction or holding of City Water Works is in Mexico frequently regarded as a highly efficient military measure. All circumstances indicate that the recapture of the Water Works at El Tejar has been thought of and planned. As a rule, newspapers in Mexico City give the first notice of contemplated military measures that are subsequently attempted. Observation indicates that newspaper information is, as a rule, announced from two weeks to a month before actual execution is attempted. However, projected military movements frequently fail to materialize, and it is regarded as highly improbable that any attempt will be made by the Mexicans to recapture the water works El Tejar.

In case any attempt should be made by the Mexicans to recapture the Water Works at El Tejar, it is most probable that the attack will be made from the direction of Tierra Blanca and along the Veracruz al Istmo Railway.

Lines of railways are almost universally followed in all Mexican military operations, on account of the small mules used for transporting artillery, the use of this branch is almost wholly confined to lines of railroad. The use of artillery has recently become so general that scarcely any expedition is attempted without artillery of either 75 or 80 millimeter caliber. The 80 millimeter artillery is universally transported on flat cars, and no attempt is ever made to use it except on such cars, or in permanent defensive positions. The 75 millimeter artillery is more mobile, but the small mules limit their transportation over roads to comparatively short distances.

The 70 millimeter mountain guns have in the past been almost universally condemned by Mexican Army officers.

W. A. Burnside
Captain Infantry.

5761-960
Headquarters U.S. Expeditionary Forces
Vera Cruz, Mexico

May 20, 1914.

Intelligence Office
Memorandum for Chief of General Staff.

1. Press reports relative to Americans who have expressed their intention of remaining in Mexico City give the number as 800. This is not believed to be correct, and reliable information all indicates that the number intending to remain it Mexico City is between 100 and 150. Press reports are probably inspired by the government for the purposes of showing the confidence supposed to be felt by Americans for the Huerta administration.

2. It is reported that Gen. Maass has been sidetracked in Mexico City; public opinion is greatly against him; he is allowed to go about the city, but is being shadowed by secret service agents. General Peña is supposed to have been sent out of Mexico City and given the command at Cordoba on account of his aspirations to become provisional president to succeed Huerta. It is also reported that Gen. Velasco will be given no further command of importance.

3. Some federals have been sent south from Mexico City to check the advance of the Zapatistas, but it is not believed there is any intention to endeavor to recapture Cuernavaca from the rebels.

4. In Mexico City, Mazatlán is reported to have fallen into the hands of the rebels.

5. Many people in Mexico City believe that Huerta is following his customary bluffing course in reporting important public buildings as having been prepared for dynamiting.

W. W. [sic] BURNSIDE Captain Infantry.

HEADQUARTERS U.S. EXPEDITIONARY FORCES
VERA CRUZ, MEXICO

June 4th 1914.

My dear Major Crawford:

Thank you very much for your letter and your thoughtfulness in connection with myself. The best possible arrangement appears to have been made and it is entirely satisfactory to me.

I have located a lithographic establishment in Veracruz, and as soon as we get through correcting the great number of errors in the existing maps of Veracruz and its vicinity, I am going to make a try for civil funds enough to reproduce this map as an experiment. The engineers are working on the basis of 10 ft. vertical interval and we will eventually arrive at something that is pretty good. As it is probable that the next two generations will be fighting battles around Veracruz, we may as well accumulate the data while we have an opportunity.

I will also keep the Jalapa plant in view, but at the present time it has a very distant appearance. Too bad we have got to stay around here and swelter with all the nice cool air going to waste just a few miles up the hill.

I have been trying to get in touch with Mexico City where I can procure plenty of the Ruhland directories, but at the present time the break in the railroad makes it very difficult to get supplies from Mexico City. I have in mind expending all my balance of the contingent fund on June 30th for these books.

I have become so accustomed to waiting patiently for big things to turn up in Mexico that the present circumstance does not trouble me in the least. Barring accidents, as I see things, we still have a very long wait before us, and I will be greatly surprised if we are suddenly withdrawn from Veracruz. Everything points in the other direction, but in a situation such as this, accidents may happen at any time and these accidents may suddenly put a new element in the situation. In addition, through the conditions constantly arising, Washington may be forced to take action on matters in connection with this unusual pacific blockade and it cannot be hoped that any of our decisions will be entirely pleasing to any class of Mexicans.

The newspaper statements that are being made about the dangerous exposure of this force here are in no sense true. Everyone continues to overestimate both the ability and the numbers of the rebel forces. The only rebel leader from whom anything can be expected is Villa. He continues to show more military talent than is combined in the federal army. However, he is a long ways from Mexico City and is liable to run into difficulties before he actually makes it dangerous for the Mexico City government. The sporting rifles and ammunition that are entering Mexico by way of Puerto Mexico do not trouble me to the extent of the shipment of high power rifles and ammunition through the west coast. My information indicates that 22,000 of the Japanese rifles have been landed. If Huerta can hold onto these he will give the northern rebels a surprise party one of these days. If he can keep his army near enough to him to exercise a sort of personal supervision over his generals, the results will be far different from those obtained when the federal army was scattered over the enormous extent of northern Mexico.

I will appreciate a note from you now and then as it will serve to give me a line on how you are all thinking in Washington. The viewpoint from Washington and Mexico is liable to be widely different.

Sincerely yours,

W. A. Burnside

[penned in at bottom of page is: "There is mighty little news here. Gen. Funston sometimes asks Wash. questions—which, of course, cannot be answered. The Navy has orders to prevent the Mex. gunboats from stopping the Tampico Rebel arms arriving on the "Atilla" (S.S.) – This step is liable to stir things. Anything U.S. may do will be interpreted by Mexs. as being unsatisfactory. WAB"]

5761-968
War Department
Office of the Chief of Staff

June 12, 1914.

Memorandum for the Chief of Staff:
Subject: Sketch of Francisco Villa
1. Herewith is biographical sketch of Francisco Villa, Mexican Constitutionalist.
John Biddle
Colonel, General Staff,
Chief of War College Division.

June 10, 1914.

Francisco Villa.

To obtain accurate biographical data about the crop of bandit revolutionary generals developed during the last four years in Mexico is a difficult task. Not only is little known of these men in the United States, but very little is known of them by the Mexicans themselves except in a very general way. The following data have been compiled from personal conversations with men of long residence in Mexico, reports, press dispatches and letters and it is believed forms as accurate a sketch as it is possible to make up from the data at hand.

Francisco (Pancho) Villa was born either in the extreme southern part of Chihuahua or the northern part of Durango about 40 years ago. One account has it that he was born at Las Nieves, Durango in 1868. This would make him 46 years of age. His parentage which is doubtful, is believed to be practically, pure Indian. While a small boy Villa's father moved his family northward into the State of Chihuahua and established a ranch near the city of that name. "Pancho" grew up as a vaquero or cowboy. Upon "Pancho's" father's death, when about 16 years old he took charge of the ranch and provided for his mother and younger sister. One story has it that the sheriff of the county eloped with Villa's sister and fled to the mountains. Villa pursued with some armed men caught the couple, forced the man to go through a marriage ceremony, made him dig his own grave and then Villa shot him and rolled his body into the grave. One account is that he was incarcerated, when 14 years of age for cattle stealing and a few months after his release was again confined for homicide at Guanavaci, Chihuahua. Whether it was these incidents or many others of a similar nature that made Villa take up the life of a bandit as a regular occupation is not definitely known, but for the last fifteen years at least he has been living the life of a bandit, roaming through the Chihuahua and Durango Mountains. A reward of $20,000 is said to have been offered for his capture dead or alive during the Diaz regime. Villa is reported to have said that during his bandit period he had forty-eight encounters with the Rurales, that he killed thirty-seven of them and was himself wounded nine times. He maintained himself by stealing horses and cattle, sometimes along the United States—Mexico border and crossing them to the opposite side from which stolen, where they were sold. He had various partners with most of whom he had trouble. One Francisco Reza is said to have been his partner in cattle stealing during 1907. Villa killed him while sitting in the plaza of Chihuahua City, then rode away to the hills. When Francisco Madero joined Pascual Orozco in the mountains of Chihuahua in the autumn of 1910, Villa attacked the factory of Mr. Soto in Allende and by threatening the daughter obtained some $11,000 which was used to arm and equip some 200 to 300 men with which he joined Orozco. Villa's partner at this time was one Jose Salgado, a butcher of Chihuahua. Salgado was accused by Villa of informing the authorities about him with the result that Villa rode over to Salgado's place of business and shot him to death. When Casas Grandes was taken in January, 1911, Villa is said to have killed Carlos Alatorre and Luis Ortez for refusing to pay a ransom. During the battle of Juarez in May, 1911, Villa distinguished himself by his personal bravery and the military disposition of his force. Orozco is said to have given Villa little credit for his part in this and other fights and thereby to have incurred Villa's undying hatred. Upon entering Juarez, Villa killed Ignacio Gomey Oyrla, a man of 60 years, because he denied having arms concealed in his house. During his service with Francisco Madero,

Villa seems to have conceived an extremely strong personal attachment for him. Upon the triumph of the Madero rebellion, Villa was given military rank in the Chihuahua Militia and obtained from the Governor of Chihuahua a monopoly for the sale of meat in the City of Chihuahua. The meat was obtained by robbing the ranches in that vicinity principally those of Luis Terrazas.

When the Orozco rebellion of 1912 broke out, Villa took the field against his old enemy, and with some 400 men captured Parral, looted the bank of $180,000, and is reported to have committed many excesses. Orozco sent one small column against Villa which was defeated. Jose Ines Salazar with a larger force was then sent which drove Villa out, but Villa kept the $180,000 and retired to Mapimí. After the defeat of General Gonzalez Salas' federal column by Orozco, and the suicide of Salas, General Huerta organized a new column at Torreon, which Villa joined with about 500 men in the summer of 1912. Villa participated in various battles and eventually commanded the 5th irregular brigade. During this time Villa is said to have committed all sorts of crimes and to have been very insubordinate. The rupture between General Huerta and Villa is said to have occurred at Jimenez on June 4, 1912, over a horse which General Huerta desired but which Villa appropriated first. One account has it that Villa thereupon deserted with some 500 followers, was pursued and captured by General Tellez with the federal cavalry, returned to Jimenez and tried for arson, rape, murder, robbery, horse stealing, etc. At any rate Villa was seized and condemned to death by General Huerta. Villa was lined up before the firing squad and the order to load had been given when Emilio Madero, who was acting as confidential agent for his brother the President, with General Huerta's army, ran out and stopped the execution producing an order from the President to send Villa to Mexico City. Villa was sent a prisoner to Mexico City and incarcerated in the penitentiary. There it was that he learned to write his name which is all he is able to write up to the present time. Villa later was allowed to escape from Mexico City and made his way via Galveston it is said to the United States, where he worked as a miner in Tucson during the winter of 1912-13. After Madero's death, Villa left Tucson in the early part of March, 1913, with seven followers proceeded to El Paso where they bought their guns from Krakouer, Zord and Moye. Mr. Krakouer sold the arms to Villa personally. Villa crossed the line near El Paso and organized his followers in the hills of Chihuahua south of Columbus, New Mexico. At this time United States officers in that vicinity were struck with the efficient manner in which Villa organized, armed, and equipped his force of some 1,000 men. He armed and mounted and men in as uniform manner as possible, organized them into companies, had wagon transportation, a paymaster, regular issues of rations and ammunition and had almost no women camp followers or "soldaderas" in camp. Since that time, March 1913, Villa has been consistently successful against the federals. In every encounter his record of killing defenseless prisoners, pillage and extortion has been added to. Among the larger places taken by him during 1913-14 have been Torreon, Juarez, Chihuahua City, Ojinaga where the remnants of Mercado's federal army was driven across the United States Mexican border, Torreon again, and Saltillo.

Villa's success has not been so much due to his own ability as to the great weakness, and incapacity to assume the offensive on the part of the federals. Most of Villa's operations against large bodies of federals have consisted in breaking their lines of communication, so that they could not be re-supplied, then by piece meal attacks making them fire away all their ammunition until an evacuation took place. In no case has he captured or entirely destroyed a federal column of any size in the open field. The evacuation of Chihuahua and Torreon being examples. In both cases the federals' retreating columns were entirely vulnerable but nothing was done to inaugurate a strong pursuit. At Torreon during the latter part of March and first part of April 1914, Villa really was whipped to a finish by Velasco. He had nothing left for defense or offense and had only a small reinforcement of ammunition and men been sent Velasco, "one hundred men and 1,000 rounds of ammunition" as one who knew the conditions said, Villa's army would have been scattered. The success at San Pedro several days after the Torreon evacuation was entirely due to dissentions [sic] among the federal generals, and the particular incompetence of General Joaquin Maass who later evacuated Saltillo. But instead of pressing his advantage and pursuing south of Saltillo, Villa has retired to Torreon, with the avowed purpose of striking in the direction of Zacatecas. The principal cause for this move is said to be due to dissention among his generals, i.e. Manuel Chao, deposed military governor of Chihuahua, Maclovio Herrera and his brother Luis, the two Arrieta generals (brothers) in Durango, General Benavides, Pánfilo Natera and other and also due to the fact that the country south of Saltillo has been stripped bare of its resources and does not afford so much prospect of loot as the vicinity of Zacatecas. This inability to enforce discipline or "get along" with his companions both inferiors and superiors has always been true of Villa, from his bandit days when he killed his partners up to the present. Villa's ability to command a party of bandits of from 200 to 500 is unquestioned. His ability to command and keep together a force of upwards of 16,000 men for any length of time is a serious question. Villa is personally brave, a good horseman, strong physically and cunning mentally. His personality is his greatest asset as he can make a very creditable appearance to strangers, and instill fear and admiration of himself into his men. His success is due to the fact that he will "butt in" where others will not and will "take a chance" personally. He obtains more of a following than others because he gives his men more plunder than others as he is more successful. The Mexican revolutionist in the ranks stays by the man who wins the most.

When not overcome by rage Villa takes advice from those he considers worthy, and follows it for a time. But there is no telling when he will break out, and do whatever his fancy for the moment dictates. One case in point is the Benton killing. He had been given a copy of the rules of civilized warfare by General Scott shortly before this. It was prescribed therein that prisoners should not be killed without proper trial. Villa announced that Benton had been regularly tried and executed. This was proved to be untrue and evidence strongly pointed to Villa as the culprit, a commission was therefore appointed by General Carranza to investigate the killing, which is said to have completely exonerated Villa and laid the blame on one General Fierro a companion of Villa. This Fierro, whose photograph is appended hereto in the group with Villa has a long list of killings to his credit and is said to be Villa's chief executioner. Nothing has yet been done to Fierro as far as known here. Since that time Villa has executed many prisoners taken in battle. Villa also told a prominent general officer of the United States Army that he was completely subservient to General Carranza's wishes, still it is a fact that during Carranza's stay in Chihuahua, there were always more of Villa's men around Carranza than Carranza brought with him, that some serious altercation took place between them, one involved the deposition of General Manuel Chao as Governor of Chihuahua another was over the expulsion of the Spaniards and the seizure of the cotton at Torreon. When the question was being agitated as to the tax on this cotton for the different foreigners Villa said "the Americans can take their cotton out free, other foreigners will have to pay 25% but the Spaniards' cotton I shall keep for myself" or words to that effect. Upon being told that General Carranza then in Durango had ordered equal treatment for all the owners' cotton Villa replied "that old ------- has nothing to do with it, I am running this" or words to that effect. The Spaniards owned nearly all the cotton, and Villa holds it up to date. An excellent press censorship in Mexico and a press bureau in the United States is maintained and little gets out that is derogatory to Villa.

The legal representatives of the Constitutionalists, Hopkins and Hopkins, Hibbs Building, Washington, D. C., are an extremely efficient firm in advancing the cause of the Constitutionalists. It is thought that they lean more towards the Carranza faction than to Villa, and that a great deal of difficulty is being experienced in justifying Villa's acts.

Whenever it has any effect Villa treats Americans, personally, very well and he has been probably more easy for the United States to deal with than any others in northern Mexico. As to Villa's carrying out any promises no dependence can be placed on them. Obregon on the West will have nothing to do with Villa, neither will Pablo Gonzalez and the insurgents in the East. That they will ever really get together seems quite impossible. Villa's whole following comes from Chihuahua and a part of Coahuila. the former has a population of 405,265 the latter 362,092, the total population of Mexico is 15,114,305. Therefore allowing Villa to have the undivided support of these two states (which in reality he has not) whose population is 767,357 Villa controls about 1/19 of the population of Mexico. His followers in great part have fought at various times under Orozco, Salazar, Rojas, Quevedo, Salas, Mercado, Huerta and others. They are chronic revolutionists largely held together because they are making a better living and amount to more as soldiers than they ever did before. They are ready to follow any leader who will bestow these advantages on them in the greatest measure. As long as a good living can be obtained and there is plenty more in sight this character of a force always keeps increasing rapidly in Mexico. With comparatively small reverses however it rapidly melts away as an organization, the men split into bands of from 50 to 200 take to the hills and exist as bandits. Villa hesitates to get far away from Chihuahua, his native heath, as he is fearful that something against his interest may occur in his absence. Villa personally is very abstemious and is said to fear being poisoned. At present he drinks no alcoholic liquors as he is very much afraid of them, having formerly had experience with them. He uses tobacco, and is said to be a pronounced raper. Where he has any ambitions to be president is a question. It is certain, however, that he brooks no opposition to his wishes by superiors. While an uneducated and ignorant man in a literary way, he has had great experience as a partisan leader of irregular troops, and lately of an organization approaching a regular army. He is extremely crafty and politic when necessary. He is learning more about all these things constantly, but due to his entire absence of early education must lean on his advisers for support on any large questions. As things stand at present Villa has more of a "punch" than any other individual military leader in Mexico. He is the embodiment of all the elements of the Northern revolutionist of Mexico.

5761-974
[Penciled in at top, "From Capt W. A. Burnside"]
May 6th, 1914.

1. My luck in landing in the centers of trouble does not appear to have escaped me up to the present time. I had information in connection with the arms and ammunition that were to arrive in Veracruz on Hamburg–American liner "Ypiranga",- somewhere along about April 8th, and telegraphed to the War Department. Then the arrest of our sailors in Tampico came along and for the first time I saw that the State Department was very strongly inclined to back up the navy ultimatum in connection with the salute and apology. This was the first instance where we had been extremely insistent on full and complete reparation. On April 20th, a day before the arms were to arrive in Veracruz, I insisted on my wife leaving Mexico City and going to Veracruz. The trip was rather disagreeable, and the people along the route were "muy bravo." The navy landed in Veracruz on April 21 about 10:00, with instructions to take the customs house. This, of course, was typical civilian order. The taking of one building in the center of a fair-sized city, without taking the remainder of the town, is naturally somewhat of a problem, if the native population does not entirely resemble a flock of goats. As might have been expected, there was another fight on the following day, and

the incident closed with the capture and occupation of Veracruz its immediate vicinity and the water works.

2. On April 22nd we had our passports handed to us by the Mexican Government. On the 23rd we left Mexico City for Veracruz aboard a special train with General Corona in command of escort, and Colonel Alberto Braniff as Assistant. Outside of some dark looks from the native population and a few "mueras" and "cabrones gringos," we made the journey to the break in the railroad at Tijerilla [Tejería] without incident. Here we transferred over two kilometers of destroyed track and took the train, operated by our own forces, to Veracruz. O'Shaughnessy finally got his instructions in the form of a telegram, informing him that it was beneath the dignity of the United States for him to remain on Mexican soil, but meanwhile we had gone through the ceremony of having the flag raised over Veracruz. A shot gun civil government, with a squad of soldiers at the Governor's door had also been established and all of the officers sworn in. A few days later this government was abolished by orders from Washington and a military government was established. This is still in the formative process, and we have had difficulties in inducing the old Mexican Government employees to return to their work, but they are gradually coming around, and within the next ten days something very closely resembling the old order of affairs ought to come about. New uniforms have been manufactured for the policemen, and today they all go back on their beats. Some of them are, of course, missing, but a great number of the old ones have returned to the only work which they understand.

3. There are scattering bunches of Mexican troops just outside of our out-posts, but they are giving no trouble; they are not in force. All of the four (4) railroads leading out of Veracruz are cut within a few miles of the City.

4. A few days ago a Mexican Major came to the Water Works at Tejar under a flag of truce and demanded that the Commanding Officer surrender within ten (10) minutes; the major stated that the Americans might leave their arms behind and retire, if they desired. The major in command of the marines told the Mexican Major that he had better make full use of his ten (10) minutes in getting as far away as possible, which the Mexican officer did. Three (3) shots, two (2) from the Mexicans, and one (1) accidentally from our forces were fired.

5. A few days ago a squad going to the water works took the wrong road, got lost and wandered into the Mexican outposts; after exchanging cigarettes and going through the usual photographic courtesies, the Mexican soldiers pointed out the proper road to our people and they went on their way and joined their command at Tejar.

6. A battalion of marines occupying one of the sandoons [sand dunes?] outside of the City has christened it "606 Meter Hill" – an appropriate name.

7. As I have told you, we had in the Embassy two (2) machine guns and two hundred and thirty (230) rifles. On April 22nd about 150 Federal soldiers came to the

Embassy with instructions to get these rifles, and we surrendered them, as might be expected. I had always opposed the idea of raising one of these friendship arms, and had advised that the arms be sent no further than Veracruz. With General Huerta as President, the matter of arming foreigners in Mexico City assumed quite a different aspect from a similar arming during the time of President Madero, and the ultimate outcome shows that I did not make a bad guess. Other foreigners in Mexico City received similar treatment in connection with the arms and ammunition that had been accumulated. I cannot feel but that this action on the part of the Mexican Government was correct and proper. However, it did have one bad effect: The newspapers extensively advised the disarming of all foreigners, and as a result the native population immediately became "muy bravo" and started the rock-throwing and window-breaking; there is nothing that pleases them more than to hear things smash. Further than a few black eyes and numerous insults, it is my opinion that Americans have had as yet suffered but little. As to what may happen no one can tell. The attitude of President Huerta when talking to the people has been that any means are proper in connection with repelling the invaders; his attitude as expressed to foreigners and through the press has been at all time that he would afford full protection to all foreigners. The practical result of these contradictory expressions has been that bad faith has been clearly evident, if one looks beneath the surface. The difficulties put in the way of Americans who have been trying to get out of Mexico have been very great. After insistent demands on the part of the British and German Ministers several train-loads of refugees have been sent out of Mexico City, every possible delay has been effected by the Mexican Government.

8. Now the A.B.C. peace conference is working, and we are advised it is to meet in Montreal, Canada. It will take some time to assemble this body; after this there will be a meeting where it will probably be discovered that some of the credentials are not in proper form; steps will be taken to obtain the necessary credentials; there will be a delay, and then this body will finally meet for the transaction of business. Then each member will want a complete set of credentials, and it will take some time to get them in the required languages. After this the proceedings will probably be blocked on the question as to the correct definition of some word. About the next step ought to be a breaking up of the peace conference with the simple explanation that the proceedings are blocked. Meanwhile Villa and Zapata will continue their bull-ring tactics and Mexico will continue yelling "mucho." General Huerta will probably remain in the ring, as he is a pretty game proposition; his fate is still uncertain; the present wabbling of his cabinet in the somewhat of a crisis is, however, a bad sign. It is believed that all of his cabinet Ministers have attempted to resign, but that he has not seen fit to accept the resignations. Some months ago he had the foresight to make them all Brigadier Generals in the Army by pinning a button and tying a sash on each of them, so that he

now has the power as Constitutional Citizen President to try these improvised Brigadier Generals by court martial, if the circumstances demand it.

9. In case the rebels succeed in routing out Huerta and his administration the situation will not be changed materially. This will be merely an incident in the final settlement - a permanent peace in Mexico, this is the big issue, and we must not permit it to be clouded by incidents that are now occurring.

10. Referring to paragraph 2, a military government appears to me much more reasonable than a civil. Mexico has had quite enough fake governments, governments purporting to be republican, but in reality a military despotism of the worst kind. A genuine just military government would probably do as much to change conditions and promote peace as any other form of government that might be put in force.

11. Referring to paragraph 9, in the event of rebel successes and the change of administration, I cannot see any promise of a permanent peace through entirely local control; permanent peace must depend on some administration which may suddenly come into power, having time and opportunity for making a sweeping change that will be required before the country settles down; another important factor is, time will be required for the new administration to organize and create an army or a police force in which dependence can be placed. Looking for material for such an organization, the rebels do not offer much promise, for conversion into a police force of a dependable nature; they are not easily reformed and will always remember the happy times when they roamed at will with stolen horses and carrying rifles. With order in the country they will miss the free enjoyment of plunder, wine, women and song. On the other hand, the majority of the present federal armed forces offer little or no future promise; they care little for the welfare of Mexico and their education has been largely confined to devising methods of graft; the present federal army can be viewed in no other light than as being suitable material for the scrap heap. If armed intervention is to be avoided, and if foreigners are to continue residing in Mexico, it would be well to take up at the present time the question of having some foreign armed force temporarily occupy important places in the republic immediately following the next change of administration. Such a course might offer a chance for holding things together for a short time and until something could be done in the way of organization; this might offer a means of starting things in the right direction; if something like this is not done, the successful revolutionary leaders will immediately start quarrelling among themselves about the division of spoils, and as a result the revolution will not cease, and once more we will be compelled to sweat through disagreeable incidents such as have taken place during the past three years. Before the revolutionists are finally successful, and while they are in doubt, something can be done in the way of influencing them; once successful, and having gained assurance, it will be as difficult to exercise a stabilizing influence over them as it was with Madero, and as it is with Huerta at the present time. A similar lack of trust that has always existed between the

Mexicans, and a similar complete failure at co-operation between the different factions will continue, and revolution will probably go on without even a temporary break. Some method must be devised of establishing peace in Mexico for a time after the next change of administration.

11. [misnumbering in original] Veracruz at the present time offers rear [sic] opportunities for gathering information as to roads and general conditions; there are assembled here many intelligent men from all over the republic; they are now enthusiastic and will devote time and labor going over items within their personal knowledge. Information with reference to roads will be as good two years from now as at present. I have all these willing workers on the go and they appear to be enjoying it. New trails not indicated on maps at the present time are being located; the work is very interesting.

12. One thing, however, must always be remembered: people in Mexico have never been closely bound to facts, and at the present time there are some very racy stories going around; in many cases it is a difficult matter to pick the substantial from the fairy tales. The origin of all news items is on a breezy basis, and as a result, we who are here, as well as the world in general have but a hazy notion of actual conditions and happenings."

<u>5761-975</u>
Notes on Mexican Constitutionalists
Or
The Northern Mexican Insurgents
By
William Mitchell,
Captain, General Staff
July – 1914
Their Personnel, Organization, Equipment, Strategy and Tactics.
With some brief notes on Mexican Federal Troops, and tactics which may be effective against both.

The sources from which these notes are compiled consist of:

Capt. J. E. Shelley, 11th Cav., eye witness, Battle of Santa Rosa

Capt. F. L. Pyle, Phil. Scouts, eye witness, Battle of Tampico

1st Lt. B. P. Disque, 3d Cav., eye witness, Battle of Nuevo Laredo

1st Lt. E. Engel, 9th Cav., eye witness, Battle of Naco, Sonora

Mr. Edwin Emerson, eye witness, Battle of Torreon.

Mr. E. V. Stoddard, eye witness, Battle of Torreon

Dr. Ryan, eye witness of the sanitary conditions obtaining with both Federal and Insurgent troops.

Reports of U.S. Marine officers participating in the capture of Vera Cruz.

U.S. Army Intelligence in the Mexican Revolution

Reports and dispatches on Mexican affairs received, particularly during the last year.

By
William Mitchell,
Captain, General Staff

Before discussing the tactical and strategical methods employed by the "Constitutionalists" or rebels, an idea should be formed as to the character of the officers, the men, their organization and equipment.

THE NORTHERN INSURGENTS

The personnel of Mexican Rebel officers is composed largely of former bandits. Their origin, ordinarily, is more or less obscure. The usual biography of one of them runs as follows: He was born of Indian or half-Indian parents at some small place in Northern Mexico; his family moved from place to place engaging in stock raising, agriculture, or in small commercial ventures. He was brought up literally in the saddle, under conditions of want and privation. He had little or no schooling, but lived in the open air; wandered over a great deal of the country, learning it well and meeting its people. Occasionally, he went to a large town, where he was carried away with the great opulence and wealth of some of the citizens and where he found amusements, liquor, and diversions to his heart's content, which he could enjoy providing he had the money. To obtain this money he committed some crime, stole cattle, or raided a store, killing its proprietor. He was pronounced an outlaw and hunted by the Rurales. He took to the mountains, gathering his friends around him. Sometimes he joined a band already in the hills. In the years which followed he had frequent brushes with the Rurales and at times with rival bodies of bandits. This work developed his military wits, as it were. He had to evolve a good system of information in order not to be caught unprepared; he had to supply himself and his followers, and he had to see that he was well mounted at all times, as his horse was his great asset both in aggression and defense. As he had no set rules or dogma to follow in these operations, the multitudinous military problems that presented themselves to him, although small, not involving, in many cases, more men or equipment than a good sized patrol, had to be solved promptly and efficiently. The efficient solution of these was rewarded by increased wealth and personal ascendancy among his followers; the poor solution meant death or incarceration in a penitentiary.

Under these conditions keenness of perception, quick decision, reliance on friends and hatred of personal enemies were developed to a high degree. The many long marches, the obtaining of food in a poor country and the exposure to the elements, developed in him a robust physique and a capability of withstanding great hardship and withal doing a great amount of military or semi-military work. The discipline of these bands was based on a mutual recognition of the strongest man as leader. His qualifications for keeping this office were his ability to devise means to accomplish the ends in view better than anyone else. A failure, especially if it involved any surreptitious dealing with an enemy, was punished by death. If any of his followers violated the unwritten code, he was macheted or shot without mercy. His ethical ideas were that "might made right", and he acted as he felt. The limit being the general opinion of his followers, because if he overdid things too greatly to the prejudice of his followers, death was the penalty. This is the discipline of the Northern Constitutionalist or rebel.

If he was more successful than most of the bandit leaders he found it convenient to associate with him persons living in the large towns, where he could take stolen articles such as cattle, gold, bullion, or produce and have it sold at a profit. In this way he made friends with people in many towns, and, if very successful, grew to know practically everything that went on all over his state and even beyond it. In places more exposed to a bandit raid than others, where ranches, mines or plantations were maintained at great profit by their owners, he made a bargain with them to "protect" these places on payment of certain sums. If these were not forthcoming, his followers "could not be held in restraint". The followers of the bandit chiefs combine these attributes of their leaders to a lesser degree, but all have had practically this training.

There are some ex-Federal officers in the Constitutionalist ranks. Most of these have had some trouble or other with their former comrades. They are looked at with suspicion, are considered lacking in initiative, and too "high-toned" by the Constitutionalists, in most cases. There are also a few educated Mexicans and soldiers of fortune of other nationalities, acting as Constitutionalist officers.

These Northern insurgents have a strongly developed individuality, and in their way are independent in thought and action. Their thought moves on a limited field and is concerned mainly with the particular object in view at a certain time. This usually is either a personal object or one relating to those immediately around them. They are strong physically, have excellent eyesight, are able to live in a sanitary environment that would ruin many white troops, are good riders, know how to take care of their horses, and how to supply themselves and their mounts. They are quick with the rifle, and fairly good shots as individuals at ranges of 500 yards or less. They have no great fear of death and withstand wounds, suffering, thirst and hunger in a remarkable manner. As they do not greatly fear death, they are brave; but this bravery is of a kind not exactly understood by the average Anglo-Saxon, and depends on circumstances. To go forward in close formation, under the direct personal command of someone else, and stand great losses, they do not relish; but if they are acting independently among a great crowd or horde of their own people in an attack against a rich town they have stood heavy losses. For individual deeds of bravery, especially if watched by their fellows, they are good. If caught at a disadvantage, they run away; but this is not necessarily due to cowardice, but may be an attempt to get into a better position or place for defense. The Northern insurgent is not particularly good in trenches where he must fight it out right there or die. He wants his horse within easy distance and his line of escape clear. With his horse within easy reach, or close behind him, he may be expected to stay to the finish and much longer than he would in a trench. This is due to his training and instinct. He is good at utilizing cover in an individual way, crafty at ambuscades and surprises.

From our standpoint, he is cruel. The ordinary punishment for any offense is death by shooting. He kills all his prisoners not willing to join his ranks. Occasionally, to make an impression on a foreigner, if he thinks it good policy; he saves a few and sends them to the rear as specimens. He has no regard for women, and those taken from his enemy he keeps as long as his fancy dictates, then gets rid of them. From his very nature, precedents, education, inclination, and the usual character of his operations, he is not inclined to the use of very large bodies of troops, or to a very definite military organization in a European sense.

He is by instinct and, in fact, a North American Indian, or nearly so; his manner of making war is more like that of the plains Indians of the United States than anyone else, although similar to that of the Russian Cossack, and to a lesser extent than that of the South African Boer. His warfare is of a "partisan" character and not that of a definite regular military organization. He has no drill regulations or field service regulations, and could not read them if he had. He has no good doctrine or dogma of war dinned into his ears since he "joined the army"; but, on the other hand, he has no bad doctrine to which he is a slave.

By hard experience, due to his adaptability, he is evolving a doctrine of war better suited to himself and his objects than an impressed foreign one would be, in much the same manner as the Americans developed a new system of tactics in the Civil War. He has now been fighting quite actively for some four years, and within the last year bodies of a comparatively large size---10,000 to 15,000 men---have been assembled on one battlefield.

The beginnings of a definite organization are commencing to appear, and it is reasonable to suppose, with such personnel, that if war kept up for three or four years longer, which required constantly increasing numbers, that a fairly efficient military system would be evolved. This is, of course, a practical impossibility for many reasons. As the group of Northern insurgents under Francisco Villa have been the most successful, therefore, having the greatest numbers, more organization and better equipment than the rest, the observations made in this paper apply particularly to them.

ORGANIZATION

Strictly speaking, there is only one kind of troops in Villa's army; that is, the mounted rifleman. It is true that he has machine guns with a pretty efficient personnel, artillery and infantry. The infantry walks because it has no horses to ride and whenever any foot man can beg, borrow, or steal a horse he becomes a mounted rifleman. The machine guns are served by men who now have had a great deal of experience with them, but if shorthanded anyone around helps out. Much store is laid by the machine guns as will be explained later.

The artillery, 75 and 80 M.M. material of excellent and up-to-date design captured from the Federals, cannot be well handled or well supplied with suitable ammunition; so its effect, except in a moral way, is small. The Northern insurgent likes to have "artillery" though, as it lends an imposing aspect to his columns, and is much talked about by the people of the cities and the country.

Villa's troops, at the Battle of Torreon, consisted of about 14,000 men; 12,000 were mounted; 2,000 were footmen. There were 32 guns and about an equal number of machine guns.

THE MOUNTED RIFLEMAN

Organization.

The captains command about thirty men. There are lieutenants, sergeants and corporals of indefinite number under them. The atmosphere of being a real commissioned officer is not very evident until the grade of captain is attained.

The brigades consist of a number of these companies or troops with a strength of from 800 to 1,600, usually about 1,000. Over these brigades is a general of brigade [Mitchell meant brigadier, Villa was the only general of brigade] with certain colonels, lieutenant colonels, and majors, of indefinite number under him. The brigade sticks pretty well together and becomes known usually by the name of its leader, such as "Ortega's Brigade," "Herrera's Brigade," etc. When two or more brigades are sent on a common mission one of the brigadiers acts as chief of all; but he consults freely with the other brigadiers and all act much according to their own ideas of the situation.

Equipment and Armament.

Horse: Mexican bred pony weighing from 700 to 900 lbs; shod in front or not shod at all. The horses are usually in pretty good condition, and there are not many sore backs.

Saddle: Mexican or American stock saddle pretty good ones such as retail from $20. to $25. in the United States. Each horse has a hackamore, a bridle usually with a spade bit (on marches the horse is handled almost entirely on the hackamore, the bit only being used when fast work is required), a long tie rope and a lasso. The saddle bags are very small, pocket affairs usually of leather or canvas. A sleeping blanket is carried on the cantle. (This is never used under the saddle; a sobrejalma being used for that). A water bottle; often a canvas water sack. A haversack; often this is a gunnie sack laid across behind the cantle.

Usually no forage, and only a little dried meat and corn as rations, are carried on the horses.

The 30-30 Winchester carbine is preferred. This has a 20" barrel, weighs loaded about 8 pounds, carries six shots in the magazine and one in the barrel, has 1900' initial velocity and shoots point blank about 200 yards. (When there are not enough of the above, Mausers and other rifles are used.)

The amount of ammunition carried by the individual varies. He will carry all he can get and pack. When it is plentiful he usually has at least 200 rounds; much of the 30-30 ammunition is soft nose. The individual soldier takes special pride in carrying large amounts of ammunition, as he considers that his prowess as a soldier is gauged to some extent by the number of rounds. Often this feeling causes him to not use it much. Many of the men carry the ammunition belts on the saddle, wound around the pommel, etc., until they go into dismounted action. The machete is carried by many, probably about 1/3. The pistol or revolver of various types is carried by all that can get them; about 1/5 have them. A khaki uniform and a large hat of various patterns is used by the men. The officers carry a saber or machete, revolver or pistol, and carbine. They are seldom equipped with field glasses, maps or compasses; otherwise their equipment is much the same as the men. Most of the officers wear Texas felt hats. There are no tents, wagons, field desks, or extra impedimenta carried. There is no sanitary service with the brigades; no first aid packages or medicines to speak of. There is no extra ammunition, forage, rations or equipment carried in wagons or on packs for the brigades. They live on the country, haciendas, and towns, or from the railroad trains.

INFANTRY
(Soldiers without horses)

The men are much the same as the mounted riflemen although possible slightly inferior in most respects and with a considerable percentage of deserted or impressed ex-federals among them.

Armament usually 7 m.m. Mauser rifles captured from the Federals.

Some carry bayonets, but not for use as such. They are used as knives, etc.

A blanket, water bottle and some small rations such as dried beef, corn meal, etc., are carried.

The number of rounds of ammunition is as many as they can get and carry, usually about 100 or more rounds.

The infantry is used principally as artillery supports near the railway; as they have little mobility their use is limited.   Their tactics do not essentially vary from those of the mounted riflemen when fighting on foot.

MACHINE GUNS

There are thirty or forty machine guns of various makes in Villa's force. These constitute one detachment, generally speaking, and are detailed with or to accompany the brigades as circumstances dictate. They are carried on pack mules, have mules for extra ammunition, and all the personnel for using them are mounted and armed.

Great use is made of the machine guns as they employ high power ammunition, much more powerful than that of the 30-30 carbines and can be availed of at greater ranges than the rifle fire. They are much more accurate in the precision and direction of their fire than that of the individual small arms fire, as a great deal of the fire of individuals, especially at close range, is delivered from the hip and in a desultory fashion. The machine guns are used everywhere and under all circumstances.

Their position in column is usually in the center of the force they accompany, and they are brought into action quickly at tactical points. Their positions are changed frequently during engagements, and in general they keep up with the advance of the skirmishers.

The use of machine guns is thoroughly understood by the personnel which operates them.

ARTILLERY

Material 75 m. m. St. Chamond-Mondragon. There are also a few 80 m.m. guns of the same make.

This equipment has all been captured from the Mexican Federals. In Villa's force there is one section (1 gun) of horse artillery, the draft animals being horses and all the personnel mounted. This section is detailed to the brigades as necessity arises, and is used actually as horse artillery.

The ammunition consists of captured Federal shrapnel, and common shell home-made in the Chihuahua Railway shops. The home-made projectiles practically gave no results. There is no high explosive shell. The amount of ammunition of all kinds depends almost entirely on captures from the Federals, and at the end of the large engagements almost, if not all, of the ammunition is expended.

The draft animals are mules (weight about 900 lbs.), six to the gun. They are shod all around and are quite efficient for their weight.

The personnel is very mediocre and know little or nothing about the handling of modern artillery. Direct laying methods are used entirely and fire by piece. At times several guns are assembled haphazard and constitute for the time being "batteries." The guns, however, continue to function separately. There is no system of the artillery's position in column, tactical distribution or massing it at any particular place or places distant from the railway. The guns are usually brought into action near the railway lines.

A few 80 m.m. guns are mounted on, and fired from, railway carriages; usually open flat cars.

It may well be imagined that the fire of this artillery is not efficient.

The personnel has no definite personal equipment or arms. None carry rifles or carbines. Some have pistols or revolvers; many have machetes.

The soldadera system is used with the artillery to quite an extent, as the women can be carried along on the caissons and limber chests.

RAILROAD DETACHMENT

The troops mentioned above are not ordinarily used for pioneering duties or the repair of railways. These things are done by a special detachment of railroad men under one Ensebio [Eusebio] Calzado, who is styled "The Superintendent of Military Railways" for Villa. He is an extremely energetic, capable and versatile railroad man, and handles the whole problem of railway and wire communications. These detachments composed of railway men, vary in strength according to the work in hand; they are supplied from a construction train. They have no particular armament, but many of them have pistols, rifles, carbines and ammunition sufficient for their own protection. At the Battle of Torreon two insulated wire field telephone lines were laid for Villa's use. No other electrical or visual system is used for transmitting information. The "soldadera" system of rationing and messing is used by the railroad detachment.

SANITARY SERVICE

There is no sanitary personnel attached to the units, or anything approaching it in the way of doctors or men of experience in this duty. When wounds are received the wounded men are cared for by their comrades as best they can, using any material available for bandages, etc. Wherever possible the wounded are carried to ranches, haciendas or huts; when near the railway they are taken to the hospital train.

The hospital railroad train under Dr. Villarreal is quite good in its way for an insurgent army such as Villa's. It contains 22 cars, one of which is an operating-room car; another a private car for Villarreal and staff; the rest boxcars. The equipment of medicines, bandages, surgical instruments, etc., is very good in quality and amount.

The personnel consists of three so-called American doctors, some Mexican doctors and male nurses. The wounded are treated in ordinary boxcars, under pretty hard, unsanitary and uncomfortable conditions. The capacity of the train is said to be the care of 1,400 wounded men, without replenishment of supply.

There were some ambulances and automobiles attached to the train. When sent out to gather wounded during an action they were usually seized and diverted to some other purpose by the troops. Occasionally wounded officers were brought in by them. The wounded usually either make their own way to the hospital train or are brought there by comrades on horseback. Ordinarily, first-aid treatment only is given in the train. The wounded are evacuated to a city as quickly as practicable.

DISEASES

The most prevalent diseases are venereal. At least 30 per cent of both insurgents and Federals suffer from gonorrhea and syphilis. One insurgent garrison of about 400 at Rio Grande, Zacatecas, inspected by an American physician were all, including officers, found to be suffering from these two venereal diseases. Some had both.

Typhus, which is transmitted by the body louse, is quite common. Heretofore this disease was more common in Southern than in Northern Mexico. The sending of southern recruits to fight in the North has greatly spread this disease.

Malaria of a pernicious form is found on the coasts, but absent in the uplands. Scarlet fever, diphtheria and dysentery are encountered in most parts of the republic. There is little typhoid fever, generally speaking, although in some localities it is more prevalent than in others.

The principal diseases to guard against in case of an invasion of Mexico will be venereal. These are being spread more and more as the fighting zones become larger, on account of the customs of the troops and the almost total absence of medical attention.

SUPPLY SYSTEM

The supplies, quite complete and of various kinds are carried in railroad trains. The water train consisted of seven tank cars, (like the large oil tank cars on the United States railways). These had long pipes along the sides arranged with about twenty spigots each, so that a number of men could come up on each side of the train and fill their canteens simultaneously. Wooden troughs were carried on the tops of the cars to be used for watering horses. The commander of this train was said to have orders to keep the tanks full to overflowing whenever possible, or he would be shot. There were three trains of rations and forage. These made issues to anyone asking who had a note from an officer authorizing the issue. (This is the only trace found of "paper work" with the units). There were also other trains carrying ammunition, equipment of various kinds for horses and men, clothing and forage. Issues of these things were made in the same manner as food, forage, etc.

The order of the trains was as follows: first, gasoline exploring cars with machine guns; second, construction train. This had 3 armored cars, carried 2 80-M. M. field guns, some machine guns and detail to operate them, which were pushed ahead of the engine: this was followed by the other cars of the construction train; next come the provision and supply trains; then the water and coal train, and last the hospital train, - a total of eight trains.

No system of wheeled animal-transportation for taking supplies to the units exists. In the first place, wagons take up too much room on the trains, are expensive, hard to get and keep in repair, and the same number of animals are well used for other purposes. Occasionally a buggy or light wagon is picked up by the units and sent back for supplies.

STAFF SYSTEM

Villa's staff system consists in having a secretary and an indefinite number (about twenty) staff officers who are especially conversant with the country, the troops, the enemy, the people and all conditions; who are individually capable and strongly attached to him personally. Information is given either by oral communication or by mounted messenger, anyone available being used for the last.

During an action Villa moves about little and is generally to be found at or near his headquarters. He holds a meeting which his principal officers attend every morning between eight and ten o'clock, where discussions occur and orders are given. He is very accessible to all, takes information and advice from anyone, and is quick in decision. A very effective press censorship is maintained, and a careful check kept on all newspaper reporters.

STRATEGY AND TACTICS

Before passing to a discussion of the constitutionalist tactics, a glance should be taken at the Mexican Federal troops to which they are opposed.

The organization of the Federal Army is contained in the Mexican Monograph, 1914. It is very good, theoretically, and along European lines. The personnel of the officers is better versed in academic and book learning, and in tactics from a European standpoint than the insurgents. There are also many officers of poor quality necessarily taken in since the large increase of the Mexican Army was made over a year ago. All these come from the better class of families; they are much more amenable to discipline than the rebel officers. On the other hand, they are much more "hide bound," lacking in initiative and slow to assume responsibility. They are physically quite as brave as the rebels, if not more so.

There is a good deal of graft evident among the upper officers (this also applies to the insurgents), and mutual jealousies between themselves, the civil authorities, and the *irregulars*. The personnel of these irregulars, led by such men as Orozco, Rojas, Caraveo and others, is of the same character as the insurgents.They constitute a very small proportion of the whole number, however. The officers, generally speaking, are very loyal to the Federal Government. The generals, although varying according to the individuals, are the exemplification of the characteristics pointed out above.

The greatest deficiency of the Mexican Army is to be found in its enlisted men. These are conscripts gathered together in the thickly settled parts of Mexico. They come from the class of agricultural laborers, town and city laborers (all peons), criminals or those accused of crimes. They are not "men on horseback" or "volunteers" like the rebels, nor do they know anything about the country in which they are operating. They are strong physically and very enduring under hardship, but lack in initiative and the most rudimentary education. They may be described as perfect "mutts." These conscripts receive their training of seldom more than 3 months, within the four walls of their cuartels [barracks]. These usually occupy about a square block in extent. The instruction consists in the school of the soldier, the squad and the company. They have no target practice and are taught only the mechanism of extended order formations before going into the field. They are herded around like sheep by their officers under whose immediate eyes they do everything, even guard duty. They are rationed by the "soldaderas" and have no wheeled transportation. Therefore, they cannot operate away from railways. Their cavalry instead of utilizing the excellent horsemen of the country, attempts to train its conscripts in this difficult branch, according to European methods. The result is that they practically have no cavalry worthy of the name, certainly none that is able to act independently. The Federal cavalry is laughed at by the northern rebel horsemen.

The Federal artillery has all the modern equipment pertaining to that arm, and while very immobile is better served, handled, and ammunitioned than the artillery of the rebels. In this arm the Federals are much stronger than the Constitutionalists; but even here, due to poor methods and deficient training of personnel, not one-fourth the efficiency of which the material is capable, even with mediocre personnel, is obtained. Few instances are cited in which artillery has been served until the instant of capture using shrapnel cut to zero. The guns seem to be abandoned rather early, or are held in masked positions from which they are incapable of delivering their fire at the assailants at close range.

With such a force, especially due to the impotency of the cavalry, and the lack of wheeled transportation, the Federals cannot assume a vigorous offensive or undertake even an extended local pursuit.

It is true that General Victoriano Huerta assumed the offensive and broke up Orozco's force in 1912. This, however, was essentially a railroad campaign, and Orozco made the mistake of committing his whole force, inferior to Huerta's, to a decisive action at Bachimba. Huerta's troops never left the vicinity of the railways in any force to pursue the rebel bands.

STRATEGY

Federal

The strategy of the Federals in the present campaign has been very simple, i.e., to hold all the railways and with these as supporting points to push mobile columns into the intervals between the railways and destroy the rebel forces. There was nothing wrong with this, except that the Federals had no mobile columns to send; so the rebels merely left the railways when whipped. There they were left unmolested, reorganized, rehorsed [remounted], and rearmed themselves.

Insurgent

As the Federals occupied practically all the railway lines soon after the beginning of the campaign, the insurgents contented themselves with catching outlying detachments or forces temporarily exposed. All this time the insurgents were gaining in strength and numbers, whereas to hold the railways against the insurgent horsemen, the Federal forces became more and more scattered in the attempt to hold all points at once. The insurgents then had their best chances. If, for instance, the railway between cities like Torreón and Chihuahua could be permanently put out of commission, and at the same time the railway from Chihuahua north could also be occupied, the Federal garrison of Chihuahua could not be resupplied with ammunition, equipment or food. By long range attacks on this place the Federals would be made to use up their ammunition and either surrender or evacuate. This is exactly what occurred at Chihuahua and the strategy of the Constitutionalists has been similar in almost all instances, much as war was made in Europe a few centuries ago. The enemy's bread line is cut where it is impossible for him to repair it in time; the hungry insurgents watch on the hills for a sign of weakness and as soon as one appears an attack is made; if successful the Federals evacuate with practically their whole force. No pursuit is made by the Constitutionalists, because, in the first place, the towns taken contain too much loot to be abandoned to anyone else, and, in the next place, the Federal's line of retreat is stripped bare of food, animals, and often of water. If the attack is unsuccessful, the Federals make no vigorous or prolonged counter-attack, and the Constitutionalists retire unmolested to the hills, resume the policy of watchful waiting and prepare for the next attack. Each time a town is taken the insurgents, unmolested by the enemy, become stronger, better equipped, and their next operation takes place against a stronger force of Federals. After each one of these successes a month or six weeks elapses before a new movement is inaugurated. This is because it takes that time to import more ammunition and supplies and because at the end of these periods most of the food, forage and pulque has been consumed in that particular locality, and more must be obtained. There is nothing complicated about this strategy; but it is hard to beat in Northern Mexico without excellent mounted troops efficient in make-up and entirely adequate in numbers.

The federals have several times held the "inner line" [interior lines] and with a mobility and a morale equal or even a little inferior to that of the Constitutionalists should have easily beaten them, even with inferior numbers. This, especially mobility, they did not have. There is only one known instance in which the insurgents availed themselves of the "inner line." In September, 1913, Villa interposed at Santa Rosalía, Chihuahua, between the Federals at Torreon and Chihuahua City. Hearing that Munguía at Torreon was scattering his force, Villa left a containing force at Santa Rosalía to guard against an advance from Chihuahua and with the rest of his force made a rapid march south. General Alvarez' [Felipe Alvírez's] column of about 400 men was encountered at Aviles about 15 miles west of Torreon, completely surprised and rapidly annihilated. Torreon had only a small garrison, which became terrified upon the reports brought by some of the soldaderas of Alvarez' column, and evacuated. Villa walked into Torreon with little or no opposition. [This is an error in intelligence, since the subsequent fight for Torreón was actually quite fierce.] Operations of this character are very unusual in Mexico, however.

There has been little or no coordinated action between the Western, Central and Eastern groups of insurgents, and none between the Zapatistas and other Southern rebels with the Northern insurgents.

TACTICS
(Mexican Insurgent, Mounted Riflemen)

The tactics of the Constitutionalists are quite simple and always adapted to circumstances. Generally speaking, on the offensive the operation is a concentric envelopment of the hostile position or force. Few reserves or supports are held out as such, but as deployments are nearly always made on the head of single columns, each unit engaging in the fight as it comes up, which takes time, some units remain formed until the attack is well developed, and are used as a reserve in a way. That is, if it is evident that they are needed at a particular place, they are sent in that direction immediately.

The habitual column of route is a single column of twos, quite elongated. This, although it does not favor quick deployments to the front, gives more time in case of surprise, and more freedom of movement for the rear elements.

MARCHES

The distance covered on a single march varies according to circumstances. On an ordinary route march the distance is usually from 15 to 20 miles, at an alternate walk and trot, with few or no halts until the objective, a watering, rationing and foraging point, such as a hacienda, village, or town, is reached. What would correspond to the ordinary route march is usually made on railway trains by Villa's men. The horses, saddles and equipment are placed in the cars; the personnel with their arms and ammunition ride on the top of the cars.

These columns can usually be heard for some distance because, in addition to the noise of the train, the men shoot at telegraph insulators, jack rabbits, cows or anything they feel like.

The marches in the presence of the enemy vary according to the objective, and in some cases are long, i.e., 30 or 40 miles and often at a trot for the whole distance with few halts unless watering places are encountered. As the men carry little food or forage a hacienda or other supply point near the hostile force is usually the objective. They always have good information about the amount of provisions these places contain.

About a week of this kind of work is all their troops can stand without rest. This they are always able to get as the Federals do not pursue them, nor do the Federals destroy all food supplies away from the railways.

An advance guard is habitually employed. This is usually a group (with a force of 2,000 men) of from 150 to 300 men, thrown out a short distance in advance, only so far that its rear element can be called to, from the head of the main body. Practically no flankers or patrols are used ordinarily. Most of the information about the enemy is picked up from the inhabitants of the country. Generally speaking, the service of security on the march is poor, but the information from the inhabitants concerning the location of the enemy is generally good, about his numbers poor. In Northern

Mexico a column of any size on the march usually raises considerable dust which can be seen for a long distance.

The commanding officers of a force usually ride at the head of the main body, the commanders of the various units at the head of their respective organizations, and some officers are detailed to bring up the tail of the   columns to prevent straggling.

Machine guns or artillery, if any, are usually placed in the center of the column.

When an enemy is met even in small force there is a decided disposition to deploy everything.

A theoretical diagram of this method of attack (Fig. 1) [not extant, but easily visualized with the mind's eye] follows:

In this diagram E represents the column of route described heretofore. What patrols are necessary to examine inhabitants, etc., are usually sent out from the advance guard, although any unit along the column sends out patrols if its commander thinks necessary.

When an enemy is encountered in position (as the Federals almost always are) a halt occurs while the commander of the force discusses the situation with his principal subordinates, who usually ride with him. After the orders are given they return to their units and a deployment, as indicated by the lines in Fig. l, occurs. The diagram may be taken to indicate two brigades, with four machine guns, one of which is deployed to the right of the line of advance, the other to the left. The intervals between skirmishers to begin with are from 20 to 30 yards, decreasing somewhat as a position is approached. Two machine guns accompany each brigade. These guns are generally used in pairs, so that it one breaks down the other can keep up the action and use the ammunition of the first.

In the advance all cover possible is availed of, and an attempt is made to get within 500 or 600 yards before the men dismount to fight on foot. No particular system of distributing, linking, or guarding the horses obtains. The horses are left by their individual riders, tied to sage brush, or otherwise secured in ditches, behind rocks, declivities, or other cover. A few men or boys not to exceed one man to ten horses are left to watch them. Each soldier, in fact, looks after his own horse. The led or rather tied horses are not mobile, and when it is desired to move them, the skirmishers come back, get them, lead or ride them forward, etc. The skirmishers by this method are able to shift easily, and when a weak place is discovered in the hostile position messengers are sent along the lines announcing it and the skirmishers get their horses and proceed to the point indicated. There are no supports or reserves, properly speaking, held out or pushed in to support the line. When a change of objective occurs skirmishers are pulled out of the line, and pushed in at another place, without regard to gaps left, continuity of fire or anything else. There is no system of fire control or direction, and sights are seldom elevated even at long range.

NIGHT ATTACKS

At the battle of Torreon in the last of March, night attacks against the Federal positions were adopted by Villa as a system. The reasons were these: The Federals occupied excellent positions with a good field of fire across the open country. Although their fire was poor in the daytime, it was much worse at night. The insurgent's fire was not effective over 500 yards, and it was very difficult to make them cross the fire-swept zone in the day time and engage the enemy at a range of 200 yards or closer. As the fire of insurgents is very haphazard anyway, the pieces being fired from the hip, held over the head when behind cover (as the large proportion of wounds in the arms and hands proves), their fire would be almost as effective at night. The Federals could be depended on not to make strong counter-assaults or pursuits, because if they did the troops executing this maneuver would get so out of hand that they would be collected again only with the greatest difficulty. The Federals, therefore, could not take the chance of losing any of their troops as they were outnumbered about 3 to 1, and had a very long position to hold.

In case units were sent in counterassault or pursuit at night the Federals' [ternary] system of supports and reserves was such that there would be no certainty of filling the gaps in the line left by the columns detached, thereby giving the insurgents a chance to get through. They were also afraid of launching supports or reserves through their front lines, due to the danger of their own men firing into each other, as actually occurred in some instances. The result was that by night attacks the insurgents crawled up to close range with few casualties, delivered their fire with almost as much effect as it had in the day time, and the Federals expended even more ammunition than they did in the day time. The insurgents had more or less regular lines of Federals to fire at while the Federals had to fire at the flashes of the insurgents' firearms scattered all over the sides of the hills and in no particular formation. The positions of the Federals' shelter trenches on the Cerro de Pila were bad, as they did not properly cover the foreground at close ranges, leaving many dead spaces. Most of these dead spaces could be covered by a slanting fire from the trenches, but with poor troops heavily engaged, especially at night, nothing can be expected except a fire straight to the front.

These night attacks were kept up for 5 successive nights, and also fire attacks during the day time at long range, the result being that although the positions were not taken by assault the Federals' ammunition ran low and they were completely worn out by night and day duty as they had not sufficient men to relieve any parts of their line with fresh troops.

Villa's troops were marked by a different manner of fastening up the side of the hat each night so as to know friend from foe. This also acted as a pass word, as it were. The manner of arranging the hat was changed each night, so that it could not be used by the Federal irregulars whose appearance is about the same as that of the insurgents.

OUTPOSTS

U.S. Army Intelligence in the Mexican Revolution

The outpost system is extremely efficient and very simple against such a force as the Federals. In the day time, they place men on watch on hills, or places commanding a good view, and leave them there all day. Sometimes only one or two men will stand watch for several thousand. These men are chosen on account of their especial qualifications for this work, such as good eye sight, knowledge of the country, alertness, etc. At night they use a system of videttes and supports, covering the principal avenues of approach, the number of these depending on the nature of the ground, proximity of the enemy, etc. A friendly population, of course, assists them. There is little trace of an efficient system of assembly points in case of attack, designation of a definite line of resistance, or good system of inspection of outposts. It is therefore to be seen that against a very mobile, aggressive enemy this system would not be especially efficient, and a surprise attack especially at night would result in a great mix-up for them, especially as each man takes care of his own horse, sleeps and eats with him. The horses are not concentrated on picket lines with guards; thereby rendering the personnel incapable of bringing the greatest number of rifles into the fight at once. (Example, Obregon's attack on Salazar, Frontera, Sonora, October, 1912) [The Battle of Hacienda de San Joaquín in September 1912 was against Colorados, not Federals, and Salazar was untrained and ignorant in military doctrine.]

COMMENTS

Against an immobile enemy, not capable of delivering counter-assaults, or against an inferior or even slightly superior force of infantry in open country equipped with about the same armament, this method of attack [Villa's tactic of "surround and pound"] is quite effective, because it develops a converging fire on the hostile position which eventually surrounds and places it in a ring of bullets, at the same time all weak places, easy approaches, etc., are found which may be quickly availed of due to the mobility of the attackers. Unless actually engaged in street fighting or something approximating it, there seems to be little difficulty in pulling the Mexicans out of the line and putting them in again. This, of course, is due to the great immobility of the Federals, that they take little advantage of faults committed by the rebels, and to the fact that no especially serious attempt is made by the Constitutionalists during the early stages of an action to close to hand to hand grips with the enemy.

Due to its thinness and great extension a line of this sort cannot develop any real efficacy of fire. The line may be pierced anywhere by a determined advance of any passably good infantry. This alone, however, does little good as the rebels merely retire with their horses and take this column in flank again, which, if the pursuit is kept up too long so that it gets out of hand, is distributed with too great intervals, or out of tactical support from the main force, will soon be destroyed, as the rebels, due to their mobility quickly concentrate on isolated groups and surround them. (These are some of the reasons why the Mexican Federal infantry can do little against the rebels). If, however, the force being attacked in the above manner has any efficient cavalry, the matter is entirely reversed.

If, in Fig. 1, the position A is held by infantry as a pivot of maneuver and a flank attack of cavalry is launched at the lead horses at C, the effect would be disastrous, and the whole line would be immediately rolled up. The Mexican rebels think first of their horses above all things and are very poor at mounted action. When once started in retreat and pursued even slightly they fly from the aggressor and do not rally until a distant water hole, town, or large body of friendly troops are met. (Example, Ojeda's counter-attack on Obregón, Naco, Sonora, June, 1913).[1] [This is not an apt example, because the forces mentioned on both sides were infantry. Obregón commanded few mounted infantry.]

If a flank attack by cavalry can be combined with a vigorous counter-assault of infantry from A, with a further pursuit by cavalry, a force of this character should be easily destroyed [a foreshadowing of Villa's defeat at Celaya].

A force of United States regular troops, say of a regiment of infantry, and a squadron of cavalry, with adequate ammunition, should have no fear of defeating and dispersing a rebel force of three or even four times its number, whether in attack, defense, or a meeting engagement in ordinary country. New or raw troops on service of this character will be at a great disadvantage, because much of this work will have to be done in very open formations and in detached groups on the open field, which is quite a different undertaking from defending a trench or making a direct attack in a shoulder to shoulder, or close formation.

Infantry alone on marches in the northern Mexico between places (garrisons) where water is available must exercise great care. Because, if struck in front (fig. 2, A) a deployment may be made in that direction; if struck, then, in flank at B, a deployment may be made in that direction, which, if extended to a rear and surrounding attack at C and D, is apt to place the defenders in a very precarious formation, as the hostile fire will sweep their whole force, front, flanks and rear. Being caught on the march in this way is especially dangerous in Northern Mexico, because if water cannot be obtained within 24 hours or less the men begin to die of thirst. (Example, Bachimba 1912, Orozco-Huerta Campaign). [Men dying of thirst has never been associated with the Battle of Bachimba, nor does a force being surrounded and "swept by fire" pertain.]

Under conditions of this kind an echelon or lozenge formation as shown in Fig. 2, E and F with wide intervals between companies or subdivisions and a distance sufficient to prevent the hostile fire from sweeping two or more echelons at one time should be adopted and the march pushed on until the objective is reached. In open

---

[1] See Janssens, *A Revolution in Military Affairs*, 1: 25-26, although this took place in April, not June, 1913.

country there need be little hesitation in prescribing intervals and distances of from 500 to 1,000 yards, according to circumstances, the nature of the ground, etc., as the power of the United States small arms and system of fire is such as to be able to sweep ranges of this length. The subdivisions, i.e., companies or platoons should be held well in hand and not allowed to scatter in any way. It must be held in mind that infantry when acting against mounted riflemen of this kind cannot pursue without detaching subdivisions, which, if they get out of tactical supporting distance from the main body, are apt at any time to find themselves completely surrounded by greatly superior numbers. The Mexican rebels can be depended on, however, under these circumstances to indulge in a protracted fire fight before closing in to short ranges. If, therefore, infantry is drawn into a protracted fire fight in position its ammunition is apt to be expended without any appreciable results being obtained, thereby being rendered helpless. Commanders of infantry columns attacked under these circumstances must push through with all vigor. The northern insurgent does not relish a rencontre or meeting engagement (hostile contact on the march). If struck in this manner his inclination is to concentrate to the rear rather than to the front if he wishes to fight, thus placing the burden of attack on the other party. In this case he holds his front with a thin fringe of dismounted riflemen, their horses close behind, and extends both of his flanks with a view to surrounding his opponent. Usually, however, he has better information of the enemy than the enemy has of him, and his attack is always a surprise or an ambuscade if it is at all practicable. Attacks of this kind at river crossings, in cañons [canyons], defiles, mountain trails and roads must be carefully guarded against by adequate reconnaissance. [see the Battle of Malpaso in December 1910] Especial care must be taken in camping to have a good outpost, definite assembly points for the units in case of attack and a definite line of resistance well understood by all.

Cavalry

Cavalry will find itself at greater tactical disadvantage than infantry in the river crossing, cañon, defile and mountain trail warfare. In the country outside of the mountains, on account of its mobility and ability to employ mounted action, it will be the arm par excellence for destroying and running down this sort of a foe, and even in the mountains, if properly handled, due to its power of rapid pursuit, will bear an almost equally important role.

The tactics of cavalry in the open country will be comparatively simple, especially due to the fact that mounted action against the rebel mounted riflemen will be more easily put through and be much more effective than against a really efficient cavalry force or against good mounted infantry armed with an excellent small arm and having a good system of fire direction and control. The Mexicans are very poor at mounted action, their horses are light in weight and they are totally unused to this kind of warfare. Individually, however, when put to it they are quick and expert, while mounted with the machete, pistol and lasso. Individual encounters between United States cavalrymen and the mounted Mexican will be by no means unequal. [a tribute to the Norteño centaurs]

The mounted charge against a dismounted line (except in prepared positions, which will be extremely rare) is perfectly feasible, as the lines so occupied will be very lightly held (not more than one man to from 10 to 20 yards), from which a fire effect equal to that from good infantry need be by no means expected. The dust raised by the charge and the fact that their led horses are almost sure to be destroyed or scattered will further influence the result. No more casualties are to be expected with tactics of this kind in Northern Mexico than were sustained by the force under General French of the British service when his cavalry division charged the Boer line of dismounted riflemen at the Modder River in South Africa. The density of the Mexican line will be no greater and their fire efficiency not as great as that of the Boers' was on that occasion.

With anything like equal numbers, more lasting effects may be expected by the mounted action of cavalry against the Northern Mexican insurgent than from any other, because if vigorously pushed it is almost sure to be successful, first against any dismounted line which should be ridden over, thereby getting into their practically unprotected and immobile led horses. Without his horse the Mexican rebel is very weak. Second, if they can be caught mounted on the march and quickly charged the great superiority of the United States Cavalry in mounted action will tell strongly. There will be little occasion in these mounted combats for a double rank formation, but a single line either boot to boot, with intervals of not more than a yard, or as foragers, combined with a rapid enveloping attack in a similar formation against a flank, will be found the best. Due to the loose and incompact formations of the adversary, compact formations are not necessary for breaking his lines, if he is fighting dismounted as much space must be covered, as practicable, to destroy, drive away or capture his led horses, and, if mounted, as much envelopment as practicable must be aimed as to prevent escape. The formation, therefore, should be extended sufficiently to obtain these results and still keep the units well in hand. It should not be drawn from those remarks that a reserve is not considered necessary. This will be found as necessary as against any troops, especially as surprise attacks on the front, flanks or rear are to be expected at any moment and as the enemy will know every foot of the country and have the assistance of the friendly inhabitants, these are apt to be embarrassing.

On the offensive against a well posted force (seldom will they be well, or even entrenched at all, except by natural cover), of northern rebels a dismounted frontal fire attack combined with a rapid flank mounted attack will often be found necessary. A dismounted frontal and flank attack will seldom yield decisive results, as the Mexicans will get to their horses and be off. It is against the adobe or masonry town prepared for defense, with incident "sniping" operations that the greatest difficulty will be encountered by cavalry. As the tactics necessary for this apply to the other arms as well, it will be treated of in another portion of this paper.

Cavalry, especially, will have to exercise great care when camping in the presence of this enemy. It has been mentioned before that the rebels are crafty at ambuscades and surprises. They are equally so at causing stampedes of horses by day or night. It is not difficult to see what would happen if one or two hand grenades were thrown into a troop grounded picket line at night, or stampeding a herd of cattle [not unheard of in the Mexican Revolution—see Campa in Durango, 1912, and the Maytorenistas at Naco in 1914] so that they will run into the camp. What was said of outpost duty and good measures for defense in case of attack by infantry applies even more so to cavalry camping in the open. Whenever corrals, towns, or places of that sort are available they should be used for the protection of the horses.

In the retreat from greatly superior numbers, especially with tired or worn-out horses, cavalry will find it necessary to adopt somewhat the same echelon formation above concerning infantry. It must be held in mind that in Northern Mexico, outside of the mountains and some swampy spots, pursuits will not be confined to roads as the whole county is available for the horsemen.

If forced to a passive defensive, with adequate ammunition, dismounted fire action will keep a greatly superior force of Mexicans off, especially if good cover or artificial protection is provided.

A vigorous counterstroke or counter-attack can be pretty effectually counted on, and a vigorous attack against greatly superior numbers will offer better results in Northern Mexico than against nearly any other troops. A commander who otherwise is hopelessly outnumbered may often extricate himself from his dilemma in this way (Example: Ojeda's attack on Obregon's Insurgents, Naco, Sonora, June, 1913). [Again, Obregón defeated Ojeda at Naco, and in June 1913 the two were engaged in fighting down south in the Guaymas Valley, so it is difficult to understand why the Captain continues to mention this single tactical move that in no way proved decisive and did not involve cavalry on either side.]

Field and Mountain Artillery

The notes made above have particular reference to operations with forces not very great in numbers; that is, from 1,000 to 4,000 Northern Mexican Insurgents and United States detachments or units of a strength from a battalion to possibly a brigade, but not larger. While the insurgents will not be able to keep forces larger than the above in the field for any considerable period of time, due to the occupation of the coasts, railways, etc., by the United States, still in the initial stages forces of the size of Villa's, Obregón's or Pablo González's, i.e., from 8,000 to 18,000 men, may be encountered. These forces may put up several good battles before they begin to scatter, especially if, in case of invasion, they make common cause with the Federals and receive ammunition from the South. Against large forces of this kind, field artillery will play a very decisive part, because there are two things that the Mexicans are not used to: first, a really efficient cavalry, capable of fighting mounted, and, second, a really efficient field artillery that can hit something. Even as it is the Federal artillery, inefficient as it is, has played a very strong part both in attack and defense. (Huerta's attack on Orozco, Bachimba, 1912, defense of Guaymas, Mazatlán, etc.) [another error, artillery played no decisive part in the Mexican Revolution.] The tactics of this arm (artillery) acting as a part of an infantry division or reinforced infantry brigade, will present nothing novel, except that an advance will be practicable to much closer range than against an efficient enemy. It will be found better to march the artillery pretty close to the heads of columns so as to bring it into action quickly [This placement of artillery in a movement to contact was Federal Army doctrine.]; the dangers incident to this position in the column will be minimized to a greater extent in Northern Mexico than if acting against an efficient enemy. Due to the absence or great inefficiency of the artillery of the Northern Mexicans, it will seldom be caught by the insurgent artillery in column of route or in close formation beyond 2,000 yards. Masked fire will not have to be used, except as a matter of concealment from view (as distinguished from its necessity as concealment from fire), to the same extent as against a more efficient enemy; neither will the small arms fire of the Northern Mexicans be as effective against it, as would be the case if acting against European troops.

Care will have to be exercised to protect it with cavalry or infantry at all times, as it will be particularly vulnerable if caught by the Mexican mounted riflemen alone and unprotected.[2] That is, the Mexican mounted riflemen will be quick to avail themselves of a temporary isolation of any field artillery which they may find, because they do not look upon an action in quite the same way that a European force would. The latter would hold as their main object the destruction of the whole force, whereas the Mexican will destroy anything he sees plainly and thinks he can do easily for the time being, without reference, particularly, to the whole operation. When the Northern insurgents break up into small groups there will be little further use for field artillery.

Horse Artillery

As long as forces of insurgents of a size of from 1,000 to 4,000 remain in the field, horse artillery will be found of great importance and assistance. Its use will present no especially new or different tactical problems. The greatest lightness for material of this kind will be of the utmost importance in Northern Mexico, on account of lack of water and forage, with the incident wear and tear on the animals, etc. Such units should have their gun carriages, limbers and caissons stripped down to the barest necessities to assure this mobility. Seldom will a force of more than a battalion of horse artillery, with a single column be used, more often the use of batteries, platoons,

---

[2] See related discussion in Janssens, *Rise of the Praetorians*, 294-95, 301.

or even sections will be found necessary [deployment of artillery in numbers lower than battery groups—i.e., multiple batteries—was specifically disallowed according to Federal Army doctrine because of the demonstrable ineffectiveness of such limited firepower, so one wonders why Mitchell would advocated such a flawed practice] according to the size of the force they accompany, the nature of the terrain and the composition of the force they will be opposed to.

Mountain Artillery

The use of mountain artillery material by the Mexican Federals has been confined mostly to the plains and along roads or in country where it could be hauled on wheels; and we have no instances where it has been used strictly as a mountain piece in Northern Mexico; that is, in a terrain, where wheeled artillery could not be taken. In many, if not the majority of instances, the Federals have pulled the guns along with draft mules and carried the ammunition on pack mules. This is a very good method with commands of infantry of the size of from a battalion to a brigade, because, when acting against the insurgents, a great expenditure of ammunition is not necessary, which therefore does not require a great many pack mules. This equipment takes up much less room when transported on railway trains, with from 30 to 50 rounds to the gun than field artillery; it takes up less road space, 2 mules pulling the gun (4 to pack it) and 1 ammunition mule to each 10 rounds, and is more easily handled in a small column than the larger field artillery. One, two, or more of these pieces are often taken with small infantry columns by the Federals. They are especially effective against adobe or masonry towns, as they throw as heavy a shell as field artillery and have a high angle of fire. In fights for a town these guns are quite convenient to use around the streets, etc. (This use will be mentioned later in connection with "sniping questions.")

While the Federals have used this material very seldom in the mountains of Northern Mexico, because they have seldom followed the insurgents that far, it may be reasonably expected that, when driven out of the plains, the insurgents will take to the hills and put up a very determined resistance; also, in such States as Guerrero and Colima almost all the warfare will be strictly "mountain" in its nature. This material, of course, will be very useful in these operations. Especially as the insurgents will avail themselves of entrenchments, which will become of a heavier and heavier character as their cause becomes more desperate. There will be plenty of use for mountain artillery in any advance from the coasts to the interior of Mexico.

There might be use for field howitzers using projectiles of from 30 to 60 lbs. on an advance from Veracruz to Mexico City; but aside from an advance on the Capital, little use will be found for this heavy material in Mexico. The Mexicans have none of it themselves.

[Comment: The U.S. Army seemed to be making the same mistake as the Federals, assuming that they would only be campaigning against small bands of insurgents, mounted infantry deployed in a widely dispersed "line of foragers" at intervals of "not more than one man to from 10 to 20 yards," and needing only to be able to destroy adobe and light masonry and other targets of opportunity. They left little or no room for the need to saturate zones of light infantry in conventional battle, except in the drive from the coast to Mexico City, it would appear; nor did the Americans consider that Federal Army artillery may have been so ineffective in set-piece battles precisely because it contained only guns and not howitzers and mortars.]

Federal Army

The tactics of the Federals are those learned from European mentors. The general characteristics of the personnel have been mentioned above. As any campaign waged by the United States against Mexico will be an offensive one, the Mexican Regular Army will be found in prepared positions on the defensive. The works so far built by the Mexican Federals have been very rudimentary from a military engineering standpoint. They have been poorly placed, with few or no covered approaches; no headcover to protect against shrapnel fire; no loopholed or crenelated parapets or cover for supports or reserves [another reference to the ternary system]. The protection for their artillery has been very poor also. Barbed wire entanglements have been used to some extent and some fougasses or mines. The entrenchment against an American invasion will undoubtedly be far better than any heretofore used against the insurgents and barbed wire entanglements of good construction will also be encountered. These works will be placed to cover the approaches to Mexico City at points which are the most difficult for the attacker to maneuver against, in a mountainous country covered with vegetation, deep barrancas, cañons [canyons] and ravines, and in some seasons of the year with swiftly flowing streams.

The tactics to be used against these positions are those ordinarily taught in the United States service to be used in a terrain of this character. The Mexican troops can be depended upon to stick to their trenches, and as they ought to have adequate supply of ammunition on account of the proximity of railway lines, a great volume of fire to the front may be expected. This, however, unless rests or frames for the small arms are provided in their entrenchments, will be poorly directed and controlled and will be high. [For this reason Federal Army doctrine specifically encouraged aiming lower rather than higher.] Machine gun fire against possible lines of approach such as ravines, ditches or stream lines will be effective and the machine guns will probably be well concealed and difficult to locate. The artillery will be well concealed, but probably not well placed. However, as the ranges will be well known, it should exert some effect. [This accurately predicts conditions faced by the Villistas at Celaya in 1915, although the Constitutionalist artillery *was* effectively placed.] In bayonet fighting or in a defense at very close quarters the Mexicans will not be as efficient as good troops. [The savage close quarter combat on La Pila at Fourth Torreón suggests otherwise.] Against a purely frontal attack, however, a pretty good account will be rendered by the Mexicans.

Against envelopments or turning movements they will be very weak, especially if at the same time they are held to their positions by a well-handled frontal attack, because due to their deficient military training a change of front, concentration of what reserves are held out on a flank or their rear is a difficult undertaking for them, and when their line of retreat is threatened they are very apt to be seized with panic. [The same could be said of any army of any country on the planet.] Once started, their retreats are made in a very poor manner and they are particularly vulnerable. If mistakes are made by the attacker, they can be covered up to a greater extent than against an efficient army, as the Mexicans will be slow to avail themselves of their opportunities and their counter-attacks will not be well supported. [This second part of the paragraph accurately described the Federals, but not the Constitutionalists, and certainly not General Obregón's army corps.]

From these characteristics it is evident that a good enveloping attack against the flank nearest to their line of retreat should be used wherever practicable, and as the terrain in these places will be very difficult to traverse, a great deal, in fact, the deciding factor ordinarily, will be an adequate reconnaissance. [These prescriptions are so basic as to be trite.] An entire or even partial reliance on maps for minute incidents of the terrain, without an excellent reconnaissance, may result in very serious complications. For instance, a barranca with precipitous sides of a depth of 300 or 400 feet is apt to be met with anywhere, which, if not known or prepared for in advance might jeopardize the whole operation. [Acquiring this intelligence was easier said than done, as Obregón's 1915 advance on Aguascalientes proved.]

The greatest familiarity possible with the localities in which operations are to be conducted should be obtained from existing data beforehand, and a very careful and efficient reconnaissance made upon actually reaching the locality. General Scott always caused very careful reconnaissance of the ground to be made in the Mexican War. While these tactics will undoubtedly succeed comparatively easy, it must not be imagined that casualties will be very small or negligible; they are apt to be quite heavy, especially during the first few fights, when a determined resistance is being made by the Mexicans and before this type of warfare [To suggest that after almost four years of combat the Mexicans were ignorant in "this type of warfare" was risible—witness the flanking maneuvers executed by the Colorados at both First and Second Rellano, and the Federal Army's favored "*martillazo.*"] is thoroughly understood by the attackers.

Upon reaching the plateau of Mexico, after the passage of the mountains, if a combination of Mexican Federal [the Federal Army only lasted for another month after the release of this paper] troops, and the Northern Mexican mounted riflemen [Completely ignored here was Obregón's Army Corps of the Northwest, the best

Mexican fighting force at the time and perhaps the best light infantry in the world] were effected, a very efficient force would be the result both for attack and defense, especially as the line of communications of the attacker would then be 200 or more miles long. The attacker must be prepared to push his advance rapidly upon gaining a foothold on the plateau to the City of Mexico. Because if the City of Mexico can be captured and occupied the bulk of the Federal infantry and artillery will be captured, scattered and the sources of their ammunition supply in the hands of the attacker. A wait of several months such as General Scott made at Puebla in the Mexican War would be much more serious at this time, because now the means of communication are so much better that more defenders can be collected in a short time, and the ammunition factories in Mexico City can turn out a considerable number of cartridges every day. The railroads will, undoubtedly, be kept in operation for moving troops and supplies up to the last minute. Upon reaching the plateau and adequate force will be very necessary to push straight ahead.

Sniping Operations

In all wars in Mexico resistance, varying in its intensity, to prevent the complete occupation of a town, has been made even after an enemy reaches the outskirts of a town and has defeated the field force in front of it. It takes the form of firing from roofs, porticos, street corners and windows, and is always fruitful of many casualties. In these operations the Mexican troops are joined by the citizens, often both men and women. Again resistance of this kind may be kept up by the civilians after the withdrawal of the troops.

The cities and towns in Mexico are of masonry or adobe brick construction. The roofs are more or less flat with copings around them, which offer footing and protection for riflemen. These may be further improved in a hasty manner by sand bags or even overhead cover, and emplacements for machine guns or even artillery. The walls of these buildings resist small arms, machine gun and shrapnel fire. Behind most of the houses in the blocks are patios and open places, separated by walls 8' to 10' high. Many of the doors of the houses and buildings are of very solid construction. Most cities and towns are laid out in quite a regular manner; that is, the streets are straight for considerable distances, especially the wide ones. If troops are caught in column in these streets, especially by the fire of machine guns, the casualties will be heavy. Even when troops have passed houses or blocks of houses, unless these are thoroughly searched, snipers appear and shoot into the rear of the columns.

The occupation of most towns will present a problem of this kind, which, although it can have little bearing on the general outcome of the war, may cause a great many casualties which probably can be avoided by good handling and management of the troops. Resistance of this kind are especially to be looked for in coast towns and ports when the first occupation takes place.

It is obvious that if the attackers advance boldly up the open streets in plain view they will offer the best targets, and have the least protection from view or fire. Also, if several parallel streets are being used as lines of advance, some columns will progress more rapidly than others. The attackers will in many cases not know where the fire against them comes from, and will begin firing not only up the streets on which they are advancing, but along streets at right angles to their advance. The result being that indiscriminate shooting into each other is very apt to result, especially if the occupation of one of these towns has to be made during the night.

Under these conditions a system varying according to circumstances something as follows should be adopted:

A certain amount of front—2, 3, 4, or more blocks—should be allotted to a subdivision of the command such as a company, battalion, or regiment until the whole front of the town is covered.

Each subdivision commander should make details for an advance between the streets, i.e., back of it, or if the intensity of the enemy's fire warrants it, over or through the house, blowing holes with explosives, or making them with crowbars through the walls for this purpose. This may be designated the attacking line. Any prisoners taken by it should be turned over to the support, so as to keep all men practicable on the line.

Another detail should be made to thoroughly search every house in the block for concealed firearms or individual snipers. (This group may be designated the support). The inhabitants should be told that no harm will come to them in any way if resistance is not made.

A third detail or reserve should be kept together for receiving the prisoners turned over to it by the support, and be utilized to reinforce the attacking line or support, as the case may be.

The attacking line when they reach a cross street should signal to adjoining groups and thoroughly clear the street; a signal should then be given either by trumpet or flag, for the whole line to cross the street and occupy the next row of blocks. Especial care should be exercised that no fire be directed along cross streets which might take other detachments in flank or reverse. The attacking line should be closely followed by the support, who, in addition to other work, should leave a few men in each block to keep it thoroughly under observation and capture or kill any snipers who may have escaped the first search and appear.

This method of advance should be kept up until the whole city is occupied.

The commanding officer should take up a central position where all reports can be sent and where a general reserve should be held. This reserve should provide a place for guarding all prisoners brought in by the various units, and as the various sections of the town are brought into subjection should attend to their police, thus relieving the support and local reserve detachments allowing them to move forward and rejoin their proper commands. Upon the complete occupation of the town the units which have worked through it should be assembled and disposed according to circumstances. The general reserve, or such part of it is necessary, having taken over the police of the city and the guarding of prisoners.

If cavalry constitutes part of the attacking force, it should occupy the approaches of the town opposite to those from which the attack is being made, to cut off reenforcements to the city and capture or destroy any forces leaving it. They should also keep all approaches and exits under observation. They should not become involved in the street fighting unless there is some very obvious necessity for it.

The character of the houses prepared for defense, barricades in the streets, and the defenses on the house tops, especially on high buildings such as churches, convents and the like, may require the use of artillery. Providing the city is a large one it may be necessary to enter the streets with this material, as the range from the outskirts may be too great or the positions unfavorable. Field artillery in transit through the streets will be particularly vulnerable. When it goes into action it will have to be used from street corners or behind sand bag, barrel or other hastily improvised parapets usually on a paved surface in the streets. Mountain artillery could be carried piece by piece, if necessary, behind the houses, through or over them or be used from the roofs, embrasurers in walls, etc. If mountain artillery is available it will be more useful than field artillery in this duty, using shell for demolition work and, possibly, shrapnel to a limited extend against personnel.

If the city is near a port or river on which gunboats are able to operate, or if artillery emplaced outside of the city can reach all parts of it, all blocks and positions in and around the city should be designated by number or otherwise, so that the fire of this artillery could be directed by signals from the forces fighting in the towns.

Hand Grenades

The Mexicans in general, due to the great number of mines in the Republic, are very used to handling explosives of all kinds and they are expert in their use. Hand grenades of all sorts and varieties are used. These are sometimes thrown by hand, sometimes by slings. With slings, they are thrown distances of from two to three hundred feet. in street fighting (and also in the attack on entrenchments), these grenades are capable of doing a great deal of damage both in defense and attack. They are very effective against personnel as well as material. For blowing down heavy or barred doors, making holes in walls or throwing onto the roofs of houses occupied by snipers, they will be especially effective. American troops campaigning in the thickly populated districts of Mexico will find considerable use for hand grenades, and should be very familiar with their use.

Engineer Troops

The duties of engineer troops accompanying troops in Mexico, aside from the railroad work, will be concerned to a great extent with reconnaissance, obtaining accurate details of roads, trails, the terrain and obstacles to the march. To make this of use it must be collected quickly and the data so drawn up that it will get to the units in the shortest time. A fine system of combined sketching methods and efficient map re-producing equipment is therefore very necessary.

Light bridges of all sorts to be put down quickly, and methods of lifting and lowering mountain or other artillery over precipices too high to be handled by their own personnel should be provided for. While there will be little use for heavy pontoon material, there will be use for light pontoon material capable of bearing a moving load of 1,500 lbs. There will not be the necessity for heavily entrenching lines, mining or sapping, demolitions and construction of military obstacles that there would be against a very efficient enemy. A great deal of pioneering work especially in the mountainous portions of Mexico will have to be done.

The finding and provision for water will be very important. In many, if not the very great majority of the wells in Mexico, water has to be pumped by steam engines. These will undoubtedly be destroyed, and possibly the pipes of the wells damaged. Efficient pumps, such as the motor fire engines used in the United States, with proper couplings for the pipes ordinarily met with in Mexico, will be of great utility. These could either be used along railways or could move under their own power away from the railways. Provision should also be made for driving wells where necessary. The water problem is one of the hardest to solve in connection with the pacification of Northern Mexico. There is little water on the surface, but plenty from 100' to 400' down.

There will be little use for searchlights or battlefield illumination. [The Constitutionalists made use of searchlights often, so it is hard to understand how the Captain arrived at this conclusion.]

Signal Troops

Aeroplane reconnaissance in all places should offer a great aid in locating the main force of the enemy; in finding out whether large bridges are destroyed, railways in operation, etc. In the North, due to the dust thrown up by columns, even small forces on the march should be easily located. Furthermore, aeroplanes have a very decided moral effect on Mexicans of all sorts. (Examples: U.S.N. flying boats at Vera Cruz, insurgent aeroplanes at Guaymas and Mazatlán).

Efficient field lines of information at all times on account of the enveloping and turning movements required against large forces, and later in the general pacification of the country, due to the great scattering of the units, will be of great use and very necessary. Radio equipment both of a mobile type for use with troops and of a semi-permanent character for use in garrisoned towns, will have great application. The personnel of Signal troops operating in Mexico should be armed with a carbine in addition to the pistol.

Documents Cited

The strategical lines of information, permanent telegraph lines along the railways, telephone systems at and near cities, submarine cables and radio stations, will require a great amount of material and personnel.

Conclusion

It must not be taken from the above remarks that the Mexican either of the North or the South is a foe to be despised in war. On the other hand, he is excellent material from which to make a soldier. What he lacks is an efficient corps of officers and adequate military training. [Not so with Obregón's Army Corps of the Northwest] The officers of the Federal Army have some good education, but little initiative; those of the insurgents have a good deal of initiative, but no education, except in an empirical way. Neither are imbued with a patriotism such as we understand that word to stand for. Could the military material of Mexico be welded into one homogenous army under a really efficient corps of officers, it would constitute a force the equal of any for fighting efficiency and the ability to withstand hardship and privation.

A Mexican constabulary organized by the United States from natives with American officers undoubtedly could be made a very effective force.

At present, the Mexicans offer a comparatively primitive military organization for the defense of their country against an efficient military force sent against them.

5761-996

SUBJECT: Federal evacuation of Mazatlán, Mexico.
From Ua3. Capt. Walter F. Martin, Cav., Military Attaché, Guatemala.
No. 428 Date October 17th, 1914.

1. The following is a copy of a letter received from A.A. Ruffo, a Mexican, Agent of the Pacific Mail Steamship Co. and Honduras Consul in Mazatlán:

2 "Mazatlán has been finally taken by the Constitucionalistas, that is, the Federal Forces evacuated the city and the Constitucionalistas immediately came in. This was on the night of the 9th. inst. On the first or second of this month the Cia. Niaviera's boats and the "Guerrero" arrived here from Manzanillo after taking the Guaymas, Federal Forces to that port, and the evacuation of this city commenced with the utmost difficulties as this had to be done at Olas Altas, in front of the Casa de Gobierno. This operation lasted until the night of the 9th, that is, practically one week as there were about four thousand men, women and children, millions of cartridges, shells and sixteen cannons to move out and you know how rough the sea is at Olas Altas at this time of the year. It was impossible to do anything at the regular wharf for the reason that the Constitucionalistas occupied Isla de la Piedra, (O'Ryan's Island) and they were constantly shooting towards the pier and the Custom House."

3. "On the afternoon of the ninth inst. everything was ready, that is, with the exception of about 400 soldiers and officers, all the Federals were on board the various

121

steamers. The 400 men were stationed at the out-posts keeping the Constitucionalistas at bar. At 8 p.m. these 400 men abandoned their posts and rushed into the city and made their way to Olas Altas, where two large lighters were ready to receive them and take them to the steamers. At about 10 p.m. about half of the men were in the lighters, which had cables running clear out to the steamers and were to be towed out as soon as the last man was on board. Unfortunately for them however at this moment the Constitucionalistas rushed into the city, and immediately attacked the Federals that were in the lighters and those that were making efforts to reach the lighters. At this critical moment the cable or hawser with which the lighters were to be hauled out to the steamers was either intentionally cut or parted with the strain and the poor federals were obliged to abandon the lighters and use the "malecon" as breast work and fight until they were overpowered as might be expected and those that were not killed were taken prisoners."

4. "During this last fight the "Guerrero" shelled the city; about twenty shots in all having been fired, fortunately without doing as much damage as might be expected. One shell dropped at H.W. Felton's residence on Calle Principal, without exploding, another passed through the second story of Mr. Claise's house at Olas Altas, the house formerly occupied by Mr. Natividad Gonzalez, and the rest, in the neighborhood of Olas Altas, wounding and killing but very few people."

5. "On the morning of the 10th. order was practically restored; at about 10 A.M. all the officers taken prisoner, about 17 (seventeen) in all, sub-lieutenants, lieutenants, captains and one colonel were shot at the Custom House and also all the volunteer soldiers, in all, as far as known at this moment, about forty. Others captured later have been shot outside of the city but the number is not known."

6. "The last few nights, preceding the foregoing, there was severe fighting between the federals stationed at Loma Atravesada, La Montuosa and the other out posts, and the Constitucionalistas who made desperate efforts to take posts held by the federals and the losses on both sides were several hundreds killed and wounded. The Casino de Mazatlán, the Circulo Comercial Benito Juarez and other large houses have been converted into hospitals which are practically filled with wounded Constitucionalistas."

7. "On the 1st. of this month the Chinese Colony was taken to Islas de Venados and practically all the other foreigners to the U.S.S. "California" where they remained until yesterday."

8. "Shortly after the foreigners left the city, nearly all the Chinese stores, including "El Palacio de Hierro" and "La Competencia" were looted by the federal soldiers and the towns-people who left nothing at the stores but the shelfs and counters, and in some cases even these were carried away and used for firewood. At a request of the Consular Corps, General Rodriguez sent soldiers to stop the looting, and sent guards to all the principal stores and order was practically restored. As you might expect

during the last two months or so the poorer people suffered very much on account of the scarcity of all the foodstuffs and during the last couple of weeks even those who had money could not buy anything with it as there was absolutely no flour, corn, beans, lard, etc. Beef was obtainable in very small quantities once or twice a week at from $3.00 to $4.00 per kilo and milk at $1.50 per litre. Fortunately, I had a pretty good supply of canned goods, flour, and corn which lasted fully up to the day the town was taken, and we, therefore, endured the siege fairly well."

9. "I was invited to spend the last week on board the U.S.S. "California" but I couldn't induce my wife to go for the reason that I wanted to remain ashore to witness the last battle and also because I was afraid that if we all went and left the house alone we would find nothing left on returning."

10. "I am living at Mr. Manuel Gomez Rubio's house on Calle Principal, near the beach. When Mr. Rubio went away he left the house in my care and as he is not in favor with the Constitucionalistas it has been the Hondurenian [sic] Flag which for the last fifteen days has been waving over the house, that has saved it from being converted into a 'cuartel' or a hospital. The very night they came in they wished to occupy the house, but fortunately they respected my flag of which I am very proud.'

11. "The foregoing, is in brief, what has happened here during the last few days. All is quiet here now and the stores are doing practically as much business as they can handle, especially in the retail line, the money in circulation being Constitutionalist paper of every colour and size and from every state occupied by them."

5761-1091/8

[This Anonymous Intervention Memorandum is inexplicably included in 5761-1091/8 along with correspondence relating to the Torreón Accords and the location of Ambrose Bierce.]

It is thought in case intervention in Mexico is necessary, that no invasion should be made except from Vera Cruz to the City of Mexico. This with a view to establishing a protectorate, and as soon as order is restored evacuate the country. It is believed that this should be done by a small but highly efficient force. The smaller the force that can accomplish the purpose, the better. A small force, say 20,000 men, can be supplied with wagon or automobile transportation sufficient to carry supplies for a month. A large force, say 50,000 men, would have to employ half of its effective strength to guard the railroad, which would be necessary for its supply, since to supply such a large force by wagons would be almost impossible.

In view of the fact that the Mexicans have no wagon transportation, and depend entirely on the railroad for transportation, it can easily be seen that our force, which goes independent of railroads, will be able to maneuver in such a manner as to place the Mexicans at great disadvantage, since we can make our choice of attacking in front, in flank, on rear, or, if need be, ignoring them altogether.

It is also apparent, I think, that such column should include a large proportion of cavalry, including an independent cavalry column, supplied with artillery and with an automobile train for carrying forage and rations, which independent cavalry column would be used for the purpose of flanking the enemy out of his chosen positions by marching around him. The army should include a large proportion of artillery to make the entrance into the City of Mexico easy, and for assuring the security of the American army while the City of Mexico is being garrisoned.

In such a movement on the City of Mexico expedition would be of the first importance. Successes gained to be followed up with the greatest promptness, the army should get out   of the unhealthy coast region and up on the high table-land, if possible within a week. The use of light automobiles for carrying infantry would be of enormous advantage, especially in the dry season. There is no question that this means of transportation will pay for itself in the results obtained.

It is claimed by many that part of our army should be utilized in such an invasion in pushing columns across the border into the interior of Mexico and taking possession of such towns as Monterey, Saltillo, Torreon, Chihuahua, Hermosillo, etc. I entirely dissent from this idea. We are told by the Mexicans themselves, and I think it true, that all the warring elements of Mexico would unite in combating our invasion. And this is what we desire, since when on arriving at Mexico City we sould [sic] be able to treat with all the Mexican commanders. But if, on the other hand, there were a number of invading columns sent in by the United States the Mexican force would be split up and a success at Mexico City would lose in importance.

Then again, the columns employed in the northern part of Mexico would necessarily be small and it is very likely that one or more of them might be unsuccessful on account of the odds employed against them. And again, when these columns reach their objective they would all be placed more or less on the defensive under very difficult circumstances, each having a long line of communication to keep up.

Again: the movement of these columns into Mexico would not necessarily protect the border of Texas, New Mexico and Arizona from Mexican bandits and brigands, who would operate in rear of the marching column and would infest the country near the border with a view to cutting off their communications.

On the other hand, it is evident that certain positions on the other side of the border should be occupied in order to prevent the towns on this side of the border from being bombarded. But I think that this is as far as an invasion of Mexico in its northern part should go.

The principal danger that confronts us in such an intervention is the destruction of small American detachments whereby public opinion may be brought to favor an occupation of the whole of Mexico. Such an occupation would require an enormous number of troops, a great expenditure of money, and might accomplish nothing in view of the fact that the American people are opposed to incorporating the Mexican race as American citizens.

Returning to the consideration of an invasion via Vera Cruz, in the advance to the City of Mexico the railroad as a means of transportation I think should be ignored. However, as soon as the enemy is beaten and dispersed, and Mexico City is occupied, the railroad should then be repaired, so that it can be used for shipping supplies from the coast.

This would require a considerable force, made up partly of civilians and partly of troops. It is not believed that any portion of the force that had reached Mexico City could be used for this purpose, since most of them will have to be a mobile force constantly in movement outside the City to prevent the enemy from concentrating against us. Part of it must of course garrison the City. To guard the railroad therefore it might be advisable to employ some Volunteers, it being assumed that the beginning of the war would be coincident with a call for volunteers.

As a matter of policy, however, it would be exceedingly advisable if the regular army were given an opportunity to show its efficiency and the fact that the great sums spent on it are not spent in vain, and that it is always ready at a moment's notice. The invading army should be entirely composed of regulars.

To constitute this force there are available in this country 23 regiments of infantry, 12 regiments of cavalry, four regiments of field artillery, 5,000 or 10,000 coast artillery, and a large proportion of United States Marines. These troops can be recruited up during the war by enlisting for them only instructed soldiers, of which a considerable number would probably apply. In addition, it should be possible to bring from Porto Rico, the Philippines and Hawaii at least a brigade of infantry and a brigade of cavalry, which would be useful in maintaining lines of communication, etc. The retention of these troops in these dependencies would not assure their safety.

It is a feature of the Mexican character that when menaced by aggression from the outside the government invariably becomes stronger, discords cease, rival parties amalgamate, and the best people come forward to take their places at the head of affairs, displacing low-down politicians.

Thus aggression on the part of the United States in the way of an intervention will of itself at once improve the state of affairs in Mexico, as far as order is concerned. The fact that Americans and American property will suffer should not be considered in this connection. As a matter of fact the damage done to American property is already so great that it is not likely that this will be a feature of much importance. Intervention then will be a blessing for Mexico for the time being. And if there be established a protectorate whereby the United States asserts its right hereafter whenever disorder occurs to intervene, it will be for Mexico a permanent blessing and the beginning of a state of prosperity unknown in the past. This prosperity can be brought about by giving the new state a slight measure of reciprocity with the United States. The peaceful character of the 500,000 Mexicans in the United States proves that the Mexican peon when prosperous is always orderly.

<u>5761-1091/19</u>

[Un-accredited letter from the First Chief written in all-caps]

CÓRDOBA, Ver. November 22, 1914. -

I HAVE LEARNED THAT GEN FUNSTON, IN EVACUATING THE PORT OF VERACRUZ, WILL DO IT BY SIMPLY LEAVING THE CITY WITHOUT MAKING PROPER DELIVERY OF THE PORT, AND THIS ACTION WOULD AMOUNT TO LEAVING THAT CITY IN THE HANDS OF ENEMIES OF THE CONSTITUTIONALIST GOVERNMENT. THE PORT OF VERACRUZ IS CONTROLLED AT THE PRESENT TIME BY THE MEXICAN POLICE, WHICH WAS MADE UP, UNDER THE AMERICANS, BY POLITICAL REFUGEES OF THE HUERTA GOVERNMENT IN THE GREATER MAJORITY, SUCH AS AGENTS VILLAVICENCIO, BOLANOS, VELEZ, ETC. BESIDES, UNDER THE PROTECTION OF THAT SAME POLICE, THERE HAVE ENTERED VERACRUZ A LARGE NUMBER OF SOLDIERS, WHO WERE THE REMAINDER OF THE FEDERAL ARMY OF HUERTA, AND THEY OCCUPY COLON AND PASAJE HOTELS, SOME HOUSES IN ARISTA AND FOURTH OF MAY STREETS, AND BEHIND THE CEMETERY, UNDER THE COMMAND OF GEN. HIGINIO AGUILAR. THESE TROOPS ARE WELL KNOWN TO BE THE REST OF HUERTA'S ARMY, FOR THEY ARE THE SAME THAT REFUSED TO TAKE LEAVE OF ABSENCE REVOLTING IN PUEBLA ON AUGUST 21ST. I AM AT A LOSS TO KNOW HOW IT IS POSSIBLE THAT THESE TROPPS [sic] ARE WELL PROVIDED WITH ARMS AND MACHINE GUNS AS THEY REALLY ARE. THE EVACUATION THAT IN SUCH CONDITIONS GEN. FUNSTON INTENDS TO CARRY OUT, AMOUNTS TO DELIVERING THE PORT TO THE REMAINDER OF THE FEDERAL ARMY WHICH WAS UNDER GEN. HUERTA. IT MAY BE SAID, THAT GEN. FUNSTON IN REFUSING TO MAKE A FORMAL DELIVERY OF THE PORT TO ANY POLITICAL PARTY, SIMPLY FULFILS [sic] HIS DUTY, BUT TAKING IN CONSIDERATION THAT BY ABANDONING THE PORT THEY WOULD PRACTICALLY DELIVER IT TO THE HUERTA ELEMENT, WHO TO-DAY PRETEND TO CALL THEMSELVES VILLA FOLLOWERS, IT WILL BE CLEARLY SEEN THAT THIS PROCEDURE WOULD BE AS OBJECTIONABLE AS ANY OTHER. ON THE OTHER HAND, GEN. CÁNDIDO AGUILAR IS CIVIL GOVERNOR AND MILITARY COMMANDANT OF VERACRUZ, AND THERE IS NO AUTHORITY HAVING OR PRETENDING TO HAVE CIVIL AND MILITARY CONTROL OF THAT STATE. THEREFORE, AS A LEGITIMATE AUTHORITY OR EVEN AS A DE FACTO AUTHORITY, HE IS THE ONE WHO SHOULD RECEIVE THE PORT TO AVOID FURTHER TROUBLES. TO DELIVER THE CITY TO ANOTHER PERSON OR BY SIMPLY ABANDONING THE PORT IN THE HANDS OF THE ENEMIES OF THE CONSTITUTIONALIST GOVERNMENT WOULD BE TO DELIBERATELY EXPOSE THE CITY TO BEAR THE CONSEQUENCES OF AN ARMED STRUGGLE DIRECTLY CAUSED BY THE CONDITIONS IN WHICH THE PORT IS EVACUATED. IT IS IN MEXICO'S INTEREST THAT THIS DO NOT HAPPEN, BUT IT IS ALSO IN THE INTEREST OF THE AMERICAN GOVERNMENT THAT THE EVACUATION DOES NOT GIVE ROOM TO DISORDERS, FOR IF SUCH AGITATIONS WERE DIRECT CONSEQUENCES OF THE MANNER IN WHICH THE AMERICAN TROOPS EVACUATE, THERE COULD BE ROOM TO QUESTION THE SINCERITY OF THE WASHINGTON GOVERNMENT TO EVACUATE VERACRUZ. IN AS MUCH AS AN ERROR ON THE PART OF GEN. FUNSTON TO EVACUATE THE PORT, COULD BE INTERPRETED AS A DELIVERED PURPOSE TO BRING ABOUT DIFFICULTIES TO THE MEXICAN AUTHORITIES. TO AVOID SUCH DIFFICULTIES, THEY

MUST TAKE IN CONSIDERATION WHAT I HAVE STATED ABOVE AS THIS COULD BE ACCOMPLISHED BY GEN. FUNSTON DELIVERING THE PORT TO GEN. CÁNDIDO AGUILAR, IN THE SAME MANNER AS HE HAS MADE DELIVERY OF CIVIL DEPARTMENTS, OR IN CASE HE DECIDES TO ABANDON THE PORT, HE SHOULD ASCERTAIN FOR HIMSELF THAT THERE ARE NO ARMED GROUPS OR BUILDINGS IN READINESS TO RESIST GEN. CÁNDIDO AGUILAR'S ARMY. GEN. FUNSTON KNOWS VERY WELL THAT BY DELIVERING THE PORT TO GEN. CÁNDIDO AGUILAR, WHO HAS FORCES SURROUNDING THE CITY, TRANQUILITY IS ASSURED.

<u>5761-1091/28</u>
War Department.
Office of the Chief of Staff.
Washington.
October 5, 1914.

MEMORANDUM FOR THE CHIEF OF STAFF:
Subject: Persons affected by the withdrawal of U.S. troops from Vera Cruz.

Refugees in Vera Cruz	267
Employees of the Department of Public Safety	242
Correctional Department	3
Charity Department	3
Mexican Employees, U.S. Postal Agency	21
Mexican Employees, Department of Public Health	64
Employees of San Sebastian Hospital	27
Employees Zamara Hospital	6
Employees Detention Hospital	4
Employees Loreto Hospital	12
Total	649.

In the 267 refugees listed above, many notations appear as Husband and Family, the number in the Family not being given so that the above total of 649 is naturally less than the real number of persons affected.

[signature indecipherable]
Captain, General Staff.

[This is a cover letter prepared by the General Staff for Veracruz Consul W. W. Canada's memo.]

No. 1038.

American Consulate
Veracruz, Mexico, September 21, 1914.

U.S. Army Intelligence in the Mexican Revolution

Subject: Persons Affected by the Disoccupation of Veracruz.
The Honorable The Secretary of State, Washington.

Sir:

I have the honor to refer to the Department's telegram of September 18th, 6 p.m. and my telegram of September 21st, 10 a.m. and to transmit enclosed herewith lists of a few of the persons, besides priest and sisters, affected by the disoccupation of this port.

The number of priests and sisters is variously estimated from a minimum of two hundred twenty upwards. The estimation is rendered the more difficult due to the fact that the priests have disguised themselves and are growing beards and the nuns are working as servant girls. All the priests with any means have already left so that those in Veracruz are destitute and have no way of leaving the port even if the Constitutionalists should begin to persecute them or order them to leave the country.

Not only, however, are the priests and nuns here in danger, but there are in this port a number of ex-federal officers and some 1500 soldiers. The officers are in danger of execution unless they can escape and the fate of the soldiers is doubtful because they were originally sent to Veracruz, it is said, by the Carrancistas in order to be gotten rid of.

In addition, there are many political refugees and wealthy men who have sought protection here. With the confiscation of their property and the cutting off of their income not a few of these have been rendered so poor that they cannot buy their steamer passage. If caught, a number will probably suffer the death penalty.

Furthermore, all the Mexicans who have served the Americans in any official capacity whatsoever are traitors according to Mexican law and may be dealt with as such unless specific provisions are made before the disoccupation of Veracruz. Only a few have the means to flee.

Moreover, it is possible that if the Americans leave within a very short time, many of the refugees who have money will be unable to get steamship passage because of the reservations made in advance and will be forced to await their fate here.

It may also be added that if all acts performed by the Americans here, such as marriages, collection of city taxes, payment of duties by merchants, receipts and disbursements by American authorities, etc., are not legalized by the Mexican Government before the troops leave, it will cause untold loss and confusion and embarrassment after the port is given up.

> I have the honor to be, Sir,
> Your obedient servant,
> Wm. W. Canada
> American Consul

[Two itemized lists are attached, with some of the more notable individuals from the "Partial List" shown below.]
17. Francisco Bulnes...Deputy, Senator and Literature.
19. Federico Gamboa...Secretary Foreign Relations.
22. Gaudencio G. de la Llave...Federal General.
26. Eduardo Cauz...[Military] Governor State of Veracruz.
27. Juvencio Robles...Federal General.
28. J. Rasgado...Federal General.
31. José Juan Tablada...Literary Man. [author of *Historia de la Campaña de la División del Norte*]
43. Carlos Rincón Gallardo...Governor Federal District.

[More notable individuals from the "Additional List" are shown below.]
8. Francisco del Toro...General
22. José Alessio Robles...
24. Enrique Goroztieta...
29. Enrique Creel Terrazas...
39. Felipe Mier...General.
40. Ignacio Bravo...General
59. Gustavo Salas...General.
60. Felix Terrazas...General.
61. Marcelo Caraveo...General.

<u>5761-1091/31</u>

From Edwin Emerson

Veracruz, May 14, 1914

My dear General Wood,

Your two recent letters arrived here while I was up in the City of Mexico on a short trip of exploration. I thank you for the remittance you sent to Mr. Kaempffert, who has written me that he has received it and deposited it for me. This settles up our account in a manner perfectly satisfactory to me.

I regret very much that I could not send you the enclosed report at an earlier date, but after the termination of my commission with you I was caught up so suddenly in my newspaper employment for the New York World, of which I advised you, that I found no time for anything except the urgent demands of each day.

To atone for this delay I send you by way of good measure some military results of my recent trip to Mexico City, which may prove of equal interest as my observations in the north, being quite up to date.

U.S. Army Intelligence in the Mexican Revolution

I was in Mexico City on May fifth, sixth, and seventh, having gone there by way of the English standard gauge Mexican Railway. On the way I saw General Joaquin Maass at Paso del Macho [Veracruz, about seventy kilometers west of the port, to oppose the American invaders], where he had his troops concentrated. General Funston's intelligence officer, Captain Thorne, and Captain Burnside had told me before I started that Maass had concentrated nearly 6,000 regular Federal troops at Paso del Macho with over 20 guns, but I feel quite certain from my own personal observations, which were made during several hours while I was left quite unmolested, that Maass has no such force. The various detachments encamped about the place looked to me like 2,000 or 2,500 men at most, of whom fully 600 were ragged irregulars, who were evidently recent levies still unused to arms. Of cavalry I saw only 300 horses, and of artillery I saw only six mountain guns placed in obviously strategic positions, four three-inch field pieces in battery park near General Maass headquarters in the town, and four three-inch pieces with their limbers and four caissons still loaded on flat cars on a siding at the railway station. General Maass impressed me as a bombastic, blustering officer, overfond of flattery and approbation. He has since then been superseded by Rubio Navarrete.

There were absolutely no other evidences of military preparation along the whole line, except the futile destruction of about four kilometers of rails and ties between Tembladora and Tejeria close to Vera Cruz. When I returned down the same line I found Maass in the act of moving his headquarters and most of his troops from Paso del Macho to Cordova [Córdoba, approximately thirty kilometers farther to the west by rail from Paso del Macho].

A German acquaintance from Jalapa, whom I met on the train, told me that Rubio Navarrete, Huerta's best artillery officer, was in command at Jalapa, where over 5,000 troops had been concentrated from Puebla and from militia and recent levies from the neighboring towns of Coatepec, Xico, Naolinco, and Huatusco. My German friend said that Navarrete four days after our occupation of Vera Cruz arrived with 3,000 troops in Jalapa, coming from Puebla over the narrow gauge Interoceanic Railway, and that he proceeded at once down to Chichicastle [Chichicaxtle, about half-way between Jalapa and the port of Veracruz] over the railroad and down to Cerro Gordo and Plan del Rio over the Camino Real, which points he proceeded to strengthen with trenches and artillery, leaving detachments at the national bridge, at Paso de Ovejas and as far down as the San Francisco bridge about thirty kilometers from Vera Cruz. For a while, during the first tension, Navarrete made his headquarters at Cerro Gordo, but afterward returned to Jalapa, where he is now, busily recruiting new levies and clamoring for reinforcements from Huerta. His second in command is General Peña [Ángel García Peña], who was in command of Jalapa when I was there last year. [Penciled in is: "An inefficient officer over fond of drink."]

130

General Navarrete enjoys a high reputation as an artillerist among the Mexicans, but I consider him much over rated. He was in command of Huerta's artillery at Bachimba, where I saw him directing the work of twenty guns in a very exposed position on the flat from a poorly chosen observing station. As you may recall from our previous reports of the Bachimba affair, the Federal artillery there accomplished little beyond frightening an enemy who had no artillery. Navarrete was again in command of the Federal artillery during the ten days street fighting in Mexico City, when the Federal artillery once more accomplished surprisingly little beyond damaging private property and killing inoffensive non-combatants. Still Navarrete is a well-educated and energetic officer, whose youth stands in his favor.

I chose May fifth for my arrival in Mexico City, partly because I hoped to be less observed during the confusion of the holiday festivities of that day and night, and partly because I knew that the customary military parades of that day would give me a better chance of estimating the present strength of the Federal garrison at the capital, in which I was not disappointed. To round up my impressions I took a taxi-cab on the following day and made a tour of all the military barracks of the capital, going as far as Guadalupe in one direction and Tlalpam in the other. From what I was able to observe I estimate the present military garrison not to exceed 3,000 or at most 4,000 men, not counting the police and gendarmes, who, so I was told by Captain Elisondo, a retire aide-de-camp of Porfirio Diaz, have been reduced from 3,000 to 900 owing to the demands of the northern campaign and the continuous troubles in Morelos, Guerrero and Puebla.

Huerta's strongest concentrations of troops in the capital, as heretofore, are at the citadel, in the Palace and neighboring War Office and Zapadores [Sappers] barracks, as well as at Chapultepec. I visited all these places, and made it a point to pass through the big patios of the War Office and to climb up Chapultepec Hill, at both of which places I found evidences of undiminished military activity. I also obtained permission to enter the Zapadores barracks for the purpose of visiting my old friend Dr. E. W. Ryan, who was kept a prisoner there, and while doing so was able to note that the barracks were quite full of their regular contingent of Zapadores troops [penciled in: "one battalion"]. On the pretext of visiting the military museum at the citadel I also tried to gain entrance there, but in vain. I was told by others that Huerta there had concentrated forty field pieces, thirty mountain guns, twenty howitzers and sixty Rexer [penciled in: "Madsen"] machine guns with large stores of ammunition and dynamite, recently confiscated from American miners. Sixty boxes of this dynamite I saw being loaded on a train destined for Puebla or further down [the line].

While I was at the capital the Mexican Congress held a joint session and by a unanimous enthusiastic ballot voted an emergency war budget of $43,258,271.48 [pesos]. The amount originally asked by war minister Blanquet was 50,000,000 but the Chambers struck out a discrepancy of $6,741,728.52 on account of the navy for the reason that the Mexican gunboat Tampico on the West coast, and two others on the Atlantic coast were apparently lost to the government, and therefore no disbursements on that score were in order. All this, of course, was in Mexican money, of which, so I was told, there was really none in sight.

Be this as it may, the budget was voted, and during the brief debate that preceded it General Blanquet, speaking for the government made public announcement that all the government's factories of arms, munitions, and powder were working at full blast, day and night, to wit: the Comisiones Tecnicas, Almacenes Generales; Maestranza Nacional; fundicion nacional; fabricas de armas, cartuchos, y polvora; laboratorio de municiones y artificios; talleras de reparacion; and parque general de artilleria y ingenieros; as well as the arsenal nacional [Technical Commissions, General Warehouses, National Armory, National foundry; factories of firearms, cartridges, and powder; munitions and artifices laboratory; repair shops; and general ammunition of artillery and engineers; as well as the national arsenal].

The war minister also made the following public statement of the present strength of the Mexican army:

Infantry: 34 battalions regulars, full war strength
  six fixed companies at Perote and in Lower California
  apparently military prison guards
  One battalion auxiliaries of Guerrero
  20 battalions federal irregulars
Cavalry: 18 regiments full war strength
  one corps (cuerpo) irregular auxiliaries
  one central remount station
  21 regiments federal irregulars equal to full peace strength of regulars
Artillery: Three regiments field artillery, full war strength
  Two regiments mountain artillery, full war strength
  One regiment of machine guns, full war strength
  Two squadrons of Madsen machine guns, full war strength
  One train of remount pack mules
    Three artillery trains
      Coast artillery, comprising batteries at Tampico, Ensenada, Mazatlán, Manzanillo, Acapulco, and Salina Cruz

[Penciled in: "The Rurales are not included in this, since they belong to the Dept of Gobernación"] [Note: This is incorrect, since the Rurales had been redesignated "Scout Corps" and transferred to the War Department in August 1913.]

During the debate one senator made some bitter remarks about batteries lost at Vera Cruz, and in the north, but he was promptly suppressed.

Documents Cited

The thing that most struck me while at the capital were the large number of volunteers enlisting in the volunteer brigade of Zaragoza, which was mustering at the old Iturbide Palace, and the large numbers of volunteers, commanded by officers in uniform, who without arms and still in civilian dress were marching up and down various avenues and streets on Cinco de Mayo day. I saw three different long columns of these men, well set up, husky looking young men, and found the largest of these columns by actual count numbering 6,000 men. They marched with military precision in double rank formation, each rank numbering twenty men with five file closers and one officer to each platoon of 25.

For reasons of personal safety I decided to leave Mexico City immediately after I had obtained a long newspaper interview with General Huerta, so my military observations while at the capital were limited to the above.

General Huerta told me, by the way, that he did not think he would have lost Torreón as well as Gómez Palacio, Lerdo, and San Pedro de las Colonias had he left old General Bravo in command. He spoke very bitterly of General Velasco's utter failure to make use of the wireless station at Torreón.

Observations in the North

As I telegraphed you from Galveston, and as I tried to express in the *incomplete* typed monograph on Torreón which I mailed to you, I found Villa's forces in northern Mexico the strongest and best field army I have yet seen in Mexico.

When I first joined Villa's army at Chihuahua City about the middle of March his forces were said to number 14,000. Knowing the Mexican tendency to exaggerate numbers I mentally corrected this figure to nine or ten thousand, but when I saw the whole army entrain, and got more opportunities at different stopping stations along the way to analyze the army's strength, I came to the final conclusion that Villa's army then numbered fully 12,000 to which must be added 3,000 from Durango who joined him at Mapimí and Lerdo.

The general composition of these forces was as follows:

Cavalry: 10,000

Infantry: 2,000

Artillery: 1,000

Miscellaneous: 2,000

By "miscellaneous" I mean railway personnel, sanitary [medical] men, packers, carters, laborers, and others employed with the army including the various staffs and their personal servants.

All of these forces were the best set up, best armed, best mounted, best equipped, best clothed, best fed, best paid and generally best cared for troops I have yet seen in Mexico. They were not, however, the best disciplined nor were they the best fighters I have seen in Mexico. From my observations within the last three years and from what I now witnessed at Gómez Palacio, Torreón, and San Pedro de las Colonias I consider the Federal Rurales and the Federal Irregulars commanded by Benjamín Argumedo better individual fighters than any of Villa's men. I also became convinced from what I saw of their mutual artillery practice in the field that the Federal artillery was better handled than that of the rebels.

Villa's greatest asset is his personality. As a former outlaw and bandit, who successfully stood his ground against Porfirio Diaz' soldiers and Rurales for over ten years, Villa is idolized by all the lower classes of Mexico. His reported exploits, true or false, such as shooting a judge on the bench, shooting a Federal colonel in the midst of his staff, shaking his fist in Huerta's face, shooting an Englishman out of hand almost in plain sight of the impotent Gringoes, and finally his amorous exploits of stealing pretty women, have brought him a renown among Mexicans like the old legends of Robin Hood or Dick Turpin. Anyone who has ever seen him on horseback knows that he is a splendid rider, a thing that all Mexicans set great store by, and he is also taking pains to spread and emphasize the universal belief that he is a "dead shot". He has further endeared himself to the lower classes and to his followers by his manifest lack of fear (never bothers about having guards around him or escorts) [except his 300 *dorados*], by his frank manners and speech, by his simple dress, simple habits, and rough unadorned speech set off by very forceful profanity and quaint obscenities. In all press interviews or at public gatherings he always makes it a point to lay stress on the fact that he is a simple, uneducated, unlettered man who never has had any advantages of culture. If he had a Machiavelli for an advisor he could not have found a surer way to the hearts of his followers, nine tenths of whom are absolutely ignorant pelados.

The American press reports that Villa is an American ex-soldier, that he is a negro, that he is the illegitimate son of a great Spaniard, that he is irreligious, or a drunkard, and above all that he is a consummate Napoleonic strategist, are all moonshine. From my almost daily contact with him during two months, I can assure you positively that Pancho Villa does not drink any alcoholic liquors whatever, neither does he smoke marijuana, he lifts his hat when he passes a church door and devoutly crosses himself when the angelus rings, although like most of his followers he has no use for Spanish bishops or sleek priests. He cannot speak a word of English except a few swear words and obscenities, and as for strategy he has merely the inborn war cunning of any Indian on the warpath, with an almost inexhaustible fund of intimate personal knowledge of the north country, gained during the strenuous years when he was chased all over the land as an outlaw, besides which he has the invaluable quality of energy, dash, and initiative, which, as you know, is exceedingly rare among Mexicans. Maps are quite lost on him, for he can no more read a map than an owl. This last also applies to most of his officers, except a few of his highest staff officers. In general most of his officers and men get along without maps. Even if they could read them they would not need them, for there is always some old or young pelado at hand who can tell them every goat path.

What glimmers of apparent strategy Villa has shown in his campaigns were either simple Indian cunning or were the result of the advice of some of the educated soldiers around him, to whom he is always willing to listen—another great military virtue. [Penciled in: "These professional soldiers include General Felipe Ángeles, Martial Poole, an American civil engineer, and young [Horst] Von der Goltz."]

Villa's strongest card with his followers is his known irreconcilable hatred for Huerta, and Huerta's known irreconcilable hatred for Villa, on most intimate personal grounds, which preclude the possibility of Villa ever selling out his men to the gobierno [government], a species of treachery otherwise so common among other Mexican rebel leaders of the past and present. [Penciled in: "Therefore, Villa's followers know they can trust him to stick."]

Next to Villa's personality, which completely overshadows that of Venustiano Carranza or any other rebel leader, his second greatest asset is the temper of the common people of Northern Mexico, who in their sentiments are thoroughly against the federal government and against the people of the center and south. This feeling extends as far down as Zacatecas, and is an inestimable advantage to Villa's army, since it brings them a constant flow of recruits, supplies and valuable information, and makes it unnecessary for them to resort to the forceful measures which make soldiers hated. During all the time that I was with them in the field I never saw an act of violence committed against natives, only against Spaniards, foreigners, or out and out federal sympathizers, who invariably belonged to the upper classes.

Villa's third great asset has been the able diplomatic and financial support he has received from the constitutionalist junta, the Madero family, the Carranza ring, and certain American financiers, resulting in the lifting of our arms embargo, and in a constant supply of good arms, ammunition and serviceable materials of all kinds, particularly good shoes, leather leggings, flannel shirts and good American stock saddles in surprising quantities. There has been no apparent stint of money, in addition to which there was never any reluctance on the part of the soldiers or populace, so far as I observed, to receive the paper money of the revolution at its face value. In as much as the troops in the field were regularly paid every five days, and in as much as all the various generals always had immense stacks of the revolutionary scrip at their disposal, there was never any lack of money. The supplies of fresh ammunition, rations, forage and clothes also were constantly forthcoming.

Apart from the financial support from the United States and from the base at Ciudad Juárez, this was due to Villa's next greatest asset, the undisputed possession of the railroad, of valuable rolling stock, comprising some fifty engines and some five hundred cars, all commandeered, and above all to the efficient management of the railroad problem by one responsible and thoroughly competent head. I mean Eusebio

González [Calzado], an unusually energetic man who knew his business thoroughly, who knew the ground and railroad men, and who enjoyed the great boon of being left practically unhampered by Villa in all the minor details of his railroad problem, which included the maintenance and repair of all telegraph and telephone communications, a highly important matter since Villa was unequipped with any signal service in our sense of the term.

I should add here that the constant railroad campaigns waged by the Mexicans in this part of the country during the last three years have resulted in making them highly expert both in the destruction and in the quick repairing of military railroads, as we shall learn to our cost whenever the time comes for us to lock horns with them on their own ground.

Villa's next great asset was the possession of the forty-two field guns abandoned to him at various points by the federals. This artillery, as I indicated above, was none too well-handled in the field, but at least it was promptly brought into action, largely through the efficiency of the railroad, and the noise it made, as well as the artillery prestige of General Felipe Angeles, who was generally believed to have personal command of it (not always the case) had much to do with breaking down the morale of the federals at Gómez Palacio, Torreón and San Pedro [de las Colonias].

The best rebel artillery officer in the field was Colonel Servín a pupil of General Angeles at [the Colegio Militar de] Chapultepec. This officer showed himself very brave and energetic every time I saw him under fire, but he allowed himself to be taken up too much with details, which resulted in a disposition on the part of the battery officers to let him do all the work. Angeles took little personal part on the firing line.

All the rebel artillery fire was direct, the trigger being pulled by a long lanyard. Positions of the guns were generally ill chosen, single pieces often getting separated from their batteries and even from their own caissons. This was because no competent artillery officer went ahead to choose the ground. When the artillery first went into action north of Gómez Palacio they did so after nightfall, each gun by itself floundering about in the uneven ground trying to find a suitable firing position. When daylight came they were found strewn all over the landscape, some in the most impossible positions, in canals, on exposed slopes, or out in open spots directly under the guns of the enemy. Luckily for them the enemy couldn't shoot straight, otherwise they would have been knocked to pieces in no time.

When guns became unmanageable they invariably withdrew from the action, instead of standing their ground, or at least remaining close to their batteries. From what I saw of the work of the battery captains I was driven to the conclusion that they were generally incapable. There was but one foreign officer among them, who professed to be an artillery captain of the regular Italian Army, but his work was no better than that of the others.

Documents Cited

The Americans who served with the rebel artillery were mere riffraff, gathered up from God knows where. Not one of them was a competent artillerist. They were used merely as drivers, mostly caisson drivers, and they were none too good at that. As soldiers I considered them almost worse than useless, for they knew no Spanish, did not know how to take care of themselves, and only bothered the Mexicans by their helplessness and inveterate grumbling. The stories published in the American press to the effect that the rebel artillery owed its successes to the presence of the American soldiers of fortune with Villa I can assure you from my own observations were totally false.

There were some more capable Americans with the rebel machine guns, where they made themselves particularly useful by their mechanical ability in fixing guns that had got out of order. These guns were always getting jammed and otherwise out of order. [Penciled in: "They were all Colt guns."]

The field pieces also gave much trouble, principally in their ammunition, which either burst high or close to the muzzle, or which refused to go off, often remaining stuck in the barrel. Another thing that gave constant trouble with almost all the guns was the pneumatic recoil mechanism.

The most satisfactory ammunition were the old French shrapnel projectiles acquired by the rebels together with the guns, but of these there were but few, about one-fifth of the total amount. The new ammunition, made in Chihuahua or in the United States, proved almost unserviceable.

After the artillery, Villa's most effective arm was the cavalry, which constituted the great bulk of his army. As you know Chihuahua and Coahuila are great horse breeding states, and the rebels were always able to supply themselves liberally with good horses from the great horse ranches of the Terrazas estates and others. The best horses came from the ranch of the Zuluagas, a Spanish family near Chihuahua, who have for many years been importing Arab stallions and mares, barbs and some English and Kentucky and California thoroughbreds. Besides these native horses, which served their purpose admirably, I also noticed a surprisingly large number of American bred horses and mules. Not a few horses bore U.S. brands [Penciled in: "without 'condemned' brands."]

Villa's great superiority in cavalry, together with his good railroad service accounts for the remarkable mobility of his troops in moments of stress. This mobility was enhanced by the fact that Villa's cavalry, when engaged in hostile operations, moves unhampered by any impedimenta, such as wagons or pack trains, the men carrying nothing but their arms, ammunition, canteen, and one blanket, their invariable practice being to make the country supply them and their horses with subsistence. This leads to much riding to and fro, from one ranch to another, in search of food and forage, but on the other hand it has the advantage of cutting them loose from bases and communications; and keeping the troops so constantly on the move makes them hardier and more contented.

During the campaign around San Pedro de las Colonias which followed immediately after the Torreón battle, I spent ten days riding around with two brigades of cavalry (4,000 horse) and during all that time we never spent more than one night in the same place. The average marches were from fifteen to twenty five miles a day. On one occasion when we had suffered the first severe reverse at San Pedro, we rode half of the night and all of the next day, covering forty five miles. Most of the horses went unshod, but I noticed very few cases of lameness or of straggling. Throughout this time we never really suffered for water, food, or forage, though it was always a question whether we would be able to get anything the next day. During this campaign not a single woman accompanied the troops, an innovation in Mexican warfare, but on the other hand we had many small boys ranging from ten to fifteen years, who, while professing to be full-fledged warriors, really were suffered to come along as mascots and body servants, or in the train of their fathers, uncles, or big brothers.

Villa's weakest arm was his infantry, of which I never saw more than 2,000. When we started from Chihuahua there were but one thousand. In a horse country like northern Mexico, people who have to go a-foot are generally scorned, therefore nobody served in the infantry by choice, but only those *poor devils* who could not get themselves mounted. During the latter part of the Torreón campaign all the Federal deserters and prisoners who were incorporated in Villa's army were stuck into the infantry, thus adding to the contumely under which this branch of service already suffered. Unlike the cavalry (who were uniformly armed with 30-30 Winchester carbines) the infantry had all kinds of heterogeneous arms, though there was a preponderance of Mauser rifles acquired from the federals. The infantry officers likewise were distinctly inferior to the general run of officers in the rest of the army.

I saw Villa's infantry in action but once. That time it marched very well, covering a distance of some ten kilometers under a broiling sun in remarkably good time and good formation with very few stragglers, but when they came under the enemy's shrapnel and rifle fire they behaved badly and drifted fully a mile to the rear, which came near having disastrous results on the whole battle line, but for the energetic action of Col. Servín of the artillery who brought up two batteries in their rear and threatened to open up on them at close range unless they returned at once to their trenches, which they did reluctantly, after Col. Servín had beaten two of their officers over the head with his automatic pistol, threatening to shoot any other infantry officer who did not at once lead his men to the front.

Of the engineer branch of Villa's army there is nothing to say, since there was none. Likewise no signal service. [In the Mexican army, signals were the purview of the artillery] Excepting some haphazard holes dug in the ground under stress of fire I never saw any concerted efforts at entrenching on the part of Villa's men. All the country of the Laguna district around Torreón and San Pedro is so thoroughly seamed with irrigation ditches, that there was scarcely any need of entrenchments. All necessary bridging or other road repairs was left to the railway repair crews, who also attended to all necessary telegraph or telephone work. The flat country of the Laguna district is such that any troops can march in any direction without reference to roads, which are mostly mere trails through the mesquite. All the main advances, as you know, followed the lines of the railways. [Penciled in: "The aeroplanes never reached the front, and his armored cars and guns mounted on cars never entered into effectual action."]

The three innovations in Villa's military organization which excited most comment among Mexicans and foreigners resident in that region, were his supply service, his commissary and quartermasters departments, and his medical department, all of which depended wholly on the railway service and never made any effectual attempt at getting away from the railway. Whenever they did so they invariably broke down, but while with the railway they did good work, in fact surprisingly good work judging by Mexican standards.

The sanitary [medical] railway trains, while I was with Villa's army, handled some eighteen hundred wounded men. The "handling" consisting mainly in transporting patients from the front to hospitals at the bases in Chihuahua City and Juarez. With the sanitary trains were many good doctors, among them several apparently competent American surgeons, but the means at their disposal were shockingly inadequate, the cars were mostly rough, ill-ventilated, dirty freight cars without cots or field beds and the sufferings of the wounded in them were horribly aggravated by the ruthless conduct of the train crews, who shunted and bumped their trains back and forth, with violent starts and stops, without any regard to the terrible consequences to patients lying on hard springless floors with bullet holes causing internal hemorrhages. I got some realization of this from a trip I took on a sanitary train from Gómez Palacio to Chihuahua City. It was the roughest railroad trip I ever took in my life, so rough that nobody in our car (we were six) could sleep a wink all night.

All their vaunted medical excellence, of which so much was made in the American press, would only have excited your pity or your indignation had you been there. Villa's medical department had some good automobiles and six mule[-drawn] ambulances (but no stretcher bearers or stretchers) but whenever they attempted to run their automobiles and ambulances into the field, which was generally many hours after the first wounded came straggling to the rear (many miles from the front) these automobiles and ambulances would be commandeered by some forceful rebel officer who preferred to use them for purposes of his own, mostly for ammunition wagons. The result was that I never saw any medical ambulances or stretcher bearers actually reach the fire line.

U.S. Army Intelligence in the Mexican Revolution

Not a man in all Villa's army carried a first aid bandage. While I was with the troops around San Pedro I never saw a doctor until the very last day, when the railway got sufficiently repaired for a sanitary train to come up to the immediate rear. During the first eighteen hours fight against San Pedro I saw many men wounded all around me, who simply had to lie where they fell with no one to attend to them.

This was not so much the case during the severer [more severe] fighting around Torreon, because the fighting lines there never got very far from the railway; still many hundred men died unattended on the steep hills and in the deep ravines immediately west of Torreon. My estimate from personal observation of the total losses on both sides was federals over 1,000 killed and 2,400 badly wounded; rebels 800 killed and 1,800 badly wounded. This does not include the San Pedro losses, of which I was unable to form a proper estimate, because your message to come out reached me on the day before the last day's battle, and I left the front immediately after your message reached me [probably sent through George Carothers].

Of sanitation in our sense of the word there was absolutely nil. Everybody was allowed to be as dirty and unsanitary as he pleased, but luckily for them the troops were kept so constantly on the move that they generally got away from their own dirt, and from the stinking cadavers of dead men and dead animals, before conditions became too intolerable. Close to the railroad stations and where there were settlements they generally tried to burn up cadavers with petroleum after the first two or three days of stench, but in the open field the sopelotes [*zopilotes*, "buzzards"] and coyotes attended to this. The strong mid-day sun and the high winds that sweep over the plains in this region also help. At all events the general health of the troops appeared unaffected by their lack of sanitation and their total neglect of cleanliness. In short the men simply lived like animals, as is their usual custom, and were as healthy as young, wild animals.

You know from your own observations in Mexico that the average Mexican pelado is a pretty sturdy creature, who can stand more privation and hardship, such as cold, heat, wetting, hunger, thirst, and vermin than any white people would put up with. I was agreeably impressed, while serving with Villa's men by their generally cheerful and contented demeanor, without any of the grumbling so usual among our own troops. During the battle days, and at other times when I came in contact with the wounded men, I was filled with admiration at the stoical way in which they bore their hurts. Men who were not seriously wounded, almost invariably continued their service in the field, and I saw many instances of men who were really quite seriously hurt—shot through the shoulders or elbows, or hands, or wounded in the head—who held their place in the firing line and stayed with their commands, refusing to go to the rear. The conduct of the women who came along on the railroad trains, and many of whom accompanied their men into the firing line around Torreón, was also notably heroic.

The thing that most impressed me about Villa's army, and his general conduct of campaign, was the swiftness of his advance from Chihuahua to Torreón. When Huerta covered the same ground in the opposite direction two years ago against Orozco it took Huerta's army of 7,500 fighting men (15,000 people) two months. Huerta, as you may recall, had 20 railroad trains. Villa, this time, had 19 trains, and two additional short trains, which he used for communication purposes and Villa carried practically the same number of people; but it took Villa only ten days. One reason for this was the fact that the federals failed to put up a more determined resistance by destroying the railroad except in a few spots; but the main reason, undoubtedly, was that Villa's advance had far more push and dash than Huerta's slow forward movement two years ago.

The most significant thing about the whole business, in my estimation, is the fact that the Mexicans, both on the rebel and on the federal side, are getting the most valuable training in the art of War, and in northern Mexico at least are getting so steeped in war conditions, that it will be very hard to get them back to peaceful pursuits. It is one thing to arm a lot of warlike Indians, like the northern Mexicans, but quite another thing to try to disarm them.

I gave your regards to General Funston, and he asked to be remembered to you.

Faithfully yours,

Edwin Emerson

Care of Ward Line

Veracruz

5761-1091/32

TRANSLATION

WESTERN UNION TELGRAM

Chihuahua, Mexico, Sept. 15, 1914

General Hugh L. Scott,

Washington, D. C.

Your kind telegram of today has been received and I assure you that my desire for the Federal prisoners at Fort Wingate to come to this State is because we have much work and can easily place all of them, but since El Senor Carranza has arranged for them to pass through Piedras Negras I believe that I ought not to insist on the matter. I salute you affectionately.

Francisco Villa,

General in Chief

5761-1091/49

HEADQUARTERS U. S. EXPEDITIONARY FORCES

U.S. Army Intelligence in the Mexican Revolution

Office of the Commanding General,
Vera Cruz, Mexico.

August 1, 1914.

General W. W. Wotherspoon,
War Department,
Washington, D.C.

My dear General Wotherspoon:

I was very glad, indeed, to receive your several very kind letters during the past two or three weeks, and shall try to give you personally some idea of the situation here, that is, as it exists today, for we have already seen that some pretty rapid changes can take place here.

We are in a pretty good position here to keep track of the general situation and of the drift of things, as we are constantly brought in contact with persons from Mexico City and other parts of the interior of the country, these including not only Americans and other foreigners, but many Mexicans themselves. The way the Mexicans come to me and unbosom themselves and tell about their troubles with their own countrymen is almost amusing. I have a hard time convincing them that I have no authority to go out from here and protect their haciendas or to occupy the City of Mexico on their invitation. I do not know what is thought in Washington, but it is my opinion as well as that of every person that I have talked with in the past few days, that the situation as regards the pacification of the country is absolutely hopeless. Even if Carranza desired to do so, he cannot grant an amnesty to those who supported Huerta for the very simple reason that his military leaders will not acquiesce in such action. Those who go to make up what is left of the Federal Army cannot be expected deliberately to put their necks in the noose, and for their own preservation will continue to resist until they [are] overwhelmed and forced into our lines here. From news received this morning we know that the forces of Jesús Carranza and Obregón are making all possible haste to occupy the Capital, the advance troops of the former having already passed Querétaro, and this while negotiations are supposed to be in progress between Carranza and the delegates appointed by Carbajal. The Federals, who are now in considerable force in and about the Capital, may either attempt to fight a decisive battle to prevent the capture of the city, or they may retire in this direction, probably to Puebla, and either try to hold out there for a time or come straight on to Vera Cruz to be interned. I hope they do not conclude to do the latter, as I have troubles enough without attempting to feed and guard thirty or forty thousand men with their "soldaderas" and other camp followers. If Carranza's people should follow them and demand entry or that the interned troops be surrendered to them, the chances for an interesting event would be excellent. I suppose the Federals would sit down on their haunches and let us fight it out. I know that I would be afraid to entrust any part of the line to them. Against four or five times our number of such troops as the rebels have shown themselves to be we would have a red-hot time of it. All talk of our being assisted by the fire of the guns of the fleet is moonshine. Those people would have too much sense to operate against our flanks where they would be exposed to fire from the fleet, but would confine themselves to an attack on our center and on cutting off the water supply. The terrain is such that if our center were being attacked the fleet would not dare fire as there would be a confused struggle in the sand hills back of the port, where friend and foe could not be distinguished. Where the fleet would be of great help would be in the men and guns that it could land to assist us, and with this help I would have no fear as to the result. I have reason to believe that it is the intention of the Federal troops if an amnesty is not arranged to retire on Vera Cruz, as the only place in the Republic where they can seek shelter. Of course, we are bound by international law to receive them.

I hope that in some way the rebels can be induced to forego their desire for more blood-letting, for in that case the Federal army, if proper guarantees were given, would probably surrender. If the attitude of those who are here in Vera Cruz is any criterion, however, that is scarcely to be hoped for, as they are absolutely uncompromising in insisting on having their revenge. It has been quite interesting here for the past two weeks, with so many prominent rebels in town at the same time with a lot of the most prominent supporters of Huerta. All they dared do was glower at each other. At one time we had three ex-cabinet ministers, five general officers, thirty or forty other officers and the families of nearly all of them. Even two sons of Huerta did not disdain to land here when they were brought up with a lot of others from Puerto Mexico by the British cruiser Bristol.

As to the remarks made by the War College apropos of my plan for saving some of the Interoceanic Railway in case of war, to which remarks you refer in one of your letters; I feel sure that in its criticism the War College overlooked several facts which were well known to us here but to which I probably did not call sufficient attention in my letter. The principal one of these was not only that we were fifty percent stronger than the Mexican force facing us on that road, but that the force referred to had very little fighting quality. After we had once got them on the run our difficulty would have been to follow them fast enough to get in another blow on them. Another thing which I think was overlooked is the fact that the rebels of the north were exerting such pressure that it would have been impracticable to reinforce the troops along the Interoceanic. For the six days before other troops from the United States could reach Vera Cruz and follow us up it would not have been necessary to patrol that part of the road that we had captured. All that was required was to get them on the run once and keep them going. Their only line of retreat was along the railroad itself or the nearby

Camino Real. As soon as we stopped they would [have] begun the systematic destruction of the track as distinguished from isolated damage here and there.

Conditions have so changed now that I acknowledge that my plan is no longer of much use, as the elimination of Huerta and the conciliatory attitude of his successor make a war with the Federals quite unlikely. If we get in a mix-up at all I feel that it will be with the present rebels, and probably under some such conditions as I have referred to earlier in this letter, in which case this little force will be thrown quickly on the defensive owing to the great numbers and the morale of the rebels. In the face of several thousands of the men who have been winning the recent victories in the north and center of Mexico it would behoove us to be pretty careful about getting too far from Vera Cruz with the troops that we have.

I do wish we could have some recruits, especially for the 4th and 28th regiments of Infantry, as those organizations are rapidly becoming mere skeletons. The health of the troops remains good, and so far, with seven thousand men on shore, we have lost but three from disease in three months – one in the Army and two, including an officer, in the Marine Corps. Of course, the vitality of the men has been considerably reduced by the hot moist climate, and a good many of them would fall out on a hard march. Really, I do not believe it is any hotter, however, than it is in most parts of the United States at this time of year. But the trouble is that but little relief will come to us in the Fall of the year. I can tell you one thing, and that is that when on clear days we gaze on the snow-covered peak of Orizaba, eighty miles to the westward, it almost makes us wish something would break.

Well, this has been a rather long and rambling letter, and I think I had better bring it to a close. I am going to take advantage of your suggestion, and from time to time write you personally about things here.

Very sincerely,

FREDERICK FUNSTON

6931-68

War Department

Office of the Chief of Staff

Washington

June 26, 1914.

Memorandum for Colonel Hodges

1. Captain Laubach, who is in charge of our monograph work on Mexico, informs me that his work can be very much facilitated if Captain Burnside, now at Vera Cruz, can be ordered to report to Washington for a few days. He also advises me that Captain Burnside's work down there will be very much improved by a personal conference with us at the War College.

2. We hesitate to make a direct application for this as coming from the War College Division, General Staff, for fear it may interfere with his usefulness in his relations with the Mexican authorities who might subject him to greater, and perhaps annoying, surveillance. We would like, therefore, it it can be arranged, to have him ordered simply to Washington, and it is suggested that on account of his relations with the State Department, that department would have to be consulted and it might be that with an explanation of the situation they would be willing to request his return temporarily.

3. It so happens that I have a personal letter from Captain Burnside this morning, in which he states:

> "I would like to visit Washington for ten days to catch up on the diplomatic dope that has slipped by me since leaving the Embassy April 25th, but of course I want to feel certain of hanging around here as all of my property that has not been stolen is in a leased house in Mexico City."

4. I have discussed this matter with General Macomb who approves of having Captain Burnside report here if it is practicable to arrange it.

Edwin F. Glenn
Colonel, General Staff.

Rec'd back,W.C.D., G.S. Jun 29 1914 with slip attached:
"Secretary does not approve at this time. W."

6931-79
War Department
Office of the Chief of Staff
Washington

December 18, 1915.

From: Secretary.
To: Captain William A. Burnside, 26th Infantry, Harlingen, Texas.
Subject: Certificate covering loss of property at Mexico City.

Under date of August 20, 1915, there was sent you for signature a certificate covering the loss of property at Mexico City, Mexico, left by you upon your departure therefrom, this with a view to having the matter disposed of. No reply having been received to that communication, I am again writing to ask that the certificate be executed by you for the purpose in question.

C. Crawford
Major, General Staff.

1st Ind.

U.S. Army Intelligence in the Mexican Revolution

Captain William A. Burnside, 26th Infantry, Kingsville, Texas, Dec. 26, 1915-To Secretary, War College, Washington, D.C.

1. Certificates returned herewith. Also certificates covering abandoned Ordnance, official and private correspondence and a letter regarding personal effects.

2. The delay in this matter has been caused by a remote hope that the American Embassy in Mexico City might be re-opened and that I might be able to have a short period in Mexico City to assemble and straighten out both Government property, correspondence and records now stored outside of the Embassy, and my household and personal effects which I cannot well afford to take further chances on losing through shipment and a disorganized customs service.

> W.A. Burnside
>
> Capt. 26 Inf

> Kingsville, Texas,
>
> December 26, 1915.

I certify that I was on duty as Military Attaché to the American Embassy at Mexico City, Mexico, and on April 25, 1915 the Embassy, with all members attached, was suddenly obliged to leave Mexico City for Vera Cruz due to the disturbed conditions arising from the landing of United States Forces at Vera Cruz. Diplomatic relations between the United States and Mexico were broken and the American Embassy Staff given passports by the Mexican Government.

Because of the haste with which the withdrawal was made, (one days' notice) and the danger to violence in which all members were, it was necessary to abandon all property, public and private."

[In addition to the normal office supplies and equipment was noted two Maxim machine guns and accoutrement, approximately eight thousand .30 caliber rounds for same, and 230 .30 caliber Model 1896 U.S. rifles with "appendages" and approximately 62,500 Ball Cartridges for same. These were surrendered to the Mexican Military authorities on April 22, 1914. The importation of these articles had been facilitated by Huerta's government for protection against rioters, etc., and the Huertistas had perfect knowledge of their existent. The Huertistas were not informed about miscellaneous ammunition, cleaning rods, and magazines left over from the time of Madero's administration and these were not turned over to the Federal Army.]

<u>8529-1</u>

> Eagle Pass, Texas.
>
> April 21, 14.

From Captain G. W. Biegler, 14th Cavalry.
To Chief War College Division, Washington, D. C.
Through Chief of Staff Southern Department.

Subject: Report on Monclova Division National Mexican Railroad.

1. Enclosed is a blue print profile of the Monclova division of the National Mexican Railroad containing the most complete available details of that line. Two changes have been marked on the blue print as "out."

A section of the Monterrey line has been cut out as shown between Anhelo and Empalme de Ixtle. There is a cut over to the Mexican Central line running from Anhelo and Paredón. A second cut off between Ramos and Ixtle brings the road back to the National line at Ixtle.

The main line from Reata to Sauceda was destroyed by the rebels over a year ago and has not been rebuilt to date. Trains to Torreón use the Anhelo-Paredón cut over and the Mexican Central line to Sauceda.

With these changes the blue print is correct and up to date.

BRIDGES: Among the list of signs a permanent bridge is meant to be a bridge of steel and stone. A wooden bridge of any kind is known as a provisional bridge.

Except at Fuente all steel bridges on the division have been badly damaged and are now solidly cribbed by means of white pine ties brought up from central Mexico.

The Hermanas bridge at kilometer 202 is a total wreck. The management has [marked out] around this bridge by means of a "shoofly."

In addition to damage done to the steel bridges most of the wooden bridges have been burned out and have been rebuilt for temporary uses by means of cribbing.

ROLLING STOCK: The normal equipment of this division consists of seventy engines, 8 switch engines, 3 yard engines, 5 passenger engines and 54 freight engines. Most of these engines are now in poor shape and a large number are now off the division, having been taken towards Torreón in the recent movement of federal troops in that direction.

It is estimated that there are about 500 freight cars remaining on the division including box cars, coal cars and a few stock cars. There are 13 passenger coaches of various types and 6 baggage cars.

FUEL SUPPLY: All locomotives are coal burners. Ample fuel supply is to be had at both Rio Escondido and Las Esperanzas. At each of these points there are large stores of coal on the ground near the track.

ROAD BED: The road is standard gauge with rails running from 54 to 75 pounds to the yard. Road bed is not ballasted except in places where the nature of the ground required it. The road bed is reported to be an excellent one.

Officials state that there are plenty of rails on hand, but no ties or bridge material.

YARDS AND SIDINGS: Every station on the line is equipped with a siding sufficient to pass full trains. There are yards containing three or more tracks at Piedras Negras, Rio Escondido, Fenix, Allende, Nava, Barroterán, Monclova and Reata.

WATER SUPPLY: Water is secured for railroad purposes as follows:

At Piedras Negras from the Eagle Pass water system.

At Nava and Leona from wells.

At Sabinas from the Sabinas river.

At Kilometer 151 from a reservoir.

At Kilometer 166 from a well.

At Kilometer 201 from a reservoir six kilometers from the track by pipe line.

At Monclova by pumping from Bocatoche a distance of 22 kilometers.

At Esperanza by a 25 kilometer gravity pipe line from the Gavia Cañón.

At La Perla from a reservoir.

At Higo from wells.

REPAIR SHOPS: In size and equipment the shop at Piedras Negras is listed as the third in Mexico. This shop is prepared to do any work required in the maintenance of rolling stock except making steel castings.

Monclova also has a small shop for repair work. For its size this is said to be the best equipped shop in the republic.

Neither of these shops have been damaged in the slightest to date.

2. Above information have been secured from officials in position to know and is accurate in every detail.

3. Request that the enclosed blue print [not included in file] be returned to me if possible at some future date as it was borrowed on my promise to return the same.

G. W. Biegler
Captain 14th Cavalry.

1 encl

To Department Engineer,
to note and return.

H.S.D. April 22, 1914.
1 Encl.
fox.

Hdqrs., Southern Dept., May 20, 1914. – To Army War College, Washington, D. C.

2d End.

The profiles referred to are being held at the office of the Department Engineer, Southern Department, during endeavor to obtain additional copies, – an arrangement which will probably be consumated within next three or four days, – at which time a copy of the profiles will be forwarded for use in connection with this communication.

Major, General Staff,
Chief of Staff.

fox.

8529-2

Report on the Monclova Division, Mexican Railway.

The Monclova Division main line runs from Piedras Negras to Saltillo, branches run from Reata to Sauceda; from Paredón to Monterrey from Monclova to Cuatro Ciénagas, also various mine branches.

Of the above the following are now in operation. Trains run on the main line from Piedras Negras to Barroterán. One train a day is now making a daily round trip over this section. The mine branch from Sabinas to Rosita 18 kilos long is in good condition. The mine branch from Barroterán to Múzquiz 40 kilos. long is being run. This is an important branch as all fuel for this Division (Coal) is obtained from the Múzquiz mines. These mines produce a high quality coal and are owned by the company, but are being worked by another corporation at present. The Monclova-Cuatro Ciénagas branch, 68 kilos long is not working as all bridges have been destroyed. Information is obtained from Constitutionalists that bridges on the main line south of Barroterán as far as Monclova have been burned and from the same source it was ascertained that the Federals are now running trains from Monclova to Mexico City. The branch from Reata to Sauceda was destroyed a year ago, all bridges having been burned. A branch from Allende in the direction of San Carlos was partly constructed some years ago. Construction progressed for 42 kilos. and further work ceased. This branch is now being destroyed to obtain material for repair.

The following information relates to the entire division.

Gauge:	All standard gauge.
Track:	Single track.
Grade:	One per cent adverse grade from Piedras Negras to Paredón. One and one-half percent grade from Paredón to Saltillo.
Curves:	The curvature (maxim.) of the main line is 7 degrees metric; branches 8 degrees metric.
Rails:	From Piedras Negras to Sabinas 61-1/2 lbs. rails. Sabinas to Saltillo 75 lbs. rails. Barroterán to Múzquiz 65 lbs. rails. The Cuatro Ciénagas and Rosita branches have 50 lb. rails; Reata branch 75 lb. rails, Paredón to Monterrey 61-1/2 lb. rails.
Ties:	Ties are of pine cut in Mexico and United States. Their condition is good.
Platforms:	There are only three platforms on this branch that have not been destroyed. One at Hermanas where two cars can unload at a time, one at Monclova for four cars, and one at Reata for three cars. All three of these platforms are constructed of masonry with dirt filling

Store-houses: There are no store-houses on this division other than the usual railroad freight house.

Watering Stations: The engines on this division are divided into two classes:- the big freight engines which carry a 9,000 gallon tank and the small engine with a 3,000 gallon tank. A large engine can travel 75 kilos without watering and a small engine 50 kilos. The railroad obtained this water from wells, dams and springs. The following are the watering stations:

Piedras Negras: Water obtained by pumping from city supply of Eagle Pass.

Nava: Water from well, wooden tank, 30,000 gallons.

Kilo. 62.7: Well steel tank 30,000 gallons.

Sabinas: Water pumped from Sabinas River into a 30,000 gallon steel tank, supply unlimited.

Kilo. 150.8: Dam. There was a wooden tank but it is now burned. Dam located 500 yards from railroad track.

Kilos. 165.7: Well. Limited supply of water especially in dry season. 33,000 gallon wooden tank.

Kilos. 200.9: Dam located 6 kilos from track. Water flows by gravity to track and pumped into a 30,000 gallon wooden tank.

Monclova: Water from Bocatoche which is located on track 22 kilos. from Monclova, a six inch pipeline parallel to track carried the water to Monclova, where it is pumped into a steel tank of 100,000 gallon capacity and a steel tank of 87,000 gallon capacity. There are two pumps, both located at Bocatoche.

Espinazo: Springs located 27 kilos. from track, at the Gavia canyon. There were two tanks here, but they have been burned. Water flows from the canyon to Espinazo by gravity.

La Perla: Dam 300 yards from track. Small wooden tank here. There is also a continuation by pipe line of the Espinazo water supply.

Higo. (Anhelo. Molino.) Well; supply limited, not exceeding 40,000 gallons in 24 hours. Drilled well 800 feet deep, supply unlimited. Water forced out by air jet, 40,000 wooden tank.

Saltillo: Wells. Unlimited supply, two steel tanks 100,000 gallon and 75,000 gallon capacity.

Water supply for branches:

Rosita Branch: Well at Rosita, 30,000 gallon wooden tank.

Múzquiz Branch: Water pumped from Polado (Palau) coal mines. 30,000 wooden tank. There is also a well with unlimited supply at Múzquiz, where there is a 30,000 gallon wooden tank.

Paredón-Monterrey Branch: At Hidalgo a four inch pipeline brings water from springs, and there is a 100,000 gallon steel tank here.

*　　*　　*　　*

Coal.

All engines on this division are coal burners. This coal is now obtained from the Palado (Palau) on the Múzquiz Branch. While the mines are the property of the railroad company they are now leased by the Coahuila Coal Co., an American corporation. It is expected that Constitutionalists will take over these mines in the near future. The Monclova Division requires about 8,000 tons of coal per month under normal conditions.

Work-shops:

Workshops on this division are located at Monclova and Piedras Negras, the latter shop is more important. The Monclova shop normally employed 175 men who are engaged in locomotive repair work. The power is electric. The Piedras Negras shop employs normally 300 men, engaged in locomotive repair work, foundry work, car shop work and brass work.

Many of the employees of these two shops enlisted in the army, but an effort is now being made by the Constitutionalists to locate them and return them to railroad shop work.

Telegraph:

The railroad has telegraph poles parallel to the track. On these poles the railroad runs three wires, and the government runs two wires. The government wires are run into government telegraph offices.

Rolling Stock:

The amount of rolling stock on this division at the present writing cannot be ascertained with any degree of accuracy. However, the following is a list of rolling stock which was obtained from the superintendent of the division. This official states that to the best of his knowledge this amount of rolling stock should be it the present time along the division; engines 15; baggage cars 2; flat cars 25; box cars 200; stock cars 30; coke cars (similar to stock cars, but without roof) 15; coal cars 125; passenger cars 5. The passenger cars are divided into first and second class and have a capacity of about 50. The freight cars have a capacity from 60,000 to 100,000 lbs.

Tunnels:

There are no tunnels on this division.

Bridges:

The Intelligence Officer of this Post submitted three weeks ago a profile which gives the construction, dimensions, etc., of the bridges, therefore, no bridges will be described in this report. The following statement can be made: all wooden bridges between

Piedras Negras and Saltillo have been burned in the course of the revolution. The reconstruction work has been very poor where bridges existed the track is now supported by cribs or by fills. In the latter case there will be wash-outs as soon as heavy rains fall. The steel bridges have had their masonry supports dynamited and at such places the dynamited portion is now supported by cribs.

Material for reconstruction: The only material on hand for reconstruction consists of 70 track kilos. of 54 lbs. rail at Allende, and 15 track kilos. of 54 lb. rails at Sabinas. There is no supply of sleepers on hand. The present method of reconstruction now in use is to destroy sidings to obtain material. Rails obtained by the destruction of sidings can be used, but ties now on sidings if dug up and replaced will be greatly injured, and of only temporary value.

Should the American Government take over this branch and obtain material and labor from the United States the road from Piedras Negras to Saltillo will be in condition for traffic in ten days.

It is understood, though not stated as a fact, that the road from Saltillo to San Luis Potosi is now being operated by the Federals. Though the bridges were destroyed, they have been cribbed or filled to support the track. The method of destruction of rail is twisting, and twisted rails cannot be straightened anywhere in Mexico.

Respectfully submitted,

John W. Wright

Captain, 17th Infantry,

Intelligence Officer.

Eagle Pass, Texas,
May 16th, 1914.

8529-3

Eagle Pass, Texas, 30 May 1914.

From: Intelligence Officer.

To: War College Division, General Staff, (through Chief of staff, southern Department).

Subject: Railroads. (Mexico).

1. I enclose duplicate copies of notes on railroad bridges in Mexico, principally on the Monclova Division.

JOHN W. WRIGHT

Capt. 17th Infty.

Intelligence Officer.

RECD AOSD May 31, 1914.

(Copy to Dept. Eng)
To Army War College, Washington, D. C.
H.S.D. June 1, 1914
1 encl.
fox.

The following notes relative to bridges on the Mexican Railway south of Piedras Negras are submitted as additional data to my railroad report.

The notes on the bridges of the Monclova Division, Mexican R.R. were furnished me by Mr. Bynum E. Nourse, a Civil Engineer who was employed by the road on bridge construction and repair during 1901 – 1907.

To insure greater accuracy, I had the notes checked by Mr. Stich, Supt. of the Monclova Division.

Neither of the two mentioned men had access to accurate data for the purpose of checking, but this data may be considered as substantially correct.

Location (Division from Piedras Negras)	Material	Base	Height of Banks (feet)	Description
Rio Escondido (Little River)	Steel	Masonry	25 and steep	5–50 ft spans
K 115. (Arroyo Blanco)	Steel	Masonry	25 steep	2 spans 90 ft each & 1 span 150 ft
K 119 (Rio Sabinas)	Steel	Masonry	35 steep	10 spans 84 ft each.
K 140 (Arroyo Barroterán)	Steel	Masonry	25 sloping	5 spans 30 ft each
K 149 (Arroyo Barroterán)	Steel	Masonry	15	2 spans 120 ft each
K 165	Steel	Masonry	15	1 span 165 ft.
K 204	Steel	Masonry	20 steep	6 spans 50 ft each.

(Near Hermanas)				
Castaños	Steel	Masonry	25	1 span 150 ft.
K 588 Matamoros de Laguna)	Wood	Pile & frame	35	15 spans 16 ft. each.

Between Matamoros and Torreón are several small unimportant bridges.

Monterrey Branch (from Reata to Monterrey) *Note*: Reata is at K 342 from Piedras Negras. (This abandoned at present; bridges in bad condition.)

Reata starting point.

Between K6 and K39 are several unimportant bridges.

K 55 (Rio Salinas)	Steel	Masonry	35	1 span 65 ft.

From K52 to Hidalgo (K78) are several unimportant bridges.

K 78 (Hidalgo)	Steel	Masonry	50 steep	2 spans of 30 ft each & 1 span of 150 ft.
Chipingue	Steel	Masonry	60	2 spans 100 ft each
Topo Grande	Steel	Masonry	45	1 span 150 ft long, 2 spans 75 ft long.

John W. Wright
Captain 17th Infantry,
Intelligence Officer.

8529-4

May 30th, 1914.

From: Intelligence Officer, Eagle Pass, Texas.
To: Chief War College Division, General Staff,
                      (Through Chief of Staff, Southern Dept.)
Subject: Railroads of Mexico.

1. I find three important errors upon the "General Map of the State of Coahuila de Zaragoza," map made by T. S. Abbott and dated 1905. This map was furnished me as Intelligence Officer. Map should be corrected as follows:

1. The railroad running west from Sauceda (10 miles south of Reata) to Hornos (Ferro Carril Internacional Mexicano) has been abandoned and the track removed.

The section from Anhelo east to Ixtle has been abandoned and removed.

The spur (about 40 miles south of Piedras Negras) from Sabinas to Hondo has been abandoned and removed.

2. From Anhelo (south of Reata) a new branch should run south and join the Monterey line at Paredón, this new section is about 8 miles long and is now the main line since the destruction of the Anhelo – Ixtle section.

3. The above data was furnished me by Mr. Stich, Supt. of the Monclova Division Mexican Railway.

John W. Wright
Captain, 17th Infantry,
Intelligence Officer.

RECD AOSD May 31, 1914.
(Copy to Dept. Eng)
To Army War College,
Washington, D. C.

H.S.D. June 1, 1914.
fox.

8529-6

886   Camp Eagle Pass, Texas
June 19, 1914.

From: Commanding Officer.
To: Commanding General, Southern Department.
Subject: Special Report.

1. Mr. Grey who is employed by the Constitutionalists as an "Agent" arrived in Eagle Pass yesterday. He came from Saltillo, having reached the place via Juarez - Torreon.

He says the line of railroad is strewn with broken cars, engines and twisted rails. He estimates Villa's force at Torreón at 15,000 men, well clothed. Villa has just received a supply of ammunition from Tampico. Mr. Grey made the trip from Saltillo to Piedras Negras in 10 hours. He says the road is in fairly good condition. He says the line south of Saltillo to San Luis Potosí is in bad shape but work is in progress. The route south of Torreón towards Zacatecas is in fair shape, but washed out in places. He estimates the forces in Saltillo at 12,000 men.

Mr. Grey says the break between Carranza and Villa was brought to a head when Natera was ordered to wait and not attack Zacatacas until he was reinforced by Villa, but Natera would not wait. He attacked, and during the attack 1,000 Federal prisoners who were serving in the ranks of Natera, revolted and went back to the Federal side. When Natera saw his attack failed, he wired to Carranza for reinforcements and Carranza ordered Villa to go to his assistance, this Villa refused to do.

Villa, according to Mr. Grey, will leave at once and attack Zacatecas, but will ignore General Natera. He will then attack Guadalajara. These operations he will undertake without reference to any instructions he may receive from Carranza. When the latter place has been taken he will go towards Mexico City.

Villa believes that an effort is being made to sidetrack him and will hereafter act alone.

2. An idea of Villa's character may be formed by the following advice which Villa gave to Mr. Grey, given half in jest, "Drink, but never get drunk; love without passion; steal, but only from the rich."

[T. W.] Griffith

Colonel, 17th Infantry

2150480 A 5   1st Ind

A. G. O., June 26, 1914 - To the Chief of Staff.

<u>8529-7</u>

Hotel Hamilton

D.J. MacDonald, Proprietor,

Laredo, Texas.

Laredo, June 27th, 1914.

Capt. H.H. Robert,

     San Antonio.

My dear Captain:

Have just had the good fortune of meeting a friend who knows all about the Soto la Marina Country Mr. Wm. Habbron of Altamira, Mexico, near Tampico. He has travelled all over this country looking into timber lands and tells me Soto la Marina is from 23 to 25 miles from the mouth of the river. Good road follows river on north entire distance. River probably 1000 to 1500 ft. wide, has usually 30' of water except on Bar at the mouth where it has only about 8 or 15. After passing Bar River is navigable for some miles beyond Soto la Marina, heavy timber & under growth on both sides. From Soto la Marina there are two roads to Victoria, one goes via Guemos or Huemos, distance about 95 miles. Good water in surface wells at Guemos 25 miles from Victoria and at another little town 25 or 30 miles further east the name of which he could not remember, but a branch of the Soto la Marina River crosses the road at this place. Says Guemos population 600 or 700 the other towns 3 to 400. The other route is over new road graded by Govt., which is only about 75 miles but a much harder and dryer route, most, all travel is by the Guemos route.

Country sparsely settled from Victoria to Soto la Marina but quite thickly settled from Soto la Marina to the mouth of River. Considerable cultivation, corn, beans, etc. Says roads are good in almost all seasons. There are other watering places along the Guemos route that he could not locate. Am writing this in a little haste as I wish to get

156

it off tonight. Haven't been able to locate Mr. Reeder, but am trying to do so and when I do will try and see him. Should the information about the Guemos route from Soto la Marina not be sufficient and in my absence you might have Col. Crane interview Mr. Hibbron here, sending him notes of such information as you wish. He was not so well informed on the other route, but says it's a much heavier route as it runs over the hills while via Guemos it runs through a more level country.

I trust this may assist you in rendering a report, Regards

Yours truly,

J.C. Gravis

Camp Eagle Pass, Tex.

June 29th, 1914.

From: Intelligence Officer, Eagle Pass, Tex.
To: Chief of Staff, Southern Department.
Subject: Road notes, Soto la Marina- Tamaulipas, Mexico.

1. Referring to your communication of June 25th, 1914 requesting information relative to the route from mouth of Rio Soto la Marina to City of Victoria, I enclose herewith in duplicate a general description of the river and adjoining country and detailed road notes of the only practicable route.

2. The Mr. "Boycott" referred to in your letter proved to be Mr. I. N. Boicourt, an American, now living in Piedras Negras. He lived 8 years in the territory referred to and the information I am sending was obtained from him.

3. I return the map furnished me. Mr. Boicourt has noted in pencil upon this map the limits of the large haciendas, and he has also marked the roads. The note "N.G." signifies bad road, and the note "G.R." signifies good road. By good road is meant one suitable for wagon transportation. Pack trains can, however, pass over the bad roads.

4. Mr. Boicourt will remain in Piedras Negras for some time, and further information can be obtained if desired.

5. Information regarding this territory can be obtained also from Mr. H. H. Reeder, whose address is The Menger Hotel, San Antonio. Mr. Reeder is the owner of El Chamal Ranch, mentioned in my road notes.

John W. Wright

Captain 17th Infantry,

Intelligence Officer.

ROAD NOTES.

State of Tamaulipas, Mexico.

Mouth of Rio Soto la Marina - El Chamal - Buena Vista - Abasola - Victoria.

Distance 118 miles.

U.S. Army Intelligence in the Mexican Revolution

General Description.

(a)   The Rio Soto la Marina is a stream that has not formed a river valley. It cuts its narrow bed through a rolling table land. Hills come down to the river bank. South of the river the hills are higher than those to the north. As a consequence the south bank is more abrupt than the northern. The country south of the river is described as a wilderness abounding in wild game, there is much timber, mostly ebony, which has never been cut. No roads or trails pass through this section and there is but little water. There is a large ranch south of and bordering the river belonging to Mr. I. O. Brictson, an American, Mr. Brictson, (who lives in Deerfield, Wisconsin) raises cattle on his ranch which contains one million acres. Oil has been found here, but has not been developed, the oil rights have been sold to a Dutch Company which has an office in Tampico. This company is known as "The Corona Oil Company."

The territory north of the river has no settlements but there is good grass. It is open country.

The average yearly rainfall in this territory is 20 inches.

(b)   Rio Soto la Marina is about 35 miles long. The stream is formed near the town of the same name by several smaller streams. It runs east to the Gulf of Mexico. Before the railroads were constructed this river was of considerable importance, vessels would ascend the river and carry supplies to the town of Soto la Marina to be shipped overland into the interior. The building of railroads caused the stream to lose its importance.

The stream is between 20 - 30 feet in depth. At its mouth is a bar over which there is 7 feet of water, at mean tide. The tide is 2½ feet. There is no harbor at the river mouth. It empties directly into the Gulf where its mouth is 300 Yds. wide. As the river is ascended it becomes less wide, the average width is 250 Yds.

The tide reaches all the way to the town of Soto la Marina and the result is that the water, east of that town cannot be used by man or beast for drinking.

There are no landing facilities of any kind on the river, such as wharves, but small boats can find many suitable places. At the mouth of the river are salt works and about 10 houses constructed of adobe. There are no boats on the river other than a few small ones.

The bed of the river is limestone; the depth and width is effected but little by the rains.

The only means of travel from the river mouth west to Victoria is by taking the route described herein. No road or trail follows along the river banks.

The Mexican Government began the construction several years ago of a Macadam road from Victoria east to Soto la Marina. This macadam road was only half completed.

All distances are from mouth of river.	Detailed Description
1. From mouth of river to El Chamal (20 miles.)	This is an old untraveled trail and has no grades of importance. It is, at places, overgrown with brush which will have to be cut. The soil is vegetable, black in color and following heavy rains becomes muddy and difficult to cross with a wagon. Heavy rains, however, are infrequent. There is plenty of grass along this route but no water until Chamal is reached.
El Chamal 20 miles.	El Chamal is a ranch property of Mr. H.H. Reeder, who can be reached at the Menger Hotel, San Antonio. About 100 people, employees, live here. There are 20 buildings of stone and adobe. Excellent grazing; wood; water from wells and tanks, the country is open. From Chamal a good road runs to Matamoros 160 miles long. Loaded wagons can pass over this road.
2. From El Chamal to Buena Vista. (15 miles.)	This section consists of a travelled road through chaparral and ebony timber. It is well watered by several small streams which are crossed by the road. No obstacles to crossing in any season. There is plenty of grass along this section. No steep grades. The soil is limestone and not effected by the rains.
Buena Vista. 35 miles.	Buena Vista has a ranch settlement with 20 houses. The ranch owner is a Spaniard. Water is obtained from wells and a dam.
3. From Buena Vista to Abasolo (18 miles.)	This section consists of a good wagon road through open country. There is some timber of the chaparral and ebony variety. There are no grades. The soil is limestone. One small stream of water is passed which goes dry in extremely dry weather only. Grazing is good.
Abasolo. 53 miles.	Abasolo is a village of 500 people. Here a Federal telegraph line crosses, connecting Tampico – Victoria and Matamoros. There is a telegraph station maintained here. The village has about 100 houses, constructed of stone and

adobe. Water from a small stream 3 miles west of town carried in carts.

From here a good wagon road runs to town of Soto la Marina.

From Abasolo to Padilla. (40 miles.)

This section is a good wagon road crossing a large range of limestone hills several hundred feet high. The road is rough in places and there are steep grades. The natives, however haul 1½ tons with 6 small mules over this road. The soil is limestone not effected by the rain.

There is plenty of wood and grass but no water until Padilla is reached.

Padilla. 93 miles.

Padilla is a village of 500 people. Houses of adobe and stone. This is a county seat. A fine river runs near the town and is the water supply. It is also used for irrigation. Considerable corn is grown in this section, and there is some stock.

From Padilla to Victoria (25 miles.)

A well-travelled wagon road (this is the road connecting Matamoros and Victoria). Along the road runs a telegraph line. The soil is limestone, but black in color. Excellent agricultural country. There are several small streams. One crossed on a small wooden bridge. All crossings can be easily effected by a wagon. Timber is abundant and much corn cultivated.

Victoria. 118 miles.

Victoria

Note----- The above data was obtained from Mr. H. H. Boicourt, who lived in this part of Mexico for 8 years.

Respectfully submitted,
John W. Wright
Captain 17th Infantry,
Intelligence Officer, Eagle Pass, Tx.

8529-10
Headquarters Southern Department
Fort Sam Houston, Texas.

August 5, 1914

From: Department Chief of Staff.
To: Commanding Officer, U.S. Troops, Eagle Pass, Texas.

Documents Cited

Subject: Water supply along certain routes.

1. Herewith memorandum compiled at direction of Department Commander by Department Engineer relative to water supply on the route between Piedras Negras and Monterrey; also copies of memorandum submitted by Department Engineer to Department Commander with this data.

2. The Department Commander desires that, as indicated in paragraph 3 of the memorandum of July 2, 1914, from the Department Engineer, you verify to the fullest extent possible commensurate with prompt action the information submitted herewith and return same with a separate communication embodying such corrections, criticisms, suggestions and recommendations as may be pertinent to the case in question.

3. The Department Commander desires that this matter be considered strictly confidential and that no one except yourself and your Intelligence Officer be acquainted therewith.

4. As the copy of water supply data furnished you herewith is the only copy on file at these Headquarters, it is desired that you preserve same carefully and return it without any change whatsoever on the copy furnished herewith.

By command of Brigadier General Bliss:

Major, General Staff.

ADDITIONAL NOTES ON WATER SUPPLY ALONG RAILROADS BETWEEN PIEDRAS NEGRAS AND MONTERREY

Fuente:	The Rio Escondido is a running stream which varies in depth from 1 to 10 feet. During the dry season there is always running water. This water is good for drinking purposes. At Fuente the stream is about 30 feet wide and the banks are not steep.
Escondida Coal Mines.	The river here has about the same width and depth as at Fuente, with good approaches. There is also a supply of water to be obtained from the mines when they are working. The water from the mines is estimated at one million gallons in 24 hours.
Fenix Coal Mines.	The river here has the same breadth and depth as at Fuente.
Maroles.	(No additional data obtainable).
Rancho De Los Muchachas.	(No additional data obtainable).

Rosita Coal Mines.	The Sabinas river has never run dry. It is from 30 to 60 feet wide. During the dry season it has 6 inches of water. After heavy rains it is often 15 feet deep. Approaches are good for watering animals. Excellent drinking water.
Alamos Arroyo.	Where the railroad crosses there is water only after heavy rains. During the dry season the bed is dry.
San Juan de Sabinas.	(No additional data obtainable).
Esperanzas.	There is no river here. There is a reservoir holding three million gallons which is filled by rain water. This reservoir has never yet been exhausted.
Barroterán.	No stream. Supply limited to wells and is uncertain.
Aura.	At the town of Aura, five miles from the station is the Rio Aura. This river never runs dry and is 10 feet wide, and 1 foot deep with good approaches.
	At the railroad station of Aura the stream runs underground in its gravel bed. There are few small wells here.
Kilo 151 on R.R.	Note on earth tank is correct. Approaches to tank good. There is 5,000 feet of water front available for watering a command.
Kilo 166 on R.R.	Note on well correct. The water in this well is the same that disappeared under ground at Aura station.
Kilo 182 on R.R.	(No additional data obtainable).
Lampacitos.	Notes on wells at mine and on Rio Salado are correct.
Hermanas.	Note on hot springs correct. This is good drinking water. A stream comes from this spring that is two feet deep and two feet wide and never runs dry.
	Note on reservoir correct. This reservoir covers about 1½ square miles. It is 18 feet deep at its deepest point and can be approached by animals on all sides.
	The note regarding the purgative effect of water in streams south of Hermanas is correct.
Rodríguez.	(No additional data obtainable).
Abasolo.	(No additional data obtainable).

Documents Cited

Monclova.	There are two places called Monclova, the town and railroad station. The town is three miles east of the station. The running stream runs through the center of the town of Monclova and is the water supply of the town. The water is good, flows the year round and is unlimited in supply. Note on artificial supply correct.
Bocatoche.	Note on artificial supply correct. There is a small stream here running under railroad bridge. It is 30 feet wide and six inches deep. It runs during all seasons.
Castaños.	The water supply referred to here is the same stream which runs under the railroad bridge at Bocatoche.
197.	This tank is known as Gloria tank as it is located on the farm named Gloria. The note is correct. This water supply never runs dry but at times gets very low.
Espinoza.	Note correct. The pipe referred to runs just below the surface of the ground and can easily be destroyed.
La Perla.	Note correct. The usual supply of water at La Perla is obtained from the overflow of the Espinoza supply. The reservoir mentioned is considered as a reserve.
Reata 8	Note correct.
Anhelo.	The note on railroad well is not correct. There is no railroad water supply here. Referring to the note on Delgado, canyon, there is no water at this canyon. There are two reservoirs at Boca de la Mula, four miles east of Anhelo. The water supply at Mula is unlimited, but these reservoirs can easily be destroyed.
Paredón.	Note correct except it should be stated that the water from this well is very bad. It is not only unfit to drink but the railroad company has never been able to use this water in boilers.
251. Kilo 593.	Rio Saltillo is so salty that it cannot be used either for drinking or for boilers.
Ramos.	The name of this stream, instead of Rio Palos, should be Rio Patos. It is dry during dry season. It empties into Rio Salinas near Ramos. Rio Salinas has water in water holes found in the stream bed the year round, but this water is salty and unfit to drink.
Icamole.	Note on Patos River is correct except it should be stated that the water is salty and unfit to drink. The note on wells cannot be verified.
Fraile.	There is a well at this place but data given in note cannot be verified.

García.   (No additional data obtainable).

Pesquería.   A stream crosses the railroad here. Water appears in the bed of the stream after heavy rains only.

All of the reservoirs along the railroad mentioned herein are built of earth, and they can very easily be destroyed. The above data was obtained from railroad officials. Cattle men do not drive their cattle along this route as it is too near the railroad. This fact limited my sources of information.

Respectfully submitted,

John W. Wright

August 13, 1914.   Captain, 17th Infantry,

Intelligence Officer, Eagle Pass, Tex.

Enc. 2

Report of Water Supply on Railroads between Monterrey and Torreón.

No information could be obtained relative to route over Coahuila and Pacific Railroad via Saltillo between Monterrey and Torreón.

The second route described in note on first sheet of accompanying memorandum of Department Engineer should be by way of Fraustro between Paredón and Sauceda Junction instead of by way of Anhelo and Reata. This distance between Paredón and Sauceda Junction being only 41 kilometers on this route. There is only one bridge of importance between Paredón and Sauceda Junction – that over the Tortigo River, while the route by way of Anhelo, Reata, from Paredón to Sauceda Junction has at least 40 bridges.

Page 1, Memo. Dept. Engr.

No information of value could be obtained concerning wagon roads along railroad.

Pesquería – No Information.

García – A garden spot, with grapes, peaches, figs, alfalfa and barley. Plenty of running water in irrigating ditches. One canal has always had water in it during last ten years. Source of water not known.

Fraile – No information.

Icamole – Water in Patos (Salinas) River is brackish but can be used by men and animals for drinking. No information of water in wells.

Ramas – Information verified. Water brackish.

Kil. 593, on Railroad.   Rio Saltillo or Salinas. Brackish water in pools the year round but supply limited in dry season.

Page 2, Memo. Dept. Engr.

Paredón – Water in well cannot be used for drinking by men or horses. All water used for drinking at Paredón is shipped on railroad or brought from Patos River about 1½ miles distant.

Amargos – Information verified.

Fraustro – Earthen tank near here can supply water for at least a brigade of infantry 6 months after the annual heavy rains. The railroad draws heavily on this tank after the rains instead of drawing on the wells in other places. The tank often goes dry in the dry season.

Hipólito – Information verified. Pipe line is near surface and can be easily destroyed. Good wagon road from iron tanks to earthen tank.

Jarral – Has pipe line connected with Hipólito line from Tulillo tank.

Tizoc – Information verified.

Ceres – Information verified. Laguna de Mayran just after rainy season reaches from Ceres to Benavides.

Benavides – Information verified. There are an abundance of irrigating canals from Nazas River between Benavides and Concordia.

Concordia – Information verified. The well water here is excellent.

Page 3, Memo. Dept. Engr. Information verified.

B Simmons
Capt. 17 Inf.
Intelligence Officer.

Eagle Pass, Tex.
Sept. 10/14.

8529-11
War Department
Headquarters Southern Department,
Fort Sam Houston, Texas.
September 18, 1914.

From: Chief of Staff.
To: Commanding Officer U.S. Troops, Eagle Pass, Texas.
Subject: Information on Mexico.

1. Herewith copy of self-explanatory communication from the Secretary of War College Division. It is directed that you have your Intelligence officer procure at once and forward without delay to these Headquarters all available information along the lines called for by said communication.

2. While it is desired that you have procured all available information on all the lines indicated herein, it is presumed that Brownsville and Rio Grande will furnish the best

sources of information for data called for in paragraph 5 of the War College communication; similarly it is believed that your station will furnish the best source for the information called for in paragraphs 2, 3 and 4.

3. You will, therefore, pay special attention to the information called for in paragraphs 2, 3 and 4 and will not delay reply to this communication awaiting information not readily available along the lines called for in paragraph 5.

4. Route No. 1 referred to in War College communication is the Eagle Pass-Monclova-Saltillo; and the route #12 is Monterey-Reata-San Pedro-Torreón.

5. A recent memorandum report from the Department Engineer at these Headquarters on water supply between Piedras Negras and Monterrey contains the following note:

> "Note: Route of wagon road from Espinazo to Monterrey is not definitely known. The following notes are compiled from such information as is here available including information from former employee of Mexican Nat. Rys. It appears that this wagon route does not pass through Reata but leaves the R.R. at Espinazo or La Perla making the distance from Espinazo to Anhelo by road less than that by rail, i.e., via Reata",

and the notes of the Intelligence officer at Eagle Pass, to whom the report of the Department Engineer was forwarded, contains the following information regarding said notation "Reata - Note correct."

6. It is believed that Mr. Chas. Stitch, former division superintendent of Nat'l. Rys. of Mexico, can give you information along the lines called for or at least can put you in touch with good sources for such information.

By command of Brigadier General Bliss:

W H Hay
Major, General Staff

1236 1st Ind.

Cmdg. Ofcr. Eagle Pass, Tex. 26 Sep.1914:- To Comdg. Gen'l. South. Dept.

1.-Returned with report of Intelligence Ofcr. this station on pars.

2, 3, & 4 of letter of Secy. Gen'l. Staff. War College Div.

2:- The available sources of information here on the subject of wagon roads is very limited, nor do I believe such information as is procurable reliable.

[illegible]
Maj. 17th Infty.

C.F. −C. Dept. Engr. 2nd Ind. HHR-WAF

Hq.So.Dept., Ft. Sam Houston, Texas, Sept. 28, 1914.- To the Secretary, War College Division, General Staff, Washington, D. C., forwarded, inviting attention to body of this communication and attached report from the Intelligence Officer, Camp Eagle

Documents Cited

Pass, Texas, dated September 25, 1914; this in connection with your communication of September 12, 1914, subject "Information on Mexico."

W H Hay
Major, General Staff
Chief of Staff, Southern Dept.

Camp Eagle Pass, Texas.
September 25, 1914.

From: Intelligence Officer.
To: Commanding Officer, 17th Infantry.
Subject: Information on Mexico.

1. I have not been able to obtain any information of value concerning the wagon roads along the routes indicated in communications herewith.

2. The following information concerning the water along the railroads between the points mentioned may be of value.

Monterrey to Paredón along Railroad.
Miles.

20	García	- Water in irrigating ditches from Monterrey to García.
25	Fraile	– Railroad well. Capacity 40,000 a day the year round. Good water.
32	Icamole	- Water in Rio Salinas. Animals can drink it but too brackish for men.
37	Ixtle	– No water.
43	Ramos	– Water in Rio Salinas, but it cannot be used for drinking.
47	Paredón	– No water.

Monterrey to Paredón along Railroad – Northern Route
Miles

22	Hidalgo	– Spring water, 80,000 gallons in 24 hours throughout the year.
27	Mina	- No railroad water. Small quantities in wells.
37	Arista	– No water.
43	Ixtle	- No water.
49	Ramos	– No water except the brackish water in Rio Salinas.
53	Paredón	– No water.

Espinazo to Saltillo along Railroad.

Note – There is a wagon road from Espinazo to Anhelo, also one from La Perla to Anhelo, which do not pass through Reata.

Miles from

Piedras Negras.

202   Espinazo – Water brought down in 4 inch pipe from springs in Gavia Canyon, 25 kilometers distant. Capacity 100,000 gallons a day. Pipe easily broken.

209   La Perla – Reservoir which contains water throughout the year except after an unusually dry season. A 3 inch pipe brings the overflow from Espinoza.

212   Reata – No water.

226   Anhelo – No water on railroad. There are 2 reservoirs at Mula, 4 miles east with unlimited supply, but reservoirs easily destroyed.

229   Paredón – No water.

250   Higo – Railroad has well producing 40,000 gallons in 24 hours throughout the year –good water.

260   Molina (1½ miles south of Ramos Arizpe) – Irrigation ditches with water throughout the year.

269   Saltillo.

B. Simmons
Captain, 17 Infantry.

8532-1
El Paso, Texas
May 26, 1914.

From: W.N. Hensley, Jr., 1st Lt., 13th Cav., Int. Officer, 8th Brigade.
To: Commanding General, Southern Department.
     (Through military channels)
Subject: Report on Mexican Central Railroad.

Enclosed is as complete a report as possible at this time on the National Railway of Mexico, known as the Mexican Central from Juarez to Torreón, including the Parral branch from Jimenez to Parral and Santa Bárbara.

1. Included in this report are complete official track charts, showing distances and directions of the railroad, with a profile showing grades, locations of sidings, location of bridges, railroad crossings, and water tanks. These track charts are not only valuable as a railroad reference but are also good topographical maps of the country for a limited distance on either side of the railroad, showing in more or less detail, wagon roads, ranch houses, corrals, rivers and mountains- in the case of mountains no heights are shown except as given by the profile. These charts were obtained by the promise of return, and it is requested that as soon as they have served their purpose that they be returned to me.

2. A time table is enclosed, which in its condensed form shows better than any report could, distance in kilometers between stations, distance of station from Mexico City and distance of station from Chihuahua. The location of sidings are shown with the capacity of each in number of cars, (each car averages forty (40) feet), where water, fuel, turn tables and Y's are located, telegraph offices and call letters for same.

3. The bridge report enclosed shows all bridges and culverts between Juarez and Jiménez and the Parral branch. In this connection it is well to state that a number of these bridges were destroyed, but it is understood that only bridges that could be burned were destroyed. The "Constitucionalistas" who are now operating this line as a military railroad, state that all bridges have been repaired or replaced by "shooflys." As soon as a Mr. Collins who is the only foreigner working on this line between Juárez and Chihuahua arrives in El Paso, a subsequent report will be rendered showing present condition of bridges and roadbed and the amount of rolling stock on this division. The civil officials of this road, now refugees in El Paso, inform me that they are entirely ignorant as to the actual condition of the road and its rolling stock. All available material for upkeep and repair has been taken to the vicinity of Torreón for use by the "Constitucionalistas" in their advance to the south. There is said to be a hundred to a hundred and fifty kilometers of fifty six (56) pound steel rails in the yards at Chihuahua, also three (3) deep well pumps complete. The rails in use on the roadbed at present are the seventy five (75) pound rail. Only enough coal is kept at coaling stations tor the actual operation of trains between Juarez and Chihuahua of which there is one every other day. The coal required is purchased in El Paso, no reserve stock being kept.

4. The pump report shows conditions at last report of water facilities, giving number, location and capacity of tanks, and all data with reference to the wells that supply these tanks.

5. It has been quite difficult to obtain any information on the present condition of this road, all employees being Mexicans who look upon Americans with suspicion. The one employee not a Mexican, mentioned above, is sick at Chihuahua. As soon as he returns arrangement has been made for an interview and it is believed a correct estimate can then be made as to present condition of road and amount of rolling stock on the Chihuahua division.

W. N. HENSLEY, Jr.
1st Lieut. 13th Cav.

To Army War College,
Washington, D.C.

H.S.D. June 5, 1914.
6 encls.

fox.

WATER FACILITIES-CHIHUAHUA DIVISION.

C. Juárez: Steel tank 33,000 gallons capacity on masonry base. Water supply taken by five pipes 7' 4", driven 23 meters and connected to a junction pump. Well 3.60 x 3.60 meters   square. Two boilers, one 15 and one 35 horse power upright. One No.7 Dean pump lever motion.

Samalayuca: Steel tank, 33,000 gallons capacity on masonry base. One well 4.50 meters in diameter, 9 meters deep. Depth of water 3.0 meters. One 25-horsepower boiler. One No. 7, Dean pump, lever motion.

Ojo Caliente: Steel tank 33,000 gallon capacity on masonry base. One well two kilometers from well on west side of track, 4.50 meters in diameter, 7.20 meters deep. Depth of water 2.10 meters. Two 25-horsepower boilers. Two No 6 Dean pumps, lever motion.

Moctezuma: Steel tank, 33,000 gallon capacity on masonry base. Open well 6.0 meters in diameter, 51.90 meters deep. Depth of water 2.10 meters. Two 25-horse power boilers. One No.7 Dean pump, lever motion.

El Sueco: Steel tank 33,000 gallon capacity on frame bents. Open well 4.50 meters in diameter, 65.40 meters deep. Depth of water 2.40 meters. One 35-horse power boiler. One No.7 Dean pump, lever motion.

Laguna: Steel tank 75,000 gallons capacity on masonry base. Open well 3.60 meters in diameter, 19.20 meters deep. Depth of water 2.70 meters. One 15 horse power boiler. One No. 7 Dean pump, lever motion.

Sauz: Steel tank 75.000 gallons capacity on masonry base. Open well 4.50 meters in diameter, 7.80 meters deep. Depth of water 3.0 meters. One 15 horse power boiler. One No.7 Dean pump.

Kilometer 672: One open well, 3.30 meters in diameter, 29 meters deep. Depth of water 1.50 meters. No. tanker pump at this place.

Chihuahua Shops: Steel tank 33,000 gallons capacity on masonry base. Open well 9.0 meters in diameter, 25.50 meters deep. Depth of water 3.0 meters. Two locomotive boilers: one, 50 and one, 35 horse power. Three pumps: one Duplex, 14 x 4½ x 12, and two No. 7 Dean, lever motion.

Documents Cited

Horcasitas: Steel tank 33,000 gallon capacity on masonry base. Open well 6.0 meters in diameter, 33.0 meters deep. Depth of water 3 meters. One 35 horse power boiler. One No. 7 Dean pump.

Bachimba: Open well 3.30 meters in diameter, 3.30 meters deep. Depth of water 2.40 meters. One 25 horse power boiler. One No. 6 Dean pump, lever motion.

Ortiz: Have no data available for water supply there, but there is a well that has an inexhaustible supply of water. A tank of about 33,000 gallon capacity is at that point.

Santa Rosalía: Kilo. 1452: Steel tank 33.000 gallon capacity on frame bents. Open well 4.50 meters in diameter, 6.60 meters deep. Depth of water 2.40 meters. One 10 horse power boiler. One No 7 Dean pump, lever motion. Understand this pumping plant was destroyed by the revolutionists last fall.

Díaz: Steel tank 33,000 gallon capacity on masonry base. Open well 4.50 meters in diameter, 9.0 meters deep. Depth of water 1.20 meters.

Jiménez: Steel tank on masonry base. Water supply taken from Jiménez River. Two boilers: one 25 and one 35 horse power. Three No 7 Dean pumps, lever motion.

PARRAL BRANCH.
Dorado: Steel tank 33,000 gallon capacity on masonry base. Open well 4.80 meters in diameter, 7.50 meters deep. Depth of water 3.0 meters. One 10 horse power boiler. One No. 6 Dean pump, lever motion.

Maturana: Wooden tank 60,000 gallon capacity on frame bents. Open well 4.50 meters in diameter, 6.0 meters deep. Depth of water 3.60 meters. One wind-mill pump: 4 x 8 Ramsey.

Parral: Steel tank 33,000 gallon capacity on masonry base. Water furnished by City of Parral.

Borjas: Steel tank 30,000 gallon capacity on masonry base. Open well 4.50 meters in diameter, 46 meters deep. Depth of water 5.40 meters. One 25 horse power boiler. One No. 6 Dean Pump, lever motion.

Rosario: Steel tank 33,000 gallon capacity on frame bent. Open well 3.60 meters in diameter, 4.80 meters deep. Depth of water 1.8 meters. One wind-mill pump: 4 x 8 Ramsey.

Deep wells at the following places, at which there are no pumping facilities and the wells are not in use:

Carrizal: Total depth 49.99 meters, water flowing. Water of bad quality. Casing 6".

Alsacia: Depth of well 83.82. Pumping level 76.20 meters. Tested 75 gallons per minute without lowering. Casing 6".

Loaeza: Depth of well 165.20 meters. Pumping level 132.59 meters. Tested 75 gallons per minute without lowering. Casing 6 1/4".

Chihuahua Shops: Depth of well 33 meters. Water comes within 15.24 meters of the surface. Tested 75 gallons per minute without lowering. Casing 8".

Bachimba: Depth of well 28.50 meters. Pumping level 21.34 meters. Water rises within 18.29 meters of surface. Could not lower water pumping at 21.34 meters. Casing 8¼".

Concho: Depth of well 91 meters. Pumping level 56.38 meters without lowering. Water shows unfavorable analysis. 10" casing.

Santa Rosalía: Depth of well 87 meters. Pumping level 7.36 meters without lowering. Water comes within .91 meters of surface. 8" drive pipe.

Jiménez: Two wells: depth of No 1 is 62 meters, pumping level 25.90 meters without lowering. 6¼" casing. Well No 2 depth 62 meters, pumping level 25.90 meters without lowering. Casing 8¼".

Understand that there is a deep well at Ranchería, but have no data at hand to give details.

EXTRACT FROM
TIME TABLE NATIONAL RAILROADS OF MEXICO.
CHIHUAHUA DIVISION.

Distance from Chihuahua Kilom.	Distance from Mexico City, Kilom.	STATIONS.	Capacity of Sidings. Cars.	
236.2	1373.3	DN Jimenez. (JN) 1.7	Yards.	ABC GMR
234.5	1375.0	Empalme Ramal de Parral, 17.4	None	Y
217.1	1392.4	La Reforma, 18.8	47	
198.3	1411.2	D Diaz, (DI) 19.2	49	A
179.1	1430.4	Bustamante, 15.7	45	
163.4	1446.1	DM Santa Rosalía (SR) 5.1	49	G
158.3	1451.2	Tanque, 10.9	None	A
147.4	1462.1	D La Cruz, (CR) 20.4	45	
127.0	1482.5	DN Concho, (C) 6.3	69	G
120.7	1388.8	D Saucillo, (SC) 9.3	10	
111.4	1498.1	Armendariz, 16.1	45	
95.3	1514.2	Las Delicias, 7.3	43	
88.0	1521.5	DN Ortiz, (OR) 12.1	78	AG
75.9	1533.6	Consuelo, 12.2	32	
63.7	1545.8	Bachimba, 1.9	33	
61.8	1547.7	Tanque, 2.3	None	A
59.5	1550.0	Carbonero, 16.2	4	C
43.3	1566.2	D Horcasitas, (HS) 19.4	33	AG
23.9	1585.6	Mapula, 11.5	50	G
12.4	1597.1	D Alberto, (BO) 2.6	24	
9.8	1599.7	Crucero F.C. Mineral, 0.3	None	
9.5	1600.0	D Morse (WS) 4.2	75	
5.3	1604.2	Crucero F.C.K.C.M. y O. 0.4	None	
4.9	1604.6	Empalme F.C.N.O. de M. 0.5	30	
4.4	1605.1	Tabalaopa, 3.4	80	
1.0	1608.5	D Chihuahua, (CH) 1.0	80	
0.0	1609.5	DN Chihuahua (Talleres) (CW)	Yards.	ABG GM R

Key: A = Agua,
         M = Turn table

R = Register
C = Fuel
Y = Y junction
D = Day telegraph station
DN = Night and Day telegraph station
( ) = Call Signs

EXTRACT FROM
TIME TABLE NATIONAL RAILROADS OF MEXICO.
CHIHUAHUA DIVISION.

Distance from Moctezuma Kilom.	Distance from Mexico City, Kilom.	STATIONS.	Capacity of Sidings. Cars.	
181.0	1609.5	DN Chihuahua (Talleres) (CW) 7.8	Yards.	ABCG RM
173.2	1617.3	Cuilty, 14.3	50	
158.9	1631.6	Corral, 3.8	50	
155.1	1635.4	Tanque, 3.9	None	A
151.2	1639.3	Molinar, 9.3	50	
141.9	1648.6	D Terrazas, (TD) 9.7	135	
132.2	1658.3	D Sauz, (SU) 10.2	50	AG
122.0	1668.5	Pinale, 9.7	50	
112.3	1678.2	Encinillas, 13.9	37	
98.4	1692.1	Agua Nueva, 13.4	25	
85.0	1705.5	D Laguna, (GN) 8.9	40	AC
76.1	1714.4	Arados 11.5	47	
64.6	1725.9	Mocho, 12.9	60	
51.7	1738.8	Loaeza, 7.3	60	Y
44.4	1746.1	D Gallego, (GI) 9.7	48	G
34.7	1755.8	El Sueco, 6.2	34	A
28.5	1762.0	Alsacia, 13.1	62	
15.4	1775.1	Chivatito, 8.5	43	
6.9	1783.6	Centauro, 6.9	50	
0.0	1790.5	DN Moctezuma, (MO)	130	ACRY

Documents Cited

Distance from El Paso Kilom.	Distance from Mexico City, Kilom.	STATIONS.	Capacity of Sidings. Cars.	
182.5	1790.5	DN Moctezuma, (MO) 13.1	130	ACRY
169.4	1803.6	Las Minas, 13.5	46	
155.9	1817.1	D Ojo Caliente, (OJ) 11.3	42	AG
144.6	1828.4	Carrizal, 11.0	44	
133.6	1839.4	D Ahumada, (MD) 11.8	34	
121.8	1851.2	San José, 3.2	45	
118.6	1854.4	Tanque, 8.7	None	A
109.9	1863.1	Lucero, 12.2	55	G
97.7	1875.3	D Ranchería, (RH) 16.7	69	
81.0	1892.0	Carbonero, 1.1	4	C
79.9	1893.1	Candelaria, 10.9	44	
69.0	1904.0	Los Medanos, 18.2	42	
50.8	1922.2	D Samalayuca, (SY) 16.1	43	A
34.7	1938.3	Tierra Blanca, 14.4	29	
20.3	1952.7	Mesa, 16.3	42	
4.0	1969.0	DN Ciudad Juárez, (P) 1.0	Yards.	ABCGRM
3.0	1970.0	Puente del Rio Grande, 3.0	None	
0.0	1973.0	El Paso.	Yards.	

Distance from Rosario Kilom.	Distance from Mexico City, Kilom.	STATIONS.	Capacity of Sidings. Cars.	
154.9	1373.3	DN Jiménez, (JN) 1.7	Yards	ABCGRM
153.2	1375.0	Empalme Ramal de Parral, 3.6	None	Y
149.6	1378.6	Orion, 24.0	5	

125.6	1402.6	D Troya, (YA) 10.3	19	G
115.3	1412.9	D Baca, (BW) 4.4	35	
110.9	417.3	Tanque, 1.7	None	
109.2	1419.0	D Dorado, (MB) 13.2	19	
96.0	1432.2	D Morita, (MG) 5.4	19	
90.6	1437.6	Adela, 11.4	19	
79.2	1449.0	Comera, 7.3	None	
71.9	1456.3	Maturana, 6.3	4	A
65.6	1462.6	D Parral, (PL) 2.2	150	ABCGRY
63.4	1646.8	Empalme F.C.P. y D. 8.9	None.	
54.5	1473.7	Zenzontle, 7.6	19	
46.9	1481.3	D Adrian, (DN) 11.7	21	RY
35.2	1493.0	Borjas, 5.5	21	A
29.7	1498.5	Peinado, 3.6	4	
26.1	1502.1	Cuevas, 5.2	30	
20.9	1507.3	Stalforth, 7.8	21	
13.1	1515.1	Escape, 0.9	5	
12.2	1516.0	Escape, 3.8	5	
8.4	1518.8	Paloma, 8.4	24	
0.0	1528.2	D Rosario, (RQ)	60	ACGRY

Number.	Type.	Number of Spans Meters and feet.	Total length of bridge in meters	Maximum height from base of rail

PARRAL BRANCH

2-A	C I P	30"
3-A	C I P	30"
3-B	C I P	30"
3-1	Cattleguard	

CIP = Cast Iron Pipe

[What follows (not reproduced) is a list of all culverts, cattleguards, and bridges for:
Orion – Kilo 5.3
Troya – Kilo. 29.3
Baca – Kilo. 39.6
Dorado – Kilo. 45.7
Morita – Kilo. 58.9

Documents Cited

Adela – Kilo. 64.3
Gomera – Kilo. 75.7
Maturana – Kilo. 83.0
Parral – Kilo. 89.3
Zenzontle – Kilo. 100.4
Adrian – Kilo. 108.0
Borjas – Kilo. 119.7
Peinado – Kilo. 125.2
Cuevas – Kilo. 128.6
Stalforth – Kilo. 134.0
Paloma – Kilo. 146.5
Rosario – Kilo 154.9
Adrian Kilo 42
Santa Bárbara – Kilo. 7.8
Jiménez – Kilo. 1373.3
La Reforma – Kilo. 1392.4
Díaz – Kilo. 1411.2
Bustamante – Kilo. 1430.4
Santa Rosalía – Kilo. 1446.1
Tanque – Kilo. 1451.2
La Cruz – Kilo. 1462.1
Concho – Kilo. 1482.5
Saucillo – Kilo. 1488.8
Armendariz – Kilo. 1498.1
Las Delicias – Kilo. 1514.2
Ortiz – Kilo. 1521.5
Consuelo – Kilo. 1533.6
Bachimba – Kilo. 1545.8
Carbonero (Coal Deck) – Kilo. 1550
Horcasitas – Kilo. 1566.2
Mápula – Kilo. 1585.6
Alberto – Kilo. 1597.1
Morse – Kilo. 1600
Tabalaopa – Kilo. 1605.1
Chihuahua – Kilo. 1608.5
Chihuahua Shops – Kilo. 1609.5
Cuilty – Kilo. 1617.3
Corral – Kilo. 1631.6
Molinar – Kilo. 1639.3
Terrazas – Kilo. 1648.6

Sauz – Kilo. 1658.3
Piñale – Kilo. 1668.5
Encinillas – Kilo. 1678.2
Agua Nueva – Kilo. 1692.1
Laguna – Kilo. 1705.5
Arados – Kilo. 1714.4
Mocho – Kilo. 1725.9
Loaeza – Kilo. 1738.8
Gallego – Kilo. 1746.1
Alsacia – Kilo. 1762
Chivatito – Kilo. 1775.1
Centauro – Kilo. 1783.6
Moctezuma – Kilo. 1790.5
Ojo Caliente – Kilo. 1818.1
Carrizal – Kilo. 1828.4
Ahumada – Kilo. 1839.4
Lucero – Kilo. 1863.1
Ranchería – Kilo. 1875.3
Candelaria – Kilo. 1893.1
Tierra Blanca – Kilo. 1938.3]

<u>8532-2</u>
El Paso, Texas,
May 26, 1914.
From: 1st Lt.W.N.Hensley, Jr., 13th Cav., Int. Officer, 8th Brigade.
To: Commanding General, Southern Department,
(Through military channels)
Subject: Report on Mexican North-Western Railroad.

Enclosed are complete pump and bridge reports to date of May 21, 1914, for the Mexican North Western Railroad from Juarez to Chihuahua, Mexico by way of Pearson and Madera. The above reports are self-explanatory. The time charts enclosed, show in their condensed form, better than a report the following:- Distance in kilometers between stations, distance of stations from Juarez and Chihuahua.

The stations where water (w), and fuel (f), are obtained and where turn tables (m) and Y's are to be found.

In column marked "Capacidad de los Loderos" is shown capacity of sidings in number of cars. The average length of a car is forty (40) feet. The mark † indicates a telegraph station and the letters in pencil to the right indicate the call letters for that station.

Documents Cited

There is at present no fuel along this road except a small quantity of coal in Juarez and some lumber at Pearson and Madera. There is at all times at Pearson and Madera enough lumber to keep the bridges along the road in repair, but it has been found more practicable to fill in and "shoofly" the less important openings for the present. All other construction material has been used up in the great amount of repair work recently done, so that at present there is no material such as rails, ties, etc., available for repair work. This road has lost a large percentage of its rolling stock on account of the revolution and at present has only forty (40) per cent of the normal amount. The stock remaining is distributed as follows:

At JUAREZ;

HEAVY FREIGHT ENGINES.
No. 132,- Mallet type,- pull 950 tons over 1½% grade.
No's. 62 and 63,- pull 900 tons over 1½ % grade. ·
SMALL ENGINES.
No's 21, 22, 40, 43 and 44, pull 400 tons over 1½ % grade.
CARS.
One hundred (100) flats, wooden deck.
Two hundred (200) Skeleton steel logging flats, can be decked.
Fifty (50) box cars.
Twenty one (21) stock cars.
Seven (7) tank cars, 6000 gallons capacity
Fifty (50) tank cars used for turpentine, can be converted in ten days,- 6000 gallons capacity.

BETWEEN MADERA AND CHIHUAHUA;

HEAVY FREIGHT ENGINES;
No 131,- Mallet type,- pull 950 tons over 1½ % grade.
MIDDLE SIZED FREIGHT ENGINES-COAL BURNING:
No's, 50, 51 and 52 pull 600 tons over 1½ % grade.
MIDDLE SIZED FREIGHT ENGINES TO BURN EITHER WOOD OR COAL:
No's. 750, 751, 754 and 756 pull 600 tons over 1½ % grade.
SMALL FREIGHT ENGINES-COAL BURNING:
No's. 28, 29, 30, 31 and 33,- pull 500 tons over 1½ % grade.
CARS:
Nineteen (19) stock cars.

AT PEARSON AND MADERA:

CARS:

One hundred fifty (150) flat cars,-wooden decked.

Three hundred (300) skeleton steel logging flats, can be decked.

Three (3) tank cars,-6000 gallons capacity.

AT MADERA:

CARS:

Thirty (30) box cars.

AT CHIHUAHUA:

CARS:

Thirty (30) box cars.

With the exception of the "Cumbre" which was destroyed by bandits some months ago this road is now capable of being operated for its entire length. "Cumbre" which pierces the mountains between the stations Cumbre on the north and Chico on the south, is a tunnel of eleven hundred sixty six (1166) meters in length. For three hundred (300) feet at north portal and two hundred (200) feet at south portal, the sides and roof of the tunnel are supported by masonry. The remaining length of the tunnel is in numerous places supported by timbering, about seventy (70) per cent of this portion being heavily timbered. This timbering was set on fire when the work of destruction took place in February of this year, by bandits.

It is estimated that there are at present twenty thousand (20,000) cubic meters of rock and earth which must be removed before traffic can be resumed through this tunnel. At present no work is being done.

It is said that troops may be transported over the mountains at Cumbre by taking the "High Line," standard gauge logging road leaving the main line north of the tunnel and mounting to the summit of the mountain that "Cumbre" pierces. At the summit troops could be detrained and marched overland by a good wagon road to the south side of the mountain, a distance of two (2) kilometers where they could again be entrained.

W. N. HENSLEY, Jr.

1st Lieut. 13th Cav.

Intelligence Officer,

8th Brigade.

LIST OF WATER STATIONS ON M. N. W. RY.

KILOM.	STATION.	CAPACITY TANK	POWER	CAP. PUMP.	SOURCE OF SUPPLY
1	Juárez	20,000 Gals.	Steam	50 Gals. Min.	40 Ft. Well. Also is connected with city.
125	Guzmán	20,000 "	"	50 " "	Spring supplies about 300 Gals. Min.
155	Sabinal	20,000 "	"	35 " "	200' well. Pump now out of commission can be repaired in 48 hours (emergency)
174	Santa Sofia	20,000 "	"	40 " "	75' Well. Supply very small, can work pump only about 20% time
218	Corralitos	20,000 "	"	35 " "	40' Well. Supply for 50% of capacity pump.
240	Casas Grandes	20,000 "	"	50 " "	65' Well. Good supply.
270*	Pearson	Connected with system of Madera Company Limited.			
300	Cuevitas	30,000 "	"	50 " "	Pump in river bed.
326	Aguaje	30,000 "	"	50 " "	" " " "
355	Chico	30,000 "	"	50 " "	" " " "
382	Babicora	30,000 "	"	45 " "	30' Well. Supply very small, can work pump only about 25% time.
420	Cebadilla	30,000 "	"	50 " "	Pump in small stream.

Note:- At Conejos, Kilom. 90, is a dug well about 20' deep, water stands in well about four feet deep, poor water, whitish in color. Carranza watered his stock at this point on his trip overland from Sonora, perhaps 150 head.

At Barreal, Kilom. 95 is a dug well 300 yds. east of track. Good Water but small supply, depth not available.

About Kilom. 101, small spring in arroyo 150 yds east of track. In April pool was perhaps 18" deep and covered superficial area of 40 Sq. ft.

Between Pearson and Aguaje, also in the neighborhood of Chico as well as north of Babicora, are running streams close to and parallel to track which will furnish water in large quantities.

WATER STATIONS, (CONTINUED)

KILOM.	STATION.	CAPACITY TANK	POWER	CAP. PUMP.	SOURCE OF SUPPLY
438*	Madera	Connected with Gravity system of Madera Company Limited.			
470	Rincón	30,000 Gals.	Gravity		Supply from wash furnishing perhaps 100 Gal. per Min. through our pipes.
517	Tejolocachic	34,000 "	Gasoline		50 Gal. Min. Pump in River bed.
575	La Junta	34,000 "	"		50 Gal. Min. Pump in River bed.
566	Rosario	34,000 "	"		50 Gal. Min. Pump in River bed.
626	San Antonio	34,000 "	"		45 Gal. Min. Pump in arroyo supply in dry season not sufficient to keep pump going over four hours per day.
650	Bustillos	34,000 "	"		45 Gal. Min. Springs. Supply fails in dry season as at San Antonio
676	San Andrés	34,000 "	"		50 Gal. Min. Pump in River bed.
705	Santa Isabel	34,000 "	"		45 Gal. Min. Pump in River bed.

| 744 | Fresno | 34,000 " | Gravity from springs two miles south. Furnishes perhaps 35 Gals. per minute steady flow. |
| 759 | Chihuahua | 60,000 " | Pump in bed of arroyo pumped by steam and gasoline. Cap. of pump about 45 gals. per minute. Supply good. |

Note:- From Rincón, Kilom. 470 to La Junta, Kilom. 575, track follows very closely to running streams.

Unless otherwise noted above we have not had any failures in supply up to railroad requirements. App pumps in Tejolocachic, La Junta, Rosario, San Andrés and Santa Isabel furnishing at all times water sufficient for our needs.

*At Madera and Pearson water supply practically unlimited.

EXTRACTS FROM
TIME TABLE MEXICAN NORTH WESTERN RAILROAD.
EL PASO DIVISION CONTINUED.

Distance from Juarez. Kilom.	STATIONS.	Distance from Chihuahua. Kilom.	Capacity of Sidings. Cars.
237.0	Colonia Dublán, 3.0	522.5	10
240.0 W	Nueva Casas Grandes, 4.0 G Y	519.5	95
244.0	Huerto, 6.0	515.5	3
250.0	Don Luis, 7.0	509.5	10
257.0	Anchondo, 5.0	502.5	74
262.0	San Diego, 8.0 D S	497.5	None
270.0 W F	Pearson, 13.5 N R T Y	489.5	Yards.
283.5	Rucio, 16.5	476.0	76
300.0 W	Cuevitas, 10.9 V I	459.5	64
310.9	Riba, 15.3	448.6	44
326.2 W	Aguaje, 9.1 W	433.3	76
335.3	Caballo, 11.4	424.2	76
346.7	Cumbre, 8.7 M U	412.8	39
355.4 W	Chico, 14.2 C K	404.1	45

369.6	Drake, 13.2 D A	389.9 Ballast Pit.	50 (B A)
382.8 W	Babicora, 14.8	376.7	84
397.6	Concha, 22.2	361.9	77
419.8 W	Cebadilla, 17.5 M D N Y	339.7	72
437.3 W F	Madera, R Y T	322.2	Yards.

(W) denotes water station. (F) denotes fuel station. (R) denotes train register. (Y) denotes "Wye" [Y junction]. (T) denotes turntable.

[What follows is a list of all bridges by number with notes on length, type, and present condition, including for Cusihuiriáchic and Miñaca branches, not reproduced here.]

EXTRACTS FROM
TIME TABLE MEXICAN NORTH WESTERN RAILROAD.
EL PASO DIVISION.

Distance from Juarez. Kilom.	STATIONS.	Distance from Chihuahua. Kilom.	Capacity of Sidings. Cars.
0.0 W F	Juarez, 10.0 R Y T	759.5	Yards.
10.0	Arena, 7.4	749.5	13
17.4	Bauche, 5.1 B H	742.1	None
22.5	Méndez, 16.5 M S	737.0	27
39.0	Sapello, 11.1	720.5	40
50.1 W	Mezquite, 18.5	709.4	12
68.6	Medanos, 6.4	690.9	34
75.0	Lena, 15.5	684.5	None
90.5	Conejos, 4.5	669.0	18
95.0 W	Barreal, 12.9 B R	664.5	12
107.9	San Blas, 17.1	651.6	34
125.0 W F	Guzmán, 15.0 G Z	634.5	50
140.0	Urrutia, 15.5	619.5	45
155.5 W	Sabinal, 3.0 S B	604.0	9
158.5	Ochoa, 15.6	601.0	36
174.1 W	Santa Sofia, 8.9	586.4	33

183.0	Empalme, 5.0	576.5	20
188.0	San Pedro, 5.4 R O	571.5	36
193.4	Summit, 7.6	566.1	18
201.0	Coyote, 17.0	558.5	None
218.0 W	Corralitos, 15.0	541.5	11
233.0	Embarcadero, 4.0	526.5	40
237.0	Colonia Dublan, 3.0	522.5	10

EXTRACTS FROM
TIME TABLE MEXICAN NORTH WESTERN RAILROAD.
CHIHUAHUA DIVISION.

Distance from Juarez. Kilom.	STATIONS.	Distance from Chihuahua. Kilom.	Capacity of Sidings. Cars.
437.3 W F	Madera, 17.8 R Y T	322.2	Yards.
455.1	Las Varas, 15.9	304.4	34
471.0 W	Rincón, 4.3	288.5	39
475.3	Yepomera, 14.0 R N	284.2	None
489.3	Temosachic, 8.2 M O	270.2	54
497.5	Loma, 6.9	262.0	5
504.4	Matachic, 13.6 M A	255.1	43
518.0 W	Tejolocachic, 21.8	241.5	31
539.8	Santo Tomas, 5.5 M X	219.7	43
545.3	Casa Blanca, 2.0	214.2	8
547.3	San Pablo, 4.7	212.2	None
552.0	Girasol, 5.8	207.5	None
557.8	San Isidro, 0.7 D O	201.7	60
558.5	Calera, 6.6	201.0	10
565.1	Basuchil, 5.6	194.4	3
570.7	Saenz, 4.6	188.8	5
575.3 W F	La Junta, 11.3 J N R Y	184.2	262
586.6 W	Rosario, 5.9	172.9	70
592.5	Paramo, 8.2	167.0	22

600.7	Pedernales, 3.2	158.8	29
603.9	Cima, 5.3	155.6	None
609.2	Mal Paso	150.3	30

EXTRACTS FROM
TIME TABLE MEXICAN NORTH WESTERN RAILROAD.
CHIHUAHUA DIVISION CONTINUED.

Distance from Juarez. Kilom.	STATIONS.	Distance from Chihuahua. Kilom.	Capacity of Sidings. Cars.
609.2	Mal Paso, 2.4	150.3	30
611.6	Casa Colorada, 6.9	147.9	6
618.5	Pampas, 7.5	141.0	21
626.0 W F	San Antonio, 13.9 FNRY	133.5	82
639.9	Llano, 4.4	119.6	21
644.3	Laguna, 7.1	115.2	31
651.4 W	Bustillos, 7.5 BU	108.1	36
658.9	Mesa, 6.6	100.6	23
665.5	Aldana, 6.4 DA	94.0	23
671.9	Sandoval, 5.0	87.6	22
676.9 W F	San Andrés, 10.7 SA	82.6	67
687.6	Chavarría, 13.1	71.9	28
700.7	La Baeza, 4.5	58.8	5
705.2 W	Santa Isabel, 3.7 SY	54.3	47
708.9	Santa Sabina, 6.4	50.6	3
715.3	Apache, 7.8	44.2	None
723.1	Palomas, 9.7	36.4	20
732.8	Salas, 11.5 Q	26.7	19
744.3 W	Fresno, 8.2	15.2	25
752.6	Ranchitos, 7.0	7.0	
759.5 W F	Chihuahua, AURYT O.O. National (DS) (CW)		Yards.

(W) denotes water station. (F) denotes fuel station. (R) denotes train register. (Y) denotes "Wye" [Y junction]. (T) denotes turntable.

Documents Cited

AGOMI Box 7473, April 20, 1914
Fort Sam Houston, Texas, April 20, 1914.
Agwar [Adjutant General, War Department], Washington, D.C.

Following abstract of latest reports from border stations communication for your information at Nuevo-Laredo twelve hundred soldiers, mostly volunteers and six or eight guns, mostly Infantry, Cavalry one or two hundred and several machine guns, about three thousand additional soldiers along railroad between Nuevo-Laredo and Monterrey. Could be brought into Nuevo-Laredo in short time. Railroad between Nuevo-Laredo and Monterrey has been interrupted for some time but is now repaired for passenger traffic from Nuevo-Laredo to within forty kilometers of Monterrey and large repair parties are at work on this section railroad bridges between Monterrey and Saltillo have been repaired. At Piedras-Negras three hundred men and about one thousand more in vicinity of railroad and at Monclova. Railroad and telegraph communication between Piedras-Negras and Saltillo interrupted all last week. Telegraphic communication with City Mexico was resumed seventeenth instant after two days interruption. Juárez garrison about four hundred fifty Infantry, two machine guns. Could probably be increased in short time to ten or fifteen thousand and large number field guns from Chihuahua and Torreón. At Palomas about seventy Cavalry at Casas Grandes seventy Cavalry at Agua Prieta less than two hundred and small garrison at Nogales. Other troops as follows: One hundred Nacozari one hundred Cumpas, two hundred Moctezuma, one hundred Morelos, one hundred fifty Tigre. Main body about ten thousand under Obregón is at Navojoa south of Guaymas and is expected to move South to attack Tepic. Attitude of Constitutionalists a few weeks ago indicated combining with Huerta in case of hostilities between this country and Mexican Federal Government. [Wrong. There is no evidence that the Constitutionalists as much as considered uniting with Huerta, and in fact, they did not join with Huerta post-U.S. intervention. This is bad intelligence.] This attitude seems to have changed recently. It appears now Constitutionalists will remain neutral so long as the territory controlled by them is not invaded or embargo on munition of war not restored.

Bliss

5:05 P.M. [Penciled in: "Copy handed to Chief of Staff Apr. 20/14 Carbon copy left by Colonel Heisland with Assistant to Chief of Staff"]

AGOMI Box 7473, April 23, 1914
110W. Vi. 235 Gov't
Fort Sam Houston, Texas, April 23, 1914.
General Wotherspoon,
Chief of Staff, Confidential, Washington, D.C.

Indications are that Federal Garrisons have gone south from Piedras-Negras and Nuevo-Laredo. Period. As Constitutionalists are strong in the vicinity of Monterrey and Saltillo the movement gives color to rumors that the two sides will join period. Villa's about to reinforce Juárez and can quickly bring ten thousand men there. Period. I can get no reliable information as to what happens across the border and must rely on Washington for it. Period. At first intimation you receive that the two sides in Mexico have united you should be prepared to immediately occupy Nogales, Juárez, Piedras Negras, Nuevo Laredo, and Matamoros. Period. The only safe course is not to wait there is no other way to protect women and children in our towns of Nogales, El Paso, Eagle Pass, Laredo and Brownsville which are in pistol range of Mexican towns opposite. Period. If Mexican towns are allowed to be occupied by unfriendly troops all women and children and other non-combatants must be ordered to leave American towns as their only safety. Period. This step must inevitably be taken if you allow Mexican towns to be strongly occupied before corresponding American towns are very strongly reinforced. Period. Mexican towns have practically no women and children and no business interests to suffer while on our side are thousands of defenseless people and great interests at stake. Period. In my opinion instant action is necessary.

Bliss

1:55 P.M. [Penciled in: "Copies to Chief of Staff Assistant to Chief of Staff April 24, 1914"]

<u>AGOMI Box 7473, April 25, 1914 #1</u>

Headquarters
Southern Department,
Fort Sam Houston, Texas.
April 25, 1914

Confidential
From: Commanding General.
To: The Adjutant General, U.S. Army.
Subject: Instructions of the Secretary of War re-Mexican situation.

1. I submit the following for the consideration of the Chief of Staff and Secretary of War.

2. Early yesterday morning April 24, 1914, I received the important code message from The Adjutant General of the Army, conveying to me the instructions of the Secretary of War to the effect that, as war has not been declared, if the United States troops at any point under my command should be attacked by armed Mexican forces, they will "resist and defend to the limit" of their powers but that they are not to

"invade" (that is to say, they are not to cross the border into Mexican territory) until they shall have received the specific orders of the Secretary of War to that end.

3. I acknowledged the receipt of three instructions and stated (after translating the code message) that I understood them. I immediately communicated the corresponding proper orders to all commanding officers of United States troops along the border and before the end of the day I received acknowledgements from them stating that they also understood their instructions. Nevertheless, I received telegrams from some of them (especially from the Commanding Officer at Laredo, where the situation at the moment seemed somewhat serious) indicating their belief that they should be permitted to cross the river in the event of certain military contingencies arising. I repeated to them by wire the previous instructions on the subject and again directed them to report whether they understood. They all so reported. Today, in order to clear up any possible doubt in the mind of any of them, I sent to all renewed instructions in writing. A copy of these instructions is by this time in the hands of the Chief of Staff to whom I have sent it for the purpose of ascertaining, beyond the peradventure of a doubt, that my original understanding of the instructions of the Secretary of War was correct.

4. The instructions will be carried out as far as it is humanly possible to do so. I desire the Department to understand (although it doubtless already does so) that these instructions are probably the most difficult of compliance that have ever been issued to troops under like conditions. Their own Government has taught them at expensive military schools and by other forms of expensive military training, ever since their entry into the service, that war is a question of fact; that although it may not have been declared by one side, it may have been declared by the other side; that, whether it be declared by either or both sides or not, the actual status may be created by warlike acts; that although there is a moral obligation on the part of nations to indicate their intention in so grave a matter, the failure to comply with this obligation merely indicates moral obliquity on the part of one or the other of them, but does not alter the fact; but last and most serious of all, this status of war which these officers have been taught at great expense by their own Government to recognize as existing through observation of actual facts, may confront one of these officers at an isolated station, perhaps beyond the reach of telegraphic communication, perhaps at midnight, perhaps when upon his instant recognition of his duty as he has been taught it through many years, will depend the lives of American women and children and other helpless non-combatants.

5. While I believe that with the officers of experience and intelligence and conscientious attention to duty and orders, that I now have under my command, nothing will happen contrary to the wishes of the Secretary of War; nevertheless, the possibility must be discounted in advance of something being done, somewhere, at

some time, contrary to the tenor of the instructions which I have received from him and which I have communicated to all others concerned.

6. In order to make the very possible, if not probable, situation that may occur at various points perfectly clear, I invite attention to the following illustrations. The Mexican towns of Matamoros, Nuevo Laredo, Piedras Negras and Juárez are practically abandoned by all their former peaceful inhabitants. All civil commercial business has entirely or almost entirely ceased. The Mexicans have practically nothing to lose by further destruction. The houses on the edges of these towns are arranged for infantry defense, and at various places, infantry trenches have been constructed. There are plenty of positions in their vicinity in which scattered field guns can be placed and out of which they could not be driven by fire from our side.

On the opposite side of the river to these towns, on American soil, are populous and rich American towns. Infantry and artillery fire from the Mexican side could not miss its target, even in the darkness of night, since they would have the whole town of ours to fire at. Colonel Crane at Laredo, wires me today that it is considered barely possible that the Mexicans who have evacuated the opposite town of Nuevo Laredo may return to do this very thing. And it is, of course, always possible even if improbable. Suppose such an incident were to happen in the middle of the night. Women and children would be killed on our side of the line, nor could any amount of fire from our troops prevent it and this would continue as long as either side had any ammunition to expend,- we doing no harm to them, they doing the greatest harm to American citizens.

In such a case as that suggested above, the art of war as it has been taught to our officers, requires them to immediately advance upon the enemy's position, drive them from it and capture their guns. Yet this is actually the thing which we are prohibited from doing under the instructions of yesterday. In my opinion, a grave responsibility is assumed in prohibiting troops from acting otherwise than as the sound principles which they have learned, teaches them that they ought to act in the war which would then be actually begun. In such a case it would be impossible for hours to obtain the instructions of the Secretary of War. I think that these are the very cases which *must* be left to the discretion of the trained officers on the spot.

Tasker Bliss,
Brigadier General, U.S. Army

thb-kls
[Penciled in: "Left with Chief of Staff April 28/14"]

AGOMI Box 7473, April 25, 1914 #2
War Department,
Office of the Chief of Staff.
April 25, 1914.

Documents Cited

Memorandum for
    The Adjutant General.

The Secretary of War directs that a telegram, in substance as follows, be sent to the Commanding General, Southern Department:
    "It has been reported to the Department that 565,000 rounds of 7 m.m. and 30-30 have been shipped on a Morgan liner to Galveston to be forwarded to Laredo to pass over the border at that point."
W.W. WOTHERSPOON,
Major-General, Chief of Staff.

Telegram to
    C. G., Southern Dept.   4/25/14
Copies to
    Chief of Staff and Asst. Chief of Staff.

AGOMI Box 7473, April 29, 1914

Received at War Department   Delivery No. 84
26-PO.R.   50- Govt.
    Fort-Sam-Houston, Texas; April 29-14
Agwar, Washington DC
Following last night from Commanding Officer Eagles-Pass Quote: Murguía's forces occupied Piedras Negras at five this afternoon. They report taking four hundred prisoners from Guajardo twelve hundred rifles, three field pieces and much small arms ammunition. My information is that prisoners are being well treated. End quote.
Bliss.
Copy to C of S
Copy to Asst. C of S.
Copy to Secretary of State 4/29 14
11:42-am.

Received at War Department   Delivery No. 108
38-PO.R.   39- Govt.
    Fort-Sam-Houston, Texas; April 29th-1914
Agwar, Washington DC
All reports from border indicate situation to-day much improved period. Constitutionalist Leaders near border apparently take conservative view period. If they can hold irresponsible bands in check, I anticipate no trouble from any present signs.
Bliss.
2:35 pm.
Sent Copy to Chief of Staff

Sent Copy to Asst. Chief of Staff.

AGOMI Box 7473, May 15, 1914

Received at War Department   Delivery No. 117

31Po. Vi. 46 Gov't.

Fort Sam Houston, Texas, May 15, 1914.

Copy for the Sec. State and Miscel Div. 5/15/14

Agwar, Washington, D.C.

Number thirty three period. Following from Commanding Officer, Eagle-Pass, repeated for information, War Department: Quote.

All Murguía's command in Piedras Negras except small detachment are leaving for the south period. Two trains have gone and last train is now loading Griffith end quote.

Bliss.

7:32 P.M.

AGOMI Box 7473, May 19, 1914

Brownsville, Texas

May 19, 1914.

From: Commanding Officer, U.S. Troops, Brownsville, Texas.

To: The Commanding General, Southern Department.

SUBJECT: Report on Mexican conditions.

Following reported on good authority from Matamoros today.

1. Third Brigade, General Elizondo, four hundred fifty men, will reach Matamoros Thursday night, now at China and Camargo.

2. Indications are there will probably be two thousand soldiers, all told, between Matamoros and Los Ramones, in next two or three weeks.

3. Carranza and Pablo González say positively they will have nothing whatever to do with any scheme for mediation.

4. Carranzistas will not, for the present, attempt to open railroad from Monterrey to Torreón. It is blocked by flank of Federal position at Saltillo, and they have not enough rolling stock. Only seven serviceable engines between Monterrey, Matamoros, Nuevo Laredo, and Tampico. Forty two freight cars stuck at Los Ramones on account no oil for bearings.

5. Villa now on his way to attack San Luis Potosí – will not attack Saltillo now. Should reach San Luis Potosí in three days.

6. Carranzistas claim they are expecting large shipments of arms and ammunition on German boats through Tampico.

7. 350,000 rounds, small arms ammunition at Monterrey, believed to represent present available supply for Carrranzistas.

8. Cipriano Castro will remain in command at Tampico, Caballero at Victoria, and Pablo González at Monterrey.

Lieut Colonel, 3rd Cavalry

1st Ind.

Hq. Southern Department, Ft. Sam Houston, Texas, May 21, 1914.
To The Adjutant General of the Army, Washington, D. C. for the information of the War Department.
The information given in paragraph 5 has not been verified and it is not known whether it is correct or not.

Tasker H. Bliss,
Brig. Gen. Comdg.

HS

AGOMI Box 7473, May 21, 1914

799

Camp Eagle Pass, Texas.
May 21, 1914.

From: Commanding Officer.
To: Commanding General, Southern Department.
Subject: Special Report.
1. General Murguía returned yesterday from Monclova to Piedras Negras, remained a few hours, and returned to Monclova. It is believed he will now make his headquarters at Monclova.
2. The garrison of Piedras Negras is now about 200, as Murguía is concentrating at Monclova.
3. A state of rivalry between Carranza and Villa is becoming more evident. Friends of Villa in conversation state that he will not submit to be dominated by Carranza as he (Villa) has accomplished everything, while Carranza has only complicated matters. It is believed this rivalry will continue to grow.
4. The Rio Grande is still high at fords.

T. W. Griffith
Colonel 17th Infantry

1st Ind.

Hq. Southern Department, Ft. Sam Houston, Texas, May 22, 1914.
To The Adjutant General of the Army, Washington, D. C.

T.H.B.

[Penciled in, "Left with Chief of Staff"]

AGOMI Box 7473, May 30, 1914

193

U.S. Army Intelligence in the Mexican Revolution

Headquarters Eighth Brigade
Fort Bliss, Texas.
May 30, 1914.

From: The Commanding General.

To: The Commanding General, Southern Department, Fort Sam Houston, Texas.

Subject: Report on border conditions, week ending May 30, 1914.

1. Nothing unusual has occurred along the patrol line this week.

2. The command has been readjusted to meet the change of station of the four troops of the 13th Cavalry.

3. Rodrigo Quevedo and a supposed member of his band were captured Friday by the city police of El Paso and turned over to military authorities under the charge of violating the neutrality laws. A recommendation has already been made that they be sent to Fort Wingate. This should have a deterring effect upon the operations of this and other bands. There is little doubt of the existence of a plan by the local junta to encourage this band, and at the proper time to launch a new revolution against the Constitutionalist Government.

The inactivity of the Constitutionalists heretofore in Northern Chihuahua has led to new accessions to these Huertista bands. Recently, however, the Constitutionalists seem to have started in to disperse these gangs. Considerable activity on the part of Villa's troops along the Southwestern railroad is reported and hopes are expressed that peaceable conditions will soon be established. There are now only about four hundred of Villa's men left in Juárez, with two hundred horses as already reported.

4. Observers returned from Paredón and Saltillo comment very favorably upon the celerity with which General Villa moved his troops from Torreón to Paredón. It is reported that he transported 18000 men, 15000 of whom were irregular Cavalry with all their horses, from Torreón to Paredón in three days. That part of the railway is level and one engine can haul from 50 to 60 cars. Horses are reported in good condition, men comfortably uniformed, and well-armed and supplied with ammunition.

John J. Pershing
Brigadier General,
Commanding.

AGOMI Box 7473, June 15, 1914

Torreón, Mexico, June 15, 1914.

It was learned today from good authority that the federal force at Zacatecas numbers between 12,000 and 15,000 men, more reinforcements, said to be part of the Saltillo garrison, having reached there during the latter part of the week. The latest information is that the force there at the beginning of June numbered no less than

6,000, that about 3,000 men arrived about June 9th and that another 4,000 or more came to the assistance of General Medina Barrón during the latter part of last week. [The most reasonable estimate of the Federal defenders was somewhere around 10,000, with about two batteries of field guns.]

In addition to a great number of machine guns, estimates run from 120 to 180, the Federals have in position several large siege pieces and a large field park.

Access to the city is possible for an army only through two narrow valleys, defended by the redoubts and entrenchments on and near the base of Veta Grande, El Grillo, and La Bufa. Of these the latter will be the hardest to take say Constitutionalist officers, and unless the Constitutionalist artillery succeed in making some of the redoubts and trenches untenable many lives are bound to be lost in the assault on the positions.

Though General Villa and his staff refused to discuss any phase of their plans, the impression generally prevails that the heavy and sustained artillery preliminary must be engaged in before attack by infantry can be made with any hope of success. In view of the fact that the supply of Artillery ammunition of the Constitutionalists is ample, no anxiety on this score is felt.

A high Constitutionalist officer expressed the opinion today that the Federals had made up their mind to defend Zacatecas to the limit of their resources. He thought that General Huerta recognized the strategic importance of preventing the Constitutionalist Army from gaining a good foot-hold in central Mexico.

The agricultural and industrial development in that part of the republic is such that the invaders could cut loose from the railroad and by virtue of their organization, cavalry brigades overrun the district and so force the entire Federal Army to concentrate at, possibly, no more than two or three points, one of them Mexico City. In this manner, thought the officer, most of central Mexico would fall into the hands of the Constitutionalists, and this General Huerta, whom he gave credit for being no mean tactician, would try to prevent at all costs. Zacatecas, virtually, he said, was the last point offering every feature favorable to this plan. Aguascalientes, seventy five miles further south and on the edge of the rich central region, being unsuited for defense.

The force which General Villa will lead against Zacatecas consists of about 29,000 men, fifty one pieces of field artillery and ninety machine guns.

<u>AGOMI Box 7473, June 17, 1914</u>
Headquarters
Southern Department,
Fort Sam Houston, Texas

June 17,1914.

From: Commanding General.

U.S. Army Intelligence in the Mexican Revolution

To: The Adjutant General of the Army, Washington, D. C.

Subject: Manuel Guerra.

1. The following extract from report commanding officer on the border at Rio Grande City is furnished as an illustration of the class of characters with whom the Military have to deal, in preventing smuggling, and shows to some extent the difficulties the Military have to contend against in the performance of border duty:

"I have a letter from Captain Lee, directing me to investigate the crossing of arms by Manuel Guerra at Roma, April 23d. This suggests a little biography of Guerra, which comes from many sources. As you know, probably, Guerra is the absolute master of this section of Texas. He is the power in the Congressional District, as well as in Starr County. Mr. Cannon of the local bank says that Guerra has made about $100,000 that he knows of, during the past year. Guerra owns property all the way from Zapata to Rio Grande City, and has in his service every outlaw and cattle rustler in this section. It is common talk that small ranchers are in constant terror of him. He seizes cattle as a matter of caprice, and the small rancher has nothing to do but submit. The other day I heard Guerra say that "he wished this damned revolution would wind up, as there was nothing doing any more; let the country rest awhile and start another one," etc. The collector of customs at Roma is a Mexican; while professedly he is against Guerra, he is thoroughly afraid of him. Withal Guerra is thoroughly equipped to get arms across most any place and in any quantity he wishes. He receives goods from Hebronville as well as Fordyce, owns his automobile, has several branch stores and scores of ranches, with a working force capable of carrying out his instructions. Guerra is the fiscal agent of the Carranzistas in this section, and I have no doubt that he is doing all he can in getting arms over. At present there are no rebels at San Pedro, opposite Roma, but about 100 at Mier, 9 miles away."

Tasker H Bliss,

Brig. Gen. Comdg.

AGOMI Box 7473, August 4, 1914

August 4, 1914.

The Secretary of War presents his compliments to the Honorable the Secretary of State, and invites his attention to the following extract from a report from the Commanding General, Southern Department, dated July 29, 1914:

"I invite especial attention to the general tenor of the reports from the El Paso District and elsewhere along that part of the border fronting the Mexican State of Chihuahua. It is evident that for some reason the Constitutionalists are making strenuous efforts to secure large quantities of arms and ammunition. There is no evidence of such efforts at any other part of the line controlled by the Constitutionalists. It seems a natural inference to be drawn from this fact, that while the Constitutionalist forces elsewhere secure all the munitions of war that they need through the port of Tampico on the East and some port on the West of Mexico, that part of their forces in the State of Chihuahua is forced, for some reason,to   rely   on smuggled munitions from the United States. [In fact, Carranza had blocked Villa from using Tampico as a port of entry for his ammunition purchases.] As has been previously reported, it is practically impossible for the Army under existing conditions, to prevent this."

AGOMI Box 7473, August 10, 1914

hfa   WWW

August 10th, 1914

The Secretary of War presents his compliments to the Honorable, the Secretary of State, and quotes for his information the following telegram, dated August 9th, 1914, which has just been received from the Commanding General, Southern Department, Fort Sam Houston, Texas.

"Adjutant General, Washington, D.C.

Number hundred fifty two. Following just received from Commanding General, Eighth Brigade, El Paso, quote: Knowing our inability to prosecute, smugglers of arms and ammunition openly defy military and customs officers. Five hundred thousand rounds arrived here Thursday. Plot to rush cars past bridge guard discovered and was frustrated by our blocking railroad track. Cars now under our guard. Attempts at bribery of soldiers and customs guard frequent. Necessary to double guards all along river front. Recommend restoration of President's embargo. Collector of Customs is making similar recommendations. End quote. In my opinion determined efforts of Villa to secure large amounts munitions of War through El Paso indicates trouble for future. I think restoration of embargo along entire frontier west from Brownsville will help maintain peace for future government of Mexico. Carranza gets all he needs through Tampico and therefore should not object. Every rifle and cartridge now going across this border may be used against new government.

BLISS."

AGOMI Box 7473, September 8, 1914 #1

hfa   WWW

September 8th, 1914.

The Secretary of War presents his compliments to the Honorable the Secretary of the Treasury, and quotes, for his information, the following telegram this day sent to the Commanding General, Southern Department, relative to the treatment of arms and ammunition passing over the Mexican border:

"Referring to cipher dispatch of April 23rd, 1914, relative to passage of arms and munitions of war going into Mexico, the Secretary of War directs that you be informed that as the conditions which induced the issuance of that order no longer exist the order is rescinded."

Mailed from A.G.O. Sept. 9.

<u>AGOMI Box 7473, September 8, 1914 #2</u>

The Secretary of the Treasury
Washington

September 8, 1914.

My dear Mr. Secretary:

I have today given Collectors of Customs along the Mexican frontier the following instructions:

"In view of the restoration of peace in Mexico, Collectors and other officers of customs may in future treat arms and ammunition as ordinary commercial shipments and permit them to go forward accordingly."

Sincerely Yours,
W. G. McAdoo

Hon. L. M. Garrison,
Secretary of War.

<u>AGOMI Box 7473, September 14, 1914</u>

FH

PC
September 14, 1914.

Winchester Bennett, Esq.,
1st Vice President,
Winchester Repeating Arms Co.,
New Haven, Conn.

My dear Mr. Bennett:

Owing to the fact that all restrictions have been removed which have heretofore governed concerning arms and ammunition which might pass from the United States across the Mexican frontier, I have the honor to inform you that the daily reports of sales of arms and ammunition will no longer be necessary.

I thank you very much for these daily reports of sales which your company has so kindly furnished the Department.

Very Truly Yours,
W. W. Wotherspoon
Major General,

Documents Cited

Chief of Staff.

AGOMI Box 7478, September 17, 1914
Western Union Special
Number: 1 GX. 0. 572/573 Collect, 2 extra. Via Galveston.
Dated VERA CRUZ, MEX. September 17, 1914
To Agwar, Washington. One fifty.

Number one sixty nine period. Replying your cablegram reference evacuation Vera Cruz comma, transportation will be required as follows: Colon: Army comma, officers two hundred thirty three comma, enlisted three thousand eight hundred eighty comma, civilian employees forty one comma, animals eight hundred seventy three comma, freight pounds six million comma, cubic feet four hundred and fifty thousand comma, ships tons eleven thousand two hundred semicolon; marine corps comma, officers one hundred and eight comma, enlisted two thousand eight hundred animals fifty six comma, freight pounds three million comma, cubic feet two hundred twenty five thousand comma, ship tons five thousand six hundred period. Included in animals of Marine Corps are thirty native mules period. Recommend they be sold here comma, as are very small semicolon; were purchased from army transportation period. Recommend date of evacuation be October tenth period. We could get out earlier comma, but I consider it desirable that we have sufficient time to familiarize with their duties new officers who will take over various departments period. If Mexican government will designate these officials and have them report to me we can employ them and pay them out of civil funds comma, they to work until evacuation under our official period. This plan will avoid confusion and will obviate probability of subsequent dispute period.

[Penciled in: "Copies to Secy State & Navy & Genls & Misc. Div. 9/17/14 WG"]

The Western Union Telegraph Company
Special
No. 1 GX. Second sheet
Dated Vera Cruz, Sept. 17, 1914
To Agwar, Washington, D.C.

Another reason for delay is that there are here a large number of Mexicans who feel that they have to leave the country period. As they have sought our protection it would seem unjust for us to leave until at least most of those who wish to go have had opportunity to do so period. Of the approximately fifteen thousand Mexican refugees here probably less than a thousand will leave comma, this number including about three hundred priests and nuns period. There is a great deal of alarm among the inhabitants of Vera Cruz because of threats that have been made by irresponsible persons to effect that when Constitutionalists get in control they will punish them severely because they submitted quietly to our rule comma, and in many cases fraternatized with us and assisted us period. I feel that most of these fears are groundless comma, and recommend that Gen. Carranza be requested to issue a statement that will reassure them period. We owe much to the people of Vera Cruz because of their exemplary conduct during our occupation period. I doubt if there is a case in history in which a people have accepted a very disagreeable situation so admirably period. Although the city was occupied by a foreign army comma, differing in race and language from its inhabitants comma, and was governed under martial law in its strictest form comma, a condition that would naturally be expected to bring about friction comma, there have been none of the incidents that would ordinarily be expected under the circumstances period. There have been no attempts at assassination comma, and not one crime of violence of any importance has been committed against any member of the occupying force period. All of our orders were carried out without question comma, and we have found the attitude of the people generally helpful period. Officers and enlisted men could go anywhere unarmed without the slightest fear period. I request that I be informed as to destinations of army organizations leaving here period. It will save much expense comma, for those who have families if this information can be given now.

Funston.

3:35 P.M.

AGOMI Box 7478, September 22, 1914

TELEGRAM SENT.

WAR DEPARTMENT,

Washington, Sept. 22, 1914.

General Funston, CONFIDENTIAL.   RUSH.

     Vera Cruz.

NUMBER ONE NINE-FOUR.

I consider it inadvisable to adopt suggestion in your Number One Fifty with respect to having new officials working in the respective offices under or with our officials period. The procedure to be followed in the turning over of the various governmental departments has not yet been determined upon period. When it is, you will be fully instructed period. Until then, extreme caution must be preserved by you in any interviews you may have with Mexican authorities period.

If, without discourtesy, you can do so, it would be well to confine your interview with General Aguilar to a statement of what is contained in my Number One Ninety-One with respect to the word that you should send to him. If, without complicating the situation or making any statements which could possibly embarrass this Government, comma, you could obtain from General Aguilar the issuance of a statement such as is suggested in your Number One Seventy-Four, comma, I approve of your doing that period.

Private inquiries made of steamship companies result in ascertaining that they will not send ships to Vera Cruz unless assured of paying passengers sufficient to make it worthwhile period. Can you ascertain number of those able and willing to pay passage, and communicate immediately here question?

GARRISON,
Secretary of War.

AGOMI Box 7478, November 7, 1914

Telegram received at the War Department, November 8, 1914, 12.19 a.m.
1W. Vera Cruz, Mexico, Nov. 7. 1914.
The Secretary of War, Washington, D.C. Nine Fifteen.

No. 223

Reference your 236. Many unsubstantiated rumors are being circulated about imminent attacks on Vera Cruz. They have *prevailed* for long time, and I think it better to discredit unsubstantiated rumors. I sustain cordial relations with General Aguilar but his *subordinate*, General Millán who is in command of troops in front of our right flank has shown bad disposition and appears inclined to make trouble. Appropriate orders have been issued to effect that if any trouble should occur Mexican authorities will be put in wrong as they will have to start it. At first we would have a hard time, but would finally win. Nothing alarming in local conditions. Jesús Carranza is believed to have gone to Tehuantepec to collect and bring North troops he left there a few weeks ago. Probably his troops are utilized fight Villa but it is possible their intention is to fight us. We have made arrangements to obtain information but cannot rely on it absolutely when received.

Funston

[Penciled in: "Original handed to Secy of War in person by H. S. Wright Cipher Clerk Nov. 8 14. This copy handed to me by Genl McBain with instructions to treat as confidential. Nov. 9. 14 The Adjutant-General"]

MGV Entry 12, File 277, May 1, 1914
HEADQUARTERS U. S. EXPEDITIONARY FORCES
Vera Cruz, Mexico, May 1, 1914.

SPECIAL ORDERS.

No. 1.

1. The verbal orders of the Commanding General of the U.S. Expeditionary Forces of the 25th ultimo, designating Colonel Benjamen Alvord, Adjutant General, as Acting Chief of Staff of the U.S. Expeditionary Forces is confirmed and made of record.

2. The verbal orders of the Commanding General of the U.S. Expeditionary Forces, of the 25th ultimo, designating Major Frederic D. Evans, Adjutant General, as Adjutant General of the U.S. Expeditionary Forces in addition to his other duties, are confirmed and made of record.

3. The verbal orders of the Commanding General of the U.S. Expeditionary Forces, of the 25th ultimo, designating Major Frederick M. Hartsock, Medical Corps, as Chief Surgeon and Sanitary Inspector of the U.S. Expeditionary Forces are confirmed and made of record.

4. Captain William H. Noble, Quartermaster Corps, is announced as Chief Quartermaster of the U.S. Expeditionary Forces, in addition to his duties as Depot Quartermaster, Vera Cruz, Mexico.

5. Captain Harry D. Blasland, Quartermaster Corps, is detailed as Assistant to the Chief Quartermaster of the U.S. Expeditionary Forces.

6. Captain George E. Thorne, Seventh Infantry, is detailed as Officer in charge of the Information Division.

7. Naval Constructor Richard D. Gatewood, U. S. Navy, is announced as Cable Censor.

8. Captain Charles W. Weeks, Twenty-eighth Infantry, is detailed as Assistant Cable Censor.

9. Captain William S. Shields, Medical Corps, is announced as Field Medical Supply Officer, and Disbursing Officer of the Medical Department.

10. The verbal orders of April 23, 1914, directing Captain Harry D. Blasland, Q.M. Corps, Pay clerks Wallace F. Baker, Harrison W. Smith, Quartermaster Sergeant Walter Ford, Quartermaster Corps, and Civil Service Clerk John W. Hitch, to proceed to Vera Cruz, Mexico, on the Transport "Kilpatrick" sailing April 24, 1914, are confirmed and made of record.

11. The verbal orders of April 27, 1914, remitting the unexecuted portion of the sentence of Private Howard Phillips, Company C, Seventh Infantry, are confirmed and made of record.

By command of Brigadier General FUNSTON:

Benj. Alvord,
Acting Chief of Staff

OFFICIAL:

F. L. Evans,
Adjutant General

Documents Cited

<u>MGV Entry 12, File 277, May 29, 1914</u>
Headquarters U.S. Expeditionary Forces,
Veracruz, Mexico, May 29, 1914.

MEMORANDUM:

1. Officers of the Services in this command will turn in to the Intelligence Officer all matter of whatever nature in their possession that possesses any military value for operations between Vera Cruz and the City of Mexico.

2. Any material turned in that is a duplicate of that already on file will be returned; reports for which officers may be responsible will be returned; if necessary after listing; all other data, including maps, employees' time tables etc., relating to the theatre of operations will be collected and filed by the Intelligence Section at these headquarters.

By command of Brigadier General FUNSTON.

F.D. Evans
Adjutant General.

<u>MGV Entry 12, File 277, November 17, 1914</u>
OFFICE OF THE MILITARY GOVERNOR
Veracruz, Mexico, November 17, 1914.

MEMORANDUM:

1. Authority has been granted for the closing of the Federal Stamp Office, the State Tax Office and the Municipal Treasury, for business at 5:00 o'clock p.m., Thursday, November 19, 1914. All accounts must be in the hands of the Chief of the Finance Department by that hour.

2. Authority is granted to pay employees on Saturday, November 21st, to include November 30th, with the understanding that they are to continue to render services until the latter date if called upon to do so by the Mexican Administration taking over the Government. [Note the generic title being given to the incoming governing authority.]

3. Such of the above employees as have made arrangements to leave this section of the country, if the number is not too great, may be exempted from the condition of holding themselves in readiness to render service to the succeeding administration.

By order of the Military Governor:
BLANTON WINSHIP
Major, Judge Advocate, U.S. Army,
In charge of Civil Affairs.

Nc

NOTE:- All accounts, except those as to employees, will be stated to include November 23d, 1914.

[written in pen at bottom:] "For Fiscal Officer, Custom House Through Administrator of Customs and Captain of the Port."

<u>MGV Entry 12, File 236, November 23, 1914</u>
OFFICE OF THE ADMINISTRATOR OF CUSTOMS
AND CAPTAIN OF THE PORT.
VERACRUZ, MEXICO.

File No. 23-24   Nov. 23, 1914

From: ADMINISTRATOR OF CUSTOMS.
To: Military Governor, C.A.
Subject: Guards for Customhouse property.

References:

1.- It is requested that military guards be posted at 5 p.m. today, or as much earlier as may be practicable to guard Customhouse officers and the warehouses. They should not permit anyone to enter these places nor remove any material unless on a written or verbal permit by me or by Paymaster Conard.

2.- There should be one post inside the Office building; a patrol of the corridors and lobby, instead of the fixed post at the Treasury, which is not necessary tonight and there should be sentries as follows outside:

(a) One at entrance to "Almacenes" (and *around* this warehouse.

(b) One in passage between warehouses 3 and 4 and patrolling around 3 and 4.

(c) One patrolling the basement of Customhouse.

(d) One patrolling around Warehouse #1 (there are arms in this house)

(e) One patrolling around #2.

(f) One patrolling around vicinity of warehouses #1 and 2, to prevent wagons or persons from removing any merchandise stored in the open air.

3.- I request that these posts remain on as long as possible, leaving only at the last moment when it is necessary for them to embark. If these sentries are not posted by 5 p.m. I fully expect the property in the Customs will be looted as it was on Apr. 21.

4.- I am unable to retain the Mexican employees at the duty; the service is now practically abandoned.

     m

<u>OQG September 18, 1914</u>

Documents Cited

TELEGRAM

received.

Fort Wingate, New Mexico, Sept. 18, 1914.

Agwar,
Washington, D.C.

Your telegram conveying approval Secretary of War of my recommendations relative release Mexican prisoners other than officers is received. Period. I note your instructions to inform prisoners that General Villa assured them full protection in territory under his control. Period. Does this mean that I can send any prisoners to Mexico via Juárez. Period. More than one hundred have asked to go this way and probably more will do so. Period. It will save much expense. Period. I expect leave for El Paso tomorrow noon. Period. Can I have decision Secretary War at once. Period.

BLISS.

TELEGRAM,
The Adjutant General's Office,
Washington,

Sept. 19, 1914.

Brigadier General Tasker H. Bliss,
Fort Wingate, New Mexico.   RUSH.

Internes cannot be sent Mexico via Juárez.

McCain.

<u>OQG September 26, 1914, #232</u>
TELEGRAM.
Ft. Sam Houston, Texas, Sept. 26, 1914.

Agwar,
Washington, D.C.
Number two three two. Period.

Following received four forty five this date from Colonel Hatfield, Douglas, Arizona. QUOTE: General Hill with about three thousand Constitutionalist troops have arrived at Naco Sonora and evidently intend to make his stand there. Period. I am ordering Col. Guilfoyle with four of his troops and machine gun platoon to proceed at once by marching to control situation there. Period. General Hill's requests to send about twenty of his wounded by train to Douglas. Period. I have refused his request but ask instructions as to entry of his wounded. Period. End of Quote.

BLISS.

QUARTERMASTER GENERAL.

205

OQG September 26, 1914, #233
TELEGRAM.

Fort Sam Houston, Texas, Sept. 26, 1914.

Agwar,
Washington, D.C.

Number two three three. Period.

Mexican Consuls along Sonora border have asked permission to bring two or more car loads wounded into United States. Period, I have replied they must get authority State Department. Period. Earnestly recommend War Department discourage their coming now or in future. Period. Mexican forces have medical attendants but don't want be embarrassed by hospitals. Period. If this not done we will soon have another large force prisoners on our hands.

BLISS/

Quartermaster General.

OQG September 27, 1914
TELEGRAM RECEIVED.

Fort Sam Houston, Texas, Sept. 27, 1914.

Staffwar,
Washington, D.C.

Reference your telegram. Mexican prisoners, Southern Department, have already been released in an orderly manner as previously directed by War Department period. Some days ago I informed War Department of exact procedure which has been carried out period. Night message from Wingate says last prisoner left there yesterday afternoon period. Prisoners Brownsville and Laredo released yesterday.

BLISS.

2.22 P.M.
Quartermaster General.

OQG September 28, 1914, #234

Sept. 28, 1914

Telegram

Ft. Sam Houston, Texas, Sept. 27, 1914.

Agwar,
Washington, D.C.

Number two three four.

Night message from Wingate. QUOTE: All Mexicans disposed of. Period. Rest of command and internes Fort Bliss left Wingate four twenty this afternoon. Period. Part

Company B. and remaining Eagle Pass prisoners on same train to El Paso. Period. End of Quote. Salazar and four others will be held at Fort Bliss.

Bliss.

Recd Sept. 28/14

D   QMG

OQG September 28, 1914, #237

TELEGRAM RECEIVED.

Fort Sam Houston, Texas, Sept. 28, 1914.

Agwar,

Washington, D.C.

Number TWO THREE SEVEN period.

Reference your Number Two Three Six. I report as follows period.

Kosterlitzky commanded Mexican Federal garrison which defended Nogales, Sonora, in March nineteen thirteen period. When defeated he and his troops sought refuge in United States and were received as Mexican Federal troops period. It is the Mexican custom to force railway guards and customs guards and other civilians into service for local defense period. Kosterlitzky is an honorable man and I should accept his certificate period. The men whom he commanded undoubtedly live in Nogales, Sonora, and vicinity period. It will be hardship to send them to Eagle Pass period. I recommend they be released at Nogales period. If this be approved direct General Murray to notify me in ample time.

BLISS.

1.00 A.M., Sept. 29.

Quartermaster General.

OQG September 29, 1914, #237

TELEGRAM.

The Adjutant General's Office,

Washington,

Sept. 29, 1914.

Commanding General.

Southern Department,

Fort Sam Houston, Texas.

Number 237 Period.

Reference your Number two three seven.   Following telegram today to commanding general, Western Department, repeated. QUOTE: Reference telegram this office yesterday this subject, Department convinced that internes now Fort Rosecrans are customs guards. Period. Send them, therefore, to Nogales and advice commanding general, Southern Department, in ample time. END QUOTE.

McCain.

Quartermaster General.

10. Civil War

<u>812.2311/195</u>

October 30, 1915.

Commanding General.
      Southern Department.
           Fort Sam Houston, Texas.

Number 689 Period.

Department State is informed by General Carranza's Agent here that he has received report that movement is on foot by Villista sympathizers to dynamite railroad in vicinity Columbus, New Mexico, in order to prevent Mexican troops from reaching Agua Prieta.

McCain.

<u>5761-999</u>

Camp Eagle Pass, Texas.
December 16, 1914

From: Commanding General.
To: Commanding General, Southern Department.
Subject: Special Report

1. An intelligent American who recently arrived from the City of Mexico was in that City when the troops of General Villa arrived. He reports that they arrived by rail and their supplies and ordnance were modern and the unloading was executed with exceptional promptness and regularity. He said that Villa was well supplied with new and modern equipment.

He also says that the men of Zapata and Villa's command were most friendly and excellent order was maintained by them in Mexico City.

He reports the sentiment in Mexico City favorable to Villa. General Angeles, Villa's Chief of Artillery, has gathered a number of ex-Federal artillery officers to serve with the artillery.

2. El Democrata, in its editions of the past week, reports as follows:

Documents Cited

The Division of the North-west, General Obregon, occupies the State of Puebla, and has 10,000 men.

The Division of the East, General Aguilar, occupies Vera Cruz State with 12,000 men.

The following is in brief, the biography of General Francisco Villa, as published in El Democrata, an unfriendly paper:

His correct name is Doroteo Arango. He was born 4th of December 1872, at a ranch called El Rodeo, near Rio Nazas, which river flows through the cotton district of La Laguna. His mother died from childbirth. His father, Guadalupe Arango, was a man servant (Mozo de Cuadro). He sacrificed a great deal to give his son a good education at the parochial school of San Francisco. During his youth Villa was a 'bad boy.' He was known as a cattle thief in the locality and was arrested by Rurales but is father attained his release and sent the boy to Sierra del Rosario.

Here he committed his first assassination, a Captain of Rurales was assassinated by Villa, who Villa said had dishonored his sister, but this was not true. He was assassinated by Villa to obtain his equipment. Villa was arrested and served a sentence in the jail at Guanaceri. When he was released, he assumed the name of Francisco Villa and went to live in Parral, Chihuahua. He arrived here well mounted and said he was a son of a rich family of Zacatecas.

While in prison, Villa met a fellow criminal named Meza, from whom he borrowed a sum of money, later when payment was insisted upon, Villa killed Meza, escaped to the country and became a bandit and highwayman. Villa became known as "El Bandito de Santa Rosalía." He fell into the hands of the law, was tried and sentenced to death, but the sentence was changed to life imprisonment. While in the prison of Chihuahua, he organized a body of prisoners, killed the jailers and fled to the mountains where he remained until Madero named him "Colonel."

His first exploit was taking a troop train. The Commanding Officer of the troop, Colonel Guerrero, and a number of men were taken and shot by Villa.

Villa, in addition to his military duties, had a private business of selling meat in towns that were captured by the Revolutionists.

Orozco recognized the character of Villa and tried to persuade Madero to take away his commission, but without success. Orozco finally went against Madero.

Villa at one time offered to go over to Orozco, but the latter refused to receive him.

Huerta, acting under Madero, had Villa arrested, and sentenced to death for crime and disobedience of orders. Villa fell upon his knees before Huerta and begged for his life, which was granted through the intervention of President Madero, the latter had him confined in a jail in Mexico City, but Villa escaped.

The Division of the North-east, General Pablo Gonzales, has in the State of Chihuahua 20,000 men.

Lieut. Colonel Jose [Fortunato] Zuazua is recruiting in Piedras Negras for a Battalion of sharpshooters (Zapadores) ["Zapadores" are "Sappers," combat engineers, not "sharpshooters"]. He offers $1.50 a day for privates.

3. The American Consul in Piedras Negras reports that Lieut. Colonel Sebastian Carranza, who was stationed in Piedras Negras, has taken a body of troops of his command to Monclova, where the Carranza forces are supposed to be concentrated from points south along the Piedras Negras – Coahuila RR.

This concentration is reported to be due to a movement of a hostile body coming from Ojinaga.

H.C. Hodges
Colonel, 17th Infantry

5761-1010

American Red Cross
National Headquarters
1624 H Street N. W.
Washington, D. C.

July 21, 1915.

From: C. A. Devol, Brigadier-General, U.S.A.
To: The Chief of Staff
Subject: Report upon conditions in Mexico as they exist today for information of the Secretary of War.

1. In compliance with instructions from the War Department I reported at Fort Sam Houston, June 9, 1915, to General Funston and remained on the Border until June 17, 1915. On my arrival in Texas arrangements were made for the entry into Mexico of relief supplies with the assistance of the military organization through the entry ports of Brownsville, Laredo, Eagle Pass, El Paso and Nogales, the Quartermaster at each of these points being designated to receive and forward Red Cross supplies. Consul General Hanna at Monterey was already in charge of the distribution of supplies at that point and the supply was continued, totaling today six cars of corn and beans and one car of packing house products from Kansas City.

2. From various sources information was received that there was destitution in the town of Monclova and also Saltillo. Acting through Vice Consul William T. Blocker a request was made for General Hernandez, commanding the Villista troops from Piedras Negras south to Monclova and Saltillo, to permit Red Cross supplies to be forwarded to Monclova. After the usual amount of delay this request was denied with the statement that no relief was required at Monclova. On a further appeal being made General Hernandez placed the matter before General Villa at Aguas Calientes who wired, stating that he would leave the matter to General Hernandez who thereupon

issued a permit for one car of corn and one car of beans to go to Monclova. He stated in his official permit that it was for these two cars only and added: "You should be convinced imperiously that there is only hunger and famine for the lazy ones and not for the people who are willing to apply their strength to honorable work." A Red Cross special agent, Mr. J. C. Weller, was appointed to take these supplies to Monclova, and in his report stated that on arrival at Monclova the visible supplies of food consisted of one bag of beans. The supplies were distributed and Mr. Weller returned in order to take down an additional consignment. This consignment consisted of one car of corn and one car of beans. He proceeded south with it about the 10th of July, but was stopped enroute owing to the fact that Carranzista troops had cut the railroad south of Barroterán and burned several bridges. A fight occurred shortly afterward between the contending forces and Weller was obliged to return to Eagle Pass with his two cars where they now are awaiting future developments.

3. I held a conference with General Garza at Eagle Pass on the 13th instant. He stated that there was no necessity for any relief supplies in Monclova and that the cars could not go forward without a further permit from General Hernandez who was south on the road conducting military operations. Consul Blocker, however, assured me that he would be able to secure a permit for these cars to go through as soon as a suitable condition of the railroad would permit it.

4. The situation, therefore, on the border after approximately six weeks effort was as follows: Five carloads of corn, one of beans, and one of packing house products had been sent to Monterrey; one carload of corn and one carload of beans had been delivered at Monclova; one carload of corn and one carload of beans on the siding at Eagle Pass awaiting a permit for transportation to Monclova.

GENERAL CONDITIONS IN MEXICO.

5.   Railroads.

I conferred with a number of railroad men, including two general managers and on down to section bosses. All of the railroads in Mexico are controlled by the different military factions. The engines are being retired one after another as they break down and robbed of spare parts to repair other locomotives. The passenger coaches and freight cars, owing to lack of repair, are steadily becoming unserviceable. The road beds are constantly interfered with owing to military operations and bridges destroyed. When repairs are made they are of the most temporary character. All over northern Mexico railroads cross many small streams which only carry water in the wet season. At many of these crossings where the bridges have been burned or destroyed the rails have been laid on what is called a shoo fly or cradle; in other words, laid in the bed of the stream. When the wet season comes, which is now due, these railroads will practically be impassable.

CONDITION OF THE PEOPLE

6. In a paper on Mexico of Yesterday and Today, written by a man who has spent a large part of his life in Mexico and whom I have known for several years, the condition of the Mexican people is summed up as follows: "Ignorance, oppression, fanaticism, lack of cohesion, political insincerity, both national and international, civic corruption and unrighteousness, drunkenness, pestilence, famine, war and anarchy, loot and bloodshed – these constitute the Mexico of today – the heritage of the mistaken policies of its former rulers and of its wealthy, responsible, classes, who, like the Bourbons, learn nothing and forget nothing and who are today striving to regain their lost dominion over Mexico."

During my stay on the border I made my headquarters at Fort Sam Houston and visited El Paso and Laredo once each and Eagle Pass twice. After conferring with General Funston I went across the river at all of these points and called on the various consuls, talked with the people introduced to me by them and also called on the Mexican Generals in command. In making these visits I impressed upon the people of Mexico that I represented entirely the Red Cross and that my visit had absolutely no military significance. [And yet here his report is, in the declassified U.S. Army Mexico file.] I also had a conference at Laredo with Secretary Mireles, Secretary for Carranza, who had just arrived from Vera Cruz. He stated that in the territory occupied by Carranza there was no destitution, only occasional shortage of food, which might have occurred as incident to any military operation. During these various visits I conferred with approximately 150 men from all parts of Mexico covering territory as far south as Oaxaca and taking in the principal cities and districts north of that point. No definite testimony could be taken and recorded for the reason that many of these men, particularly those having property in Mexico, urgently requested that their names be not mentioned and in some instances, very particularly that of a man who had left his family in Mexico, a conference had to be held at a place free from observation from anyone connected with the Mexican forces. I can, therefore, only generalize in regard to this testimony.

Conditions are undoubtedly most acute in sections of the country that are constantly being occupied and reoccupied by contending forces. This applies in great force to Mexico City and the surrounding country to San Luis Potosí, the country in the vicinity of Saltillo, Monclova and Paredón. In El Paso it was stated that the State of Chihuahua was normal no food supplies were needed. I, however, talked with a man right from the city of Chihuahua and he stated that there would be a decided shortage of food in a very short time.

There have been large exports of food supplies to the United States and to Cuba, corn, beans and cattle. I was offered Mexican exported beans for sale by various dealers all the way from Arizona to the Eastern part of Texas. One estimate of the amount of beans exported was two thousand carloads. I secured and have on file a definite list of the exportations of food supplies through Laredo and Eagle Pass since the first of January.

Both General Hernandez and General Villa prohibited the purchase of food supplies in Mexico for relief of the Mexican people, stating that any such supplies must be purchased outside of Mexico. The reason given is that the food in Mexico is required to feed the Mexican soldiers. The real reason, in my opinion, is that there is a military export tax, or, in other words, that a charge on all exportations amounting on the northern border to $400 to $500 for each car that crosses the border. This money is paid direct to the military leaders. For instance, at the time supplies were being purchased for Monclova three carloads of corn could have been purchased on the Mexican side for $.56 per bushel; the actual supply cost the Red Cross $.94 per bushel. The difference in the price per bushel indicates the amount of revenue accruing to the military authorities from export products. Relatively the same difference applies as regards beans.

There are still large quantities of food supplies assembled in various parts of Mexico and held by the military authorities and not permitted to reach the people. In the province of Michoacán there is a large deposit of corn and wheat, said to be ten thousand tons of corn and two thousand tons of wheat under control of Carranza. Northwest of Mexico in the State of Guanajuato there is another large deposit, the owner, several weeks ago, not being able to obtain information as to which military faction was in control. In the vicinity of Saltillo there is a large wheat crop variously estimated at from one thousand to two thousand carloads. I was informed by a man who had just returned from there that he had entered into negotiations for exporting this entire crop, stating that 25 cents per bushel would have to be paid to Carranza. There is also a thousand tons of corn opposite Del Rio which was offered to me by the owner stating that he was holding it and also explaining that his high price was due to the fact that a portion of the money received would have to be paid to the military authorities. As this place had just been captured by Carranza it is presumed that the tax would go to that faction.

SUMMARY.

7. At the present time there may not be any actual cases of starvation in Mexico. The Mexican people can live on little or nothing and when driven to it can subsist for some time on cactus and other plants. There is in nearly every part of Mexico destitution and constant misery. People outside the military factions subsist only by sufferance and only when permitted to do so by the military leaders. It is evident that not 50 per cent of a crop has been planted generally throughout Mexico. Continual suffering obtains and is growing worse from day to day. The methods of exaction used by the leaders of all factions appear to be identical. All commercial supplies and industries are exploited for the benefit of the military. Owing to the political and physical conditions in Mexico there appears to be no future for adequate relief even if

the American people would contribute for this purpose. This, however, they have only done to a very limited extend to this date.

Very respectfully,

C. A. Devol

Brigadier-General, U.S.A.

5761-1023

Fort Sam Houston, Texas, Sept. 28, 1915.

Captain H. H. Robert. Memorandum.

A telegram to the Carrancista daily, La Raza of San Antonio, Tex. reads as follows:

"Special. Laredo, Texas, Sept. 27, General of Division Alvaro Obregon in a message sent today to some of his friends in this city, says that he has re-established telegraph and railway communication as far as San Pedro de las Colonias, which place has been occupied by his forces without meeting any resistance. From other sources it is known that as the Villistas have left Torreon, the occupation of that place is a question of days only, as the column of operations, in its formidable advance, is ready to overcome all obstacles."

The following correspondence from La Raza gives some information as to the movement upon Torreon.

"Saltillo, Coahuila; Sept. 20 (1915). Advices received in this city say that communication, at least telegraphic, has been reestablished between this city and Piedras Negras, Coahuila, as I have seen various telegrams dated in that frontier city.

On the 18th of this month the mobilization of strong units of troops began from this place upon the stronghold of Torreon, in combination with other columns that are marching, against the Laguna region. Yesterday the headquarters of the forces were at Paredón station, but according to the latest news, they will be transferred to Hipólito, in order that the advance will be carried out without interruption. A few moments before leaving General of Division Alvaro Obregon called to say farewell to Senor Gustavo Espinosa Mireles, Provisional Governor of the State, who accompanied "The Maimed One of Leon" as far as the station where the troops were embarking. In order to protect official secrets it is impossible to tell you the exact strength of the troops under the command of the vanquisher of Villa and Angeles. (Speaks of the railway under construction to the Sierra of Arteaga.)

An interview with General Obregon.

The reporter of a local daily has had an interview with General Obregon, of which I send you the essential part.

Reporter--- "How then, do you believe that the question of peace in Mexico will be solved?"

214

Documents Cited

Obregon--- "Just as we are solving it, at the point of the bayonet and by the mouths of cannon. Beasts do not understand caresses they only understand a club or the blow of an axe. For snakes you have to smash their heads, and that is just what we are doing.

Reporter. "Do you consider the broken up Division of the North as a military problem anymore?"

Obregon.— "No, Pancho Villa is now what he was before, a highway robber.

Reporter. "What is the time you calculate for the definite triumph of Constitutionalism?"

Obregon. "The military triumph is already gained, but, as I said to you before, Villa is not a military problem any more and now what is lacking is the pacification of the country. As to the time; I must tell you that it will be the time it will take for my forces to get to Ciudad Juarez; in my march I will finish up all these rascals."

A special telegram to La Raza says:

"Monterey, September 21. Today General Jacinto B. Trevino arrived at this city after his short visit to the port of Matamoros, where he talked with General Nafarrate about the latest occurrences reported on the Texas frontier. Although General Trevino has not made any declarations in relation to his trip to the frontier port, it is thought here that it was of great importance for the stopping of difficulties between the military authorities of Mexico and those of the United States.

Yesterday a great mobilization of troops took place by the orders of General Treviño for an advance to Paredón for the purpose of taking part in the campaign begun against the last bands of Villistas that have remained behind in Torreon, Gomez Palacio and some other points and above all in the state of Chihuahua.

The soldiers when they were in the trains to take them to the front were shouting with enthusiasm "on to Ciudad Juárez."

I can confirm the arrest of former General Lucio Blanco in the neighborhood of the city of Saltillo, by forces of General Fortunato Maycotte. Lucio Blanco declared publicly that he would not be shot because he had been given amnesty by General of Division Pablo Gonzales in the pact of Galeana. In the same sense General Adolfo Huerta Vargas, commanding officer of Saltillo, has also addressed General Trevino, but up to now it is not known what has been decided upon in the matter.

Yesterday the telegraph lines between this city and Paredón, Saltillo and Piedras Negras were re-established with great pleasure on the part of the public.

With the evacuation of Cuatro Ciénagas by the last Villista forces and the arrival at that place of the column of General Elizondo, the coal district of Coahuila has come under the absolute domination of the Constitutionalist forces.

It has been learned by entirely trustworthy account received in this city that Villa recovered the sum of six million pesos after assassinating his old friend Tomas Urbina at the ranch of Las Nieves, in the state of Chihuahua.

From other sources it is sure that "Doroteo Arango" will not make any resistance in Torreon, a place which without doubt will be occupied by our forces at almost any moment, as the advance, as I have learned, will be rapid and efficient.

(General Obregon was at Hipólito, Coahuila, September 21st.)

News received in this city says that General Miguel Zapata who some time ago rebelled against Carrera Torres has defeated that leader in the region of Tula, Tamaulipas. Details of the encounter have not yet been received, but I will send them to you as soon as I learn them."

(Sgd) Luis F. Bustamante,

Chief of the Office of

Information."

Special correspondence to La Prenza, a daily of San Antonio, says:

"Laredo, Texas, September 26. Lieut. Colonel Gustavo Salinas of the Carranza forces, who is said by a local paper to be the chief of artillery of General Álvaro Obregon, left Nuevo Laredo yesterday on a special car for Saltillo. In the same car a foreigner named Solano also left, having come to this city to buy a linotype machine and presses for Governor Espinosa Mireles of Coahuila. This foreigner will edit a paper in the city of Saltillo and will be assisted by Eduardo Guerra, a newspaper man who is for Carranza. This paper will be a personal organ of Governor Espinosa Mireles.

Eduardo Guerra came to Nuevo Laredo as inspector of the custom house there. The Carranza newspapers have given him the title of Doctor to which he is not entitled. Guerra was at one time Municipal Treasurer when Alfredo Perez was first Alcalde."

Special correspondence for La Prensa printed September 28th says:

"El Paso, Texas, September 25th. Since the execution of Tomas Urbina and the mobilization of Villista troops towards Ciudad Juárez, nothing else of importance has happened in the region dominated by the dying Villista faction. The Chief of the Villistas is in the capital of the state of Chihuahua, in a wavering attitude and it would seem in hope of developments in political affairs in the American capital, the solution of which he is waiting for in order to guide his future acts as a revolutionist. It is said insistently that there has been a real struggle between Villa and his advisors. In order to convince him that he must still struggle on diplomatic grounds against Carranza because the "Chief of the Operations" wished to take a violent stand in view of the marked tendency of the American government to leave him out of all further combinations. The famous evacuation of Torreon which turns out to be one of the famous bluffs invented by the Carranzistas has not been carried out yet, and it seems more that the Villistas have decided to play their last card at the Laguna city. A fact that calls attention is that so much time has passed without neither Alvaro Obregon nor Hill have thought of beginning the attack on Torreon, and acting as if they find themselves confronted by the enemy.

The advices that are coming in, referring to the capture of Raoul Madero, say that this chief saw himself obliged to cross the frontier line because he was completely defeated by the Carranzistas, who were pursuing him very closely. The Villistas of this city are making great efforts to have Madero set at liberty, stating to the military authorities that Madero came to American territory to carry out a commission given him by Villa, with the Pan-American conferees."

The following article from Justicia, a daily of El Paso, Texas, of September 25, says:

"Villaism in its flight is showing all its wounds and exhibiting throughout its whole body all its terrible physical and moral condition. With good information we are placing in the knowledge of our readers the news that demonstrates it.

We have already announced that long convoys of railway trains are constantly arriving from Chihuahua carrying people, horses, goods, and every class of revolutionary booty. Yesterday there arrived in Juarez twenty seven cars with the regular number of men but all in a state of lamentable demoralization but for Villaism a natural condition. Above all more than nine-tenths of these men were in a terrible condition. There are not enough lurid colors in order to paint the state of these people. More than nine-tenths of them, we repeat, came in terrible condition; the men barefooted, hungry, almost naked; which gives an idea of the treatment that they have received from their so-called Chiefs. As may be presumed the majority of them are ready to desert and if they have taken the determination to cross the line, it is in order to come over and find protection in El Paso, Texas, and if not, it is due to the fact that the line is so well guarded because as we said in Justicia, it is in the plans of Villa to stop the flight of his unhappy soldiers. It seems that this was one of the things that was discussed in the Villa-Scott conference, whether this be certain or not, the facts make it seem true. Many are of the soldiers and "Generals" who have escaped to the mountains, in order not to be taken along in this current of deaths, which is emptying itself into Sonora, and many others have succeeded in getting over the line in order to save themselves from the wrath of the insatiable jackal. The soldiery tired out by the fatigue of a constant and unskilled campaign, which in place of bettering their condition has made it worse, submerging them in a misery worse than death itself; thirsty and hungry are knocking at the doors of private houses of Juarez, begging, demanding better, food and water.

These facts have been seen by persons who, upon telling them to us, said that it gave them pain and made them indignant against Villa on account of the state into which the ignorant and unthinking soldiers had been brought.

There are very well founded presumptions that something very grave may happen in Ciudad Juarez for the reasons which we have just set forth and because people are all becoming convinced that Villa in his craziness does not know any more what he is doing. The execution of Urbina and other leaders of the same class, has begun to open the eyes of some, because they are saying that if Villa pays this way to those who served him best yesterday and who he made generals, what can the rest expect.

A mutiny, a barrack rising, the sacking of the city, which really has very little to be sacked, but which would end in a bloody struggle, is very much feared, very much indeed. It is well for persons who are accustomed to go to Juarez or who go there because their business may oblige them to, should stay away for a while to avoid death in a moment of popular vengeance.

The destination of these troops as we have said, is Casas Grandes, but it is not true that troops are going through American territory to Nogales. What did pass recently to Maytorena were some cannon of the few that the jackal has saved. From Casas Grandes the exhausted forces of Villa have begun to march for Sonora."

W.L. Gibson

1st Ind.   HHR-vve

Office Dept. Engr., So. Dept., Ft. Sam Houston, Texas, Sept. 28 ,1915. – To Chief of Staff, So.Dept., For forwarding to Chief, War College Division, Washington, D.C.
Henry H. Robert
Captain, Corps of Engineers,
Department Engineer.

2d Ind.

Hq. Southern Department, Fort Sam Houston, Texas, Sept. 29, 1915. –
To Chief, War College Division, General Staff, Army War College,
Washington, D.C.

W. H. Hay
Major, General Staff,
Chief of Staff.

WHH-nc

5761-1025
Fort Sam Houston, Texas Sept. 30, 1915

Special correspondence to La Prensa of San Antonio, says:

El Paso, Texas, Sep. 29, 1915. Telegrams to this city from Chihuahua say that Lic. Miguel Diaz Lombardo, Minister of Foreign Affairs of Villa, is ill, and for this reason will not be able to make his announced trip until the end of this week. The Villistas say that the object of the trip of Diaz Lombardo is to confer with some functionaries of the Convention government who are now here and later to go to Washington on a mission confided to him by Villa.

Chihuahua news also says that Francisco Villa arrived on Tuesday (Sep. 28) at Casas Grandes via the Mexico Northwestern Railway direct, crossing the summit at Cumbre over the new branch. It is not known in Ciudad Juarez why he made the trip by that line, as it was believed that he would first come to our neighboring city in order to make the trip.

The trip of Villa to the District of Galeana has been made for the purpose of watching the movement of his forces by marching into the State of Sonora. This movement will begin at Casas Grandes. Ten trains carrying about six thousand troops and plenty of artillery have already arrived at that place. These troops are under command of Generals Jose E. Rodriguez, Santibañez, Jurado, San Roman, Bracamonte, and some others.

It is not known whether Villa, on his return, will pass through Ciudad Juarez, en route from Casas Grandes to Chihuahua, for the reason that as yet he has made no statements as to the route he may take. Nevertheless, everything causes it to be believed that he will come to Ciudad Juarez in order to see Diaz Lombardo, who, as I said, will arrive at the end of the week at the neighboring city.

Special telegram to La Prensa.

El Paso, Sep. 30, 1915. The first passenger train arrived at Ciudad Juarez this morning from Chihuahua. During the present week and last week passenger traffic between the two cities has been suspended in order to allow the mobilization of the Villista troops for the occupation of Sonora. Also, today a passenger train left Ciudad Juarez, and according to public notice, will make the trip to Jimenez. It is believed that the people waiting to leave Chihuahua will arrive by tomorrows train.

The Mexico Northwestern will cease operations tomorrow (Oct. 1). All the employees have received the notice of the resolution of the Directors. All employees will have all necessary facilities to come to the frontier, if they desire to do so.

It is announced this morning that the Mormons who live in Colonia Morelos, Chihuahua have made up their minds to leave Mexican soil in view of the circumstances at present. A train bringing the Mormon families is expected tomorrow, and these families will probably locate near El Paso.

The advanced parties of the Carrancistas are now thirty miles to the North of Torreon and are engaged in repairing the railroad which is torn up as far as Jimenez.

The cotton crop abandoned by Villa in the Laguna is estimated to have a value of eleven million dollars.

On account of the discontent reigning in the garrison of Ciudad Juarez the authorities have decided to pay the men in American gold. It is also said that this measure is taken for fear that the garrison will mutiny.

Only eight hundred (800) men have been left in the City of Chihuahua, as a garrison. This force has orders to tear up the railway line when it evacuates the place, retiring to the North by order of Villa.

Correspondence to the Prensa says:
Monterrey, Sep. 28. According to the news given, out by the Carrancistas in this city, the forces of Cesáreo Castro took possession today of Torreon, as it had been evacuated by the Villistas. It is reported that the cavalry corps headed by Castro made its entrance amidst the best of order, as it had been four days since the place had been evacuated. It is said that the reason that the troops of Alvaro Obregon remained so long in San Pedro de las Colonias, was to repair all the railroad lines destroyed by the Villistas in their retreat towards the Laguna. The Carrancistas will leave Torreon in a few days for the North, for the purpose of keeping up their pursuit of the Villistas, and are only waiting to replenish their ammunition and rations in the Laguna City, before beginning the advance, as they have news that the Villistas have left not a trace of anything usable for food, and that the line of advance will be practically over a desert.

In the advance upon Chihuahua all the forces of Obregon and Trevino, amounting to more than twenty thousand (20,000) men, will be used, for the leaders are determined to take the frontier states away from Villa, and think that the campaign will be from now on, harder than that just ended, as the terrain is very advantageous for their foes.

A special telegram to El Presente of San Antonio says:
El Paso, Texas, Sep. 30. Information received in this city from Chihuahua says that yesterday (Sep. 29) the people rose, and when the Villista soldiers tried to suppress the disorder, they killed more than fifteen people, including one American and one French citizen. Part of the Villista forces also mutinied, uniting with the people. It has not been possible to obtain more details as to the condition of things there, as the censorship in Juarez is rigorous and no information in the matter is allowed to go through.

A special telegram to La Raza of San Antonio says:
Laredo, Texas, Sep. 30. On Sep. 29 at 1 P.M. the Constitutionalist forces under the command of General of Division Francisco Murguía and General of Brigade Luis Gutierrez made their triumphal entrance into the city where Villa had dominated more than a year. A great reception was given the victors, and the people of Torreon seemed sincerely pleased to see the entrance of the troops. All those addicted to the Villista cause, together with those who served the Villistas, left with the enemy.

The line of the railroad from Zacatecas to Torreon, along which General Murguía advanced, has been repaired. Telegraph communication has been restored and the trains will be able to run from now on. Many people are getting ready to make trips to that region.

The Carrancistas captured three hundred and sixty eight (368) freight cars in good condition, five cabooses, eleven oil tank cars, twenty one passenger cars, and fourteen locomotives. A large amount of material for track repairing was also found.

General Obregon has announced that he intends to remain in Torreon but a short time, in order to give the necessary orders and will advance towards Chihuahua with the object of fighting Villa wherever he may encounter him.

W.L. Gibson

1st Ind.   HHR-vve

Office Dept. Engr., So. Dept., Ft. Sam Houston, Tex., Oct. 2, 1915. – To Chief of Staff for forwarding to Chief, War College Division, Washington, D.C.

Henry H. Robert
Captain Corps of Engineers

2d Ind.

Hq. Southern Department, Fort Sam Houston, Texas, October 2, 1915. – To Chief, War College Division, General Staff, Army War College, Washington, D. C.

W. H. Hay
Major, General Staff,
Chief of Staff.

WHH-ne

5761-1027
Fort Sam Houston, Texas, Sept. 30, 1915.

Memo. for Captain H. H. Robert.

A special telegram to Justicia of El Paso, Texas, in its issue of Sept. 29 says:

Douglas, Ariz. Sept. 27. The reporter of the "Douglas Daily International" has just arrived at this place, and says that Plutarco Elias Calles entered "The Republic of Agua Prieta" as some of our "cousins" have nicknamed it on last Saturday (Sep.25) and this confirms my former advices.

The forces that arrived at Agua Prieta, counted carefully by the reporter, came to two thousand one hundred and forty six (2146) men, with practically his artillery intact with which he left the place. When Calles entered Agua Prieta in an auto, he seemed very well satisfied and unvanquished giving himself the airs of a conqueror at the head of his followers who came along in their march as if they were at a funeral.

Calles has ordered the concentration of the garrisons of El Tigre, Moctezuma, Cumpas, Nacozari, and Fronteras at Agua Prieta in order to make himself as strong as possible and to repulse the coming attack that the Villistas are preparing, who, as I said in my former message, are beginning to invade the State of Sonora.

As soon as it was learned in Cananea that the mining camp of "El Zorrillo" had been left without a garrison, on the instant a gang of ten bandits was organized who went into the camp to rob it, sacking some Chinese shops, and committed all classes of depredations. These bandits had been taking refuge previously in the hills near the mining camp.

It has been said that the police of Cananea have been witnessing almost with pleasure, similar bandit attacks and on learning the fact Calles sent Colonel Aguirre with three hundred (300) men, who came suddenly into the mining town, and killed some of the bandits, upon whom the troops opened fire at sight, and who did not offer any resistance.

In his official report Calles states that in his battle with the Maytorena forces he killed about one hundred of the latter, and wounded a great number.

Yesterday, under bond, numerous families and Carrancista employees passed through American territory in a special train, evacuating completely the town of Naco, where formerly Calles had an idea of fortifying himself.

The customs collector of Naco, who is already employed by the Villistas, has received a telegram from Maytorena, saying that a force of employees is leaving Nogales today to take possession of the respective offices in Naco.

From two hundred to three hundred persons who fled from Cumpas when the town was left without a garrison, and who fear the depredations that may be committed by the Villistas, have arrived at Nacozari en route to this place (Douglas, Ariz.)

W. L. Gibson

Confidential

B35-16   1st Ind.   HHR-vve

Office Dept. Engr., So. Dept., Ft. Sam Houston, Texas, October 1, 1915. – To Chief of Staff, for forwarding to Chief, War College Division, Washington, D.C.

Henry H. Robert
Captain, Corps of Engineers,
Department Engineer.

5761-1028
Fort Sam Houston, Texas, October 1, 1915.

MEMORANDUM FOR CAPTAIN ROBERT:

A special telegram for La Prensa, San Antonio, says:

Documents Cited

"Laredo, Texas, September 29 – General Jacinto Treviño, Chief of Arms, Corps of the Northeast, with all the troops which he has been able to dispose of for the campaign, left yesterday morning, (September 28) from the station of Hipólito, proceeding in the direction of San Pedro de las Colonias, Coahuila.

It is said that General Cesáreo Castro, Chief of Cavalry of General Alvaro Obregón, entered the city of Torreon yesterday (September 28) at 4 P.M. without firing a single shot, as the place was found to be completely evacuated. It is said that the Villista troops left Torreon on Monday (September 27) and that only a small body of neutral police remained to keep order. Nevertheless, no absolutely accurate report has been received to confirm the news spread by the Carranzista agents that the city of Torreón was occupied yesterday afternoon by the cavalry commanded by General Castro."

Another telegram to the same paper says:

"El Paso, Texas, September 29 – The authorities in Ciudad Juárez have admitted to us that the convention forces have evacuated the city of Torreón and also all of the rich district of the Laguna which they have had in their power since April of last year. His Villista Generals who were the last to leave the city of Torreón were Juan N. Medina and General Madinaveytia, who left that place with 2,000 men of the infantry and cavalry, leaving according to all probabilities for Chihuahua, although this has not been confirmed.

The Consuls of Germany, France and England who are stationed at El Paso telegraphed today to General Álvaro Obregón asking him if their nationals who may be in the city of Torreón would have every kind of protection upon the arrival of the Carranzista troops. As yet no reply has been received to the message but they believe that Obregón will answer in a favorable manner.

I wish to ratify my previous advices to the effect that Francisco Villa is now in the city of Casas Grandes. Telegraphic news received today at 6 P.M. (September 29) confirm my information and add that Villa has been directing the sending of his troops toward the state of Sonora. It is said that Villa will return to Chihuahua as soon as he finishes instructing the troops that will go to Sonora. Nevertheless, information from a reliable source indicates that he will go directly to Sonora after all his men have crossed in that state and that he will go to Nogales, in which place he will establish his general headquarters and where he will install a cartridge factory and the works to make equipment for his army.

It is also known in some circles here that all the cars and in general all the rolling material that Villa has used to transport his troops towards Casas Grandes will be destroyed along the north-western line by orders of the revolutionary chief, in order to avoid the falling of this material into the hands of the Carranzistas. It is said that Villa has at present fifty locomotives and 2200 passenger and freight cars in his hands and that all of these will be destroyed.

Special correspondence to La Prensa says:

"El Paso, Texas, September 27 – The following message has just been received by the Villa Consular Agent in El Paso from General Villa: 'Headquarters of General Villa in Chihuahua, September 26 – The forces of General Hilario Rodríguez have been fighting in San Juan del Mezquital, state of Aguascalientes (should be Zacatecas) having defeated the enemy after capturing some arms and commissaries from them.

The column of General Margarito Salinas which has been operating between the station of Catarina and San Miguel del Mezquital, Zacatecas, has defeated various hostile bands and capturing much war material from them and more than 100,000 goats.

Forces of General Canuto Reyes have fought a band of Carranzistas near San Juan de Guadalupe and defeated them, inflicting some losses and taking fifteen prisoners. (Signed) Francisco Villa.'

In effect, according to information which has been received through passengers arriving from the south, it seems that the Villistas are displaying some activity in the state of Zacatecas, toward which state the forces of Salinas, Reyes, and Fierro were sent a few days ago (September 21, about). The fact that these forces are remaining to the south of Torreón causes it to be understood that Villa as yet does not propose to evacuate that place (Torreón evacuated September 27).

General Juan Banderas, formerly the Zapatista General, who was recently appointed Governor and Military Commander of the state of Sinaloa left the city of Chihuahua on September 26th, at the head of a column claimed to be five thousand strong. This force left over the Mexican Northwestern and at some point towards the northern frontier of the state will leave the railroad in order to march to their destination. Banderas has orders for all the Villista leaders to put themselves under his orders, as Villa has put the campaign on the west coast into the hands of Banderas. It seems that the plan of the Villistas is to take the rear guard of General Manuel M. Vieguez [Diéguez], who is reported to be advancing northward from Mazatlán and to cut him off from the troops remaining in that city. The troops that passed through Ciudad Juárez for Casas Grandes will co-operate in the campaign to be begun in the states of Sonora and Sinaloa. The Carranzistas say that the concentration of the troops of Plutarco Elías Calles at Agua Prieta has for its first object the preparation of adequate resistance against the Villa troops, who are expected to invade the state of Sonora."

Since the above was received, news has been received that J. M. Maytorena, commanding the anti-Carranza forces in Sonora, has crossed the United States line and that General Francisco Leyva, the Villista commander of Guaymas, Sonora, has also crossed the line. It is further reported that bandits have occupied the mining town of Cananea.

Documents Cited

W. L. Gibson

<u>5761-1029</u>

Fort Sam Houston, Texas,
October 2, 1915.

MEMORANDUM FOR CAPTAIN ROBERT:

This morning at 9.30 A.M. as a representative of the San Antonio Light, the undersigned had an interview with Carranza Consul J. Z. Garza of Brownsville with whom I have been acquainted for about one month. Mr. Garza said in the course of our conversation following the formal interview that Major Cavazos, of General Nafarrate's Staff, had investigated the reports as to bandits on the Mexican side and that yesterday Major Cavazos had reported the presence of numerous bands of bandits on the Las Flores Ranch and that General Nafarrate had telegraphed to General Jacinto B. Treviño at Torreón asking for instructions in the matter and that he supposed that the Mexican troops would attack the bandits. The only ranch known as Las Flores in the vicinity of Progreso that can be located is a place one half mile southeast of Las Cruces Ranch and one half mile south west of San Joaquín Ranch in the southwest corner of Sheet No. 20 of the topographical map of the Rio Grande survey. It is not marked on this map but is found on the map issued by the War College giving the route from Matamoros to Monterrey.

Mr. Garza further stated that he was trying to get the true statement of affairs on the border before First Chief Carranza and that he had been called to San Antonio by Consul-General Beltrán for this purpose. Mr. Garza intends returning to Brownsville tonight but is somewhat "under the weather" and may not return until tomorrow night.

Mr. Agras, Vice-Consul at Brownsville has left the service and I think on account of the fact that he and Consul Garza were unable to get along together. Mr. Garza does not visit Matamoros except on official business and I do not think the relations between him and General Nafarrate are at all cordial. Mr. Agras did reside in Matamoros and I think that his separation from the service is due to political reasons more than personal ones. Mr. Agras was formerly a reporter on the Mexican Herald in Mexico City.

Major Cavazos, who investigated the rumors and located the bands at Las Flores, is a member of the family which owns the Cavazos Ranch, giving its name to Cavazos crossing south of Mission where Captain McCoy had his encounter with the organization commanded by Captain Porras some time ago. Whether this Major Cavazos is the same man who was until recently the first captain in the garrison at Reynosa under Lieutenant Colonel Alonzo Velazco, is not yet known. From another source it has been learned that Nafarrate is said to have asked permission to attack the bandits. Why he should do this unless he is uncertain of the temper of his troops who are local men largely of the border, is not clear, unless he desires reinforcements from another part of the country who are not affected by local conditions. It is hard to see why he should await orders before attacking the force which is nominally hostile to him within the limits of his own command, which extends from the mouth of the Rio Grande to and including Guerrero, Tamaulipas, opposite Zapata, Texas. General Nafarrate is now the General of Brigade and his title is "Chief of the line of the frontier of the Bravo (Rio Grande)."

W. L. Gibson

Confidential
B35-20   1st Ind.   HHR-vve

Office Dept. Engr., So. Dept., Ft. Sam Houston, Texas, October 2, 1915. –
To Chief of Staff, Southern Department, Fort Sam Houston, for forwarding to Chief, War College Division, Washington, D. C.

Henry H. Robert
Captain, Corps of Engineers,
Department Engineer.

B35-20   2 Ind.
Hq. Southern Dept., Fort Sam Houston, Texas, October 4, 1915.

To Chief, War College Division, General Staff, Army War College, Washington, D. C.

W. H. Hay
Major, General Staff
WHH-hs   Chief of Staff.

5761-1033
Fort Sam Houston, Texas, October 7, 1915.
Memorandum for Captain Robert: Situation in Chihuahua State.

Reports state that Villa took eight hundred men with him and proceeded on Oct. 3 to Santa Rosalía, (Ciudad Camargo) one hundred miles towards the southeast, along the line of the Central, and after a conference with his former subordinate, an encounter took place Oct. 4 between forces of Villa and Gen. Rosalío ("Chalío") Hernandez, the subordinate referred to. The result was the withdrawal of Villa to Chihuahua. Many soldiers are said to have deserted and others were killed and wounded in the fight.

Villa is said to have made up his mind to evacuate Chihuahua City, and to be due in Ciudad Juarez Oct. 7 or Oct. 8.

Rodolfo Fierro, general in Villa's forces, is reported with five hundred picked men to be at Ciudad Juarez, waiting orders from Villa.

General Tomas Ornelas is the local commander of Ciudad Juarez. 600 men.

The bulk of Villa forces is reported to be at Casas Grandes, 150 miles S.W. of Ciudad Juarez on the Mexico Northwestern Railway. This force is said to number 12000 to 15000 men, with 18 pieces of artillery. Gen. Jose Rodriguez is the principal leader. Gen. Juan Banderas is also probably with this force, having come over the Mex. N.W. Ry. but he may be at La Junta or Guerrero. His force is claimed to be 5000 men, but this is very doubtful. Another report from New Mexico says that Villa's forces are about due this date at Carretas, west of Casas Grandes, on the Sonora line.

Gen. Ochoa is the local commander at Casas Grandes, with 500 men. He has garrisons at Janos and Ascención to the Northwest, Colonia Juarez nearby Galeana to the southeast, and Pearson and Cumbre to the south.

There is a small Villista garrison at Ojinaga on the Rio Grande opposite Presidio, Texas and another at Guadalupe, opposite Fabens, Tex.

Chihuahua is also said to be still held by Villista troops, probably under Gen. Juan Medina, who was the last Villista commander to leave Torreon.

There were three raiding columns of Villista troops to the south of Torreon about Sept. 25, one under Gen. Hilario Rodriguez near San Juan del Mezquital, Zacatecas, another under Gen. Margarito Salinas between Catarina, Durango and San Miguel del Mezquital, Zacatecas, and a third under Gen. Canuto Reyes in the vicinity of San Juan de Guadalupe, Zacatecas.

The Carrancistas seem to have no troops as yet in Chihuahua except two small commands. The first under Gen. Luis Herrera, is composed of local men, and is reported to have occupied Hidalgo del Parral, west of Jimenez, on Oct.4. This place was formerly garrisoned by Villista troops of the brigades of Tomas Urbina and Manuel Chao, but these have probably dispersed. The other band is a small one of 50 to 200 men at the Gavilondo ranch on the New Mexico line, near Laings Ranch, N. M. Gen. Rosalío Hernandez is probably now a Carrancista, as he is said to have been negotiating with the enemy for some time. He is at or near Santa Rosalía. W. L. Gibson.

Confidential

B35-25 C.F.   1st Ind.   HHR-vve

Office Dept. Engr., So. Dept., Ft. Sam Houston, Texas, Oct. 7, 1915. – To Chief of Staff, for forwarding to Chief, War College Division, Washington, D.C.

Henry H. Robert
Captain, Corps of Engineers,
Department Engineer.

5761-1039

U.S. Army Intelligence in the Mexican Revolution

Fort Sam Houston, Texas, October 12, 1915.

Memorandum for Captain Robert: Movements of V. Carranza.

A special telegram to La Prensa of San Antonio, Texas, says:

Veracruz, Oct. 11. The First Chief of the Constitutionalist Army, Venustiano Carranza, accompanied by his staff, sailed today from this port for that of Tampico, Tamaulipas, on board the gunboat "Bravo", being escorted by the "Zaragoza" and the "Atlanta." Carranza is going to Tampico, from which place he will later make a trip to the capital of Nuevo Leon (Monterrey), and other cities that he announced a short time ago that he would visit.

From another source it is learned that General Álvaro Obregón will meet General [sic] Carranza at Tampico.

The rumors spread by a group at El Paso, another at Monterrey and later repeated from Washington by the correspondent of the Los Angeles Times, that Obregón was working with Jacinto B. Treviño to declare themselves against Carranza, seem to be discredited by the facts, as far as known at this date.

W. L. Gibson

Confidential

B35-30 C.F.   1st Ind.   HHR-vve

Office Dept. Engr., So. Dept., Ft. Sam Houston, Texas, Oct. 13, 1915. – To Chief of Staff, Southern Department, for forwarding to Chief, War College Division, General Staff, Washington, D.C.

Henry H. Robert
Captain, Corps of Engineers,
Department Engineer.

5761-1040

Fort Sam Houston, Texas, October 12, 1915.

Memorandum for Captain Robert: Military movements in Sonora.

Special correspondence to La Prensa of San Antonio dated Nogales, Ariz. Oct. 9, 1915 says:

Since yesterday (Oct.8) the railroad (Southern Pacific of Mexico) has been in good repair between Nogales, Sonora and Guaymas, Sonora (265 miles). The line will be used at once for military purposes, as Governor Carlos Randall (at Nogales) has, in accord with General Francisco Urbalejo, ordered that troops shall be sent at once from the neighboring city for the port of Guaymas, for the purpose of re-enforcing the garrison of that place, so that Guaymas may be defended, in case that the advance of the forces of General Manuel M. Diéguez (Carrancista) continues.

Documents Cited

Yesterday the rumor was in circulation that a band of hostiles had burned a small bridge between Nogales and Magdalena (54 miles) but the rumor has been discredited by the fact that a military convoy left over the railway, and arrived without incident at Hermosillo (175 miles) today.

Tomorrow, (Oct. 10) the first passenger train will leave for the south from Nogales, it is stated. Yesterday coaches were sent from the U.S. side to Nogales, Sonora, for the renewal of public service. All coaches that the railway company had in service in Sonora were brought to this city when traffic was suspended at the time the forces of Calles were mobilized and came quite near Nogales, Sonora.

A special telegram to La Prensa, dated, San Diego, California, Oct. 11, 1915, says:

Wireless telegrams received here today say that a military train with more than one thousand Villista soldiers, four pieces of artillery and some machine guns, arrived today at the port of Guaymas, to reinforce the garrison of that place. It is believed that these forces will leave shortly for the southern part of the state, with the object of opposing the advance of the Carrancista forces commanded by General Diéguez.

In other advices received from Topolobampo, Sinaloa (on coast about 200 miles S. E. of Guaymas), the arrival is reported of the gunboat "Korrigan II," belonging to the naval forces of the Carrancista government, with a shipload of troops.

From other sources it is learned that on Oct. 10 Governor Carlos Randall approved the appointment by General Francisco Urbalejo, commander of the Villista forces in Sonora, of Colonel Adolfo Islas as military commander of the District of Ures, of Colonel Fortunato Tenorio as military commander of Guaymas, and of Lieutenant Colonel Gaytan as military commander of Magdalena.

Tenorio has been commander at Guaymas on various occasions before. He came to the frontier at the time General Plutarco Elias Calles of the Carrancista forces advanced against Nogales in September.

A special telegram to La Prensa from Washington, D. C. dated Oct. 12, 1915 says:

In addition to the authorization given by the Secretary of State today to U.S. consuls to leave their posts if they deemed it best, in territory dominated by the Villista faction, and the advices given to American citizens urging them to leave that territory; the Secretary of the Navy has sent peremptory instructions to the commander of the American men of war that are in the port of Guaymas, Sonora, to be ready with their vessels to protect American citizens in case that the Villista soldiers attempt to commit any outrages against Americans. The instructions have been given as precautionary, as it is known in this capital that the forces that left Nogales for Guaymas have not arrived at that port, and no hostile demonstrations up to the present have been made by the garrison there.

A special telegram to the Express of San Antonio, dated Douglas, Ariz., Oct. 12, says:

An attack upon Guaymas, in Southern Sonora, the only west coast port controlled by Villa, was reported imminent in a message received by General P. E. Calles in Agua Prieta today from General M. M. Diéguez, Carranza commander at the head of 7000 attacking troops near Guaymas. The message said Diéguez had the gunboats "Guerrero" and "Pacific" and five armed transports (one of these is the Korrigan II).

Two thousand Villa troops were reported today by a cattleman to be moving westward along the border opposite South Columbus N.M. Another Villa force estimated at more than a thousand men was reported near Bavispe, Sonora, last night. In their vicinity were nine hundred cavalry under Colonel Carranza. There has been outpost skirmishing for three days, it was reported.

An Associated Press wire to the Express from Guaymas, Sonora, dated Oct. 12 (by radio to San Diego, Calif.) says:

The Carrancista gunboat "Guerrero" arrived off here last night, loaded with troops under General Diéguez, who has demanded the surrender of the city, which is held by Villa forces. The commanding officer of the United States cruiser "Chattanooga" informed General Diéguez that sufficient notice should be given the towns of Guaymas and Empalme before commencing bombardment to permit non-combatants to reach a place of safety. Diéguez is said to have replied that he did not think it would be necessary to bombard but that if surrender was refused he would establish a neutral zone, where non-combatants would be safe. A special train with forty Americans, mostly women and children left here yesterday for Nogales, but the train was held up at Carbo, 130 miles south of Nogales, because of nine burned bridges between Carbo and Nogales. Railroad officials said the train may return to Guaymas. There are 110 Americans remaining in Guaymas, ten of them women and children. The United States cruiser Chattanooga and the United States supply steamer Glacier are here.

W. L. Gibson

Confidential

B35-31 C.F.   1st Ind.   HHR-vve

Office Dept. Engr., So. Dept., Ft. Sam Houston, Texas, Oct. 13, 1915. – To Chief of Staff, Southern Department, for forwarding to Chief, War College Division, General Staff, Washington, D.C.

Henry H. Robert
Captain, Corps of Engineers,
Department Engineer.

5761-1043

Documents Cited

Fort Sam Houston, Texas, October 21, 1915.

Memorandum for Captain Robert: Movements of Villistas.

The Associated Press sent a message from El Paso last night which is amplified by a wire to La Prensa of San Antonio, of the same date. This last message says:

El Paso, Texas, Oct. 20. 1915. Passengers arriving here today from Chihuahua say that when they left the state capital, the railway stations were in flames, as Villa had ordered that they should be burned with all the rolling stock in the yards. The conflagration, according to the passengers (or travelers in this case), included the round house and the yards of the freight stations, for which reason it is thought that not one element that is usable remains of what was in the place. In the railroad yards there were fourteen (14) locomotives that were also destroyed.

Those who brought the foregoing news added that the Villistas are destroying, to the North as well as to the South of Chihuahua City, all bridges on the railway line, with the purpose of obstructing the advance of the Carrancistas as much as possible when they approach from the direction of Torreón. The news brought by the travelers caused a great sensation in this city, where it is believed that Villa has begun his work of destruction that he has been announcing in a dissembling way, since he got the news of the recognition of Carranza. It is known that the Villistas are trying to do all the damage possible in order to make complications for the Carrancistas. Villa declared a short time ago, that in future he would not be responsible to any foreign government, for what might happen in the territory that might be controlled by his forces.

The travelers of whom we have been speaking made a very distressing trip from the state capital (Chihuahua City) as they came in coaches, carts, and on horseback, suffering all kinds of privations and molestation.

The Associated Press reports from El Paso under date of Oct. 21 as follows:

Reports are current here that General Manuel Madinaveytia (Villista) in command at Ciudad Juarez, will remain there until the troops from Chihuahua City (under the command of General Eduardo Ocaranza) arrive, and then proceed with them to Casas Grandes.

Reports from Casas Grandes today were that General Villa had departed for Sonora and that the Casas Grandes district is virtually bare of troops. A slaughter of more than two thousand (2000) head of cattle belonging to an American cattle company was reported to have provisioned the troops for the overland march to Sonora. A delegation of Mexican and American cattlemen heavily interested in herds in Northern Mexico today urged Consul Andres Garcia of the Carranza government to make representations to General Carranza in connection with the shipment from Juárez of confiscated cattle and hides. W. L. Gibson

U.S. Army Intelligence in the Mexican Revolution

Confidential
B35-35 C.F.   1st Ind.   HHR-vve
Office Dept. Engr., So. Dept., Ft. Sam Houston, Texas, Oct. 21, 1915. – To Chief of Staff, So. Dept. for forwarding to Chief, War College Division, General Staff, Washington, D.C.

Henry H. Robert
Captain, Corps of Engineers,
Department Engineer.

<u>5761-1048</u>

Cover: Capt Clayton This is the letter from MacKinlay I phoned you about. Please return when you are through with it. Crawford.

Maj. Crawford Have read your friends section carefully but saw nothing of value in it. Clayton

McAllen, Texas. Nov. 8. 1915

Body: Dear Friend:-

I have intended to write to you ever since I landed down here, on Aug. 18th last, but have not done so. I did not get to the school on the "Big Muddy" except to look in Aug. 15th, when a telegram was handed me to get to San Antonio as quickly as possible, or there was a job for me on the "banks of the Rio Grande". So I came and am down here for the "San Antonio Light" the leading daily of San Antonio. You may know the owners + editors. Mr. Diehl formerly Lt. Col. of the 1st Illinois Inf. and Mr. H. L. Beach formerly Capt. in the 2nd Illinois. Both were on the Associated Press in Chicago for many years and both were in Cuba in '98. Diehl was in charge of the A. P. boat and Beach was ashore with the army. My work has also been in cooperation with Major W. H. Hay, our friend.

There are many factors in this so called "bandit problem," but I am convinced that it originated in Los Angeles, California, with the Mexican branch of the I.W.W. This Mexican branch is known as Magonistas, and their leader is Ricardo Flores Magón, of Los Angeles, who publishes an anarchist paper there, called "Regeneración," which has 3 pages in Spanish and 1 in English, the latter edited by Wm Owen.

About last Feb. or March, there was a Mexican arrested here at McAllen by Deputy Sheriff Desdoro Guerra as an agitator for the "Plan de San Diego" a famous document on the anarchist style, drawn up by a group of Magonistas, at San Diego, Texas. The name of this agitator was Belisario Ramos, who was taken to the U.S. Court at Corpus Christi and released by the judge who said Ramos was fitter for the lunatic asylum than for the penitentiary. Guerra then resigned as deputy sheriff in disgust + Ramos spread the news that the Americans were *afraid* to punish him. then the attack of the Villistas on Matamoros came and all the Mexicans + a lot of Americans on the Texas side did all they could to help Rodriguez, the Villista commander, take Matamoros and defeat Gen. Emiliano P. Nafarrate, the Carrancista commander there. The Villistas were defeated, however, and retreated, leaving Nafarrate bitter towards the Americans + ready to wink at injuries to them.

Then Luis de la Rosa, formerly a deputy sheriff of Cameron Co., (of which Brownsville is the county seat) took a hand, along with Aniceto Pesaña (or Pizaño) of Los Tulitos ranch east of Brownsville. They got a gang together, probably in Matamoros of "Tex-Mex" and "straight Mex." 'bad men' and commenced depredations, first coming into conflict with U.S. troops on Aug. 3d at Los Tulitos. The subsequent operations are probably well known to you on account of the feeling on the U.S. side that Nafarrate at Matamoros was conniving at the bandit's use of Mexican territory, Consul José Z. Garza of Brownsville, with whom I am well acquainted, recommended to the Minister of Foreign Affairs, Acuña, that Nafarrate be relieved and he was *promoted* on Sept 22 last to be "general of brigade" and relieved a week or so later by his subordinate "general brigadier" Eugenio López de Lara, formerly a milkman at Matamoros and now only 24 years old. Nafarrate is now at Ciudad Victoria, the capital of Tamaulipas, where he is temporary commander of the Fifth or Tamaulipas Division of the Army Corps of the Northeast, of which General Jacinto B. Treviño is the commander.

Nafarrate and López alike are very "Mexicano" and like all ignorant men, are trying to show that they don't fear or need to respect "foreigners." In other words, they are "playing to the gallery" as good as any American politician can do it.

Since Carranza sent out orders for the local garrisons to get busy after "bandits" the love feast has begun. Colonel Eutemio Cantu, the officer in charge at Reynosa, just across the Rio Grande from Hidalgo, Tex., sent an officer over Saturday night to see Lt. Waiver of the 28 Inf. at Hidalgo, and I think a lot of good will come of it. And it would be of great advantage to the U.S. officers if they were allowed to return these calls, now that the Carranza government has been recognized as the *de facto* one in Mexico. Men like F. R. McCoy at Mission, Waiver at Hidalgo, Major J. W. Carter at Ft. Ringgold and a lot of others up + down the river, can get up a system of joint patrols and co-operation, that cannot be arranged without personal contact. You can do more in a half hour over a bottle wine + a cigar with these Mexican officers than by a month of *notes*. If it can be done, an order rescinding the prohibition for U.S. officers to cross the line would be of great effect and much good can be accomplished. Of course this is just a personal letter to you for your information. I think López has 1700 in his "brigade" all told- divided into 5 "regiments" 1 corps of "fiscal guards" and 1 corps of "Rurales." The stations are Matamoros, Lt Col. E. Benavides, Comdg. 2. *Reynosa* (Hidalgo, Tex.) Lt Col. Eutemio Cantú. 3 Camargo (Rio Grande City + Ft. Ringgold Tex) Col. Marciano González 4 Mier (Roma Tex) small garrison 5 Guerrero Tamps. (Zapata Tex) small garrison

U.S. Army Intelligence in the Mexican Revolution

It is hard to figure out how the internal situation of Mexico is. If Carranza is loyally supported by Obregón Treviño and González, he will probably win, although Villa has not been finished as yet. The financial problem is the big one for Carranza to get over. The possession of Vera Cruz and of Yucatán are two big factors in his favor. I will try to keep you in touch with any new things that develop. Remember me to Major Howze + Major Rowell. I inclose an addressed envelope so you can reach me by mail quickly. Hope someday to see you and have a talk over things as ever.

Sincerely yours,

W. E. W. MacKinlay

Don't forget to use my *nomme de plume* (William L. Gibson)

5761-1052

Note: The following translation was received at Office of Department engineer, November 26, 1915, from Mr. W. L. Gibson, envelope in which it came being postmarked "Brownsville, Texas."

The following is a translation of the report of General Alvaro Obregón to First Chief Venustiano Carranza, omitting formal address. Agua Prieta, Son. Via

Gen. Hqrs. "Torreón, 17th. November, 1915.

The bandit Villa, after his defeat in this place (Agua Prieta) has concentrated his forces in Naco and Cananea where he is daily losing about two hundred men by desertion.   November 10 one hundred and sixty-one (161) Villistas presented themselves to our forces, turning in their arms. Two of them were colonels, the rest enlisted men. The Maytorena forces that are in Nogales have refused to obey the orders of Villa and I have reason to believe that trouble between them may take place.   I am getting a column in shape to engage the reactionaries, in case that they do not surrender.   November 6 Hermosillo was occupied by the forces of General Manuel M. Diéguez, who captured war material and rolling stock of the railroad.   Telegraphic communication with Hermosillo has been opened via Tepic and Mazatlán. November 4 General Plutarco Elias Calles sent me the following telegram: The enemy has made a general retreat with his forces in front of this place (Agua Prieta). His present place of concentration is Naco.   There have been picked up three hundred and sixty dead (360), one hundred and ten prisoners (110) have been made, and I calculate that his losses from desertion have been considerable, as numerous bands have left, some of them presenting themselves to the authorities on the American side of the line.   At this moment, cavalry forces are going out for reconnaissance and to pick up stragglers, and also to get the wounded men left by Villa, abandoned, at Gallardo.   According to advices of the consul at Naco, two hundred (200) wounded Villistas have arrived here. The Zapatista general, San Román, was found dead on the battlefield, and I am informed that other generals were wounded, among them Orozco and Arroyo.   I can

234

assure you that the attack of Villa on Agua Prieta has been a real disaster for him. The victory is yours, because your pupils have been the victors."

Durango State—The forces of General Domingo Arrieta, on November 3 defeated the enemy, commanded by the rebel leaders, Maximo Garcia, Gandara, Liramontes and Sarabia, at the hacienda Los Pinos causing them losses of more than a thousand, and capturing eight machine guns, arms, ammunition and two loads of provisions from them. The enemy came from Canatlán (on the Tepehuanes branch of the Mexican International R.R.), where he burned the railroad station, and cars filled with goods stolen from the merchants of Durango. General of Division Francisco Murguía has organized various flying columns with which to pursue resolutely the various bands of rebels marauding in different parts of the state, until they are exterminated. A band of rebels under Calixto Contreras having come down to the railroad line between Torreon and Durango with the design of breaking it at Pedriceña, a force of infantry and cavalry composed of twelve hundred (1200) men commanded by General Miguel de la Vega left Gomez Palacio in order to engage the reactionaries in their strongholds of Cuencamé, Nazas and Peñón Blanco. Forces from this column occupied Ocuila and Cuencamé, after defeating the bandits who were found in those places. The line of the railroad was again opened on November 13. The repairs on the telegraph line are being hastened. Preparations are being made for energetic operations against the bandits who may be at Nazas, by forces from this column. General Jacinto B. Treviño, commanding the Army Corps of the Northeast, has made the following report of the occurrences in the states where forces of his command are operating:

November 3—Four field officers, twenty-three officers and one hundred and thirteen men of the Villista forces surrendered to the forces of General Domingo Arrieta, with twenty carts, 105 Mauser rifles, and other war materiel.

State of Sinaloa—Nothing to report

Territory of Tepic—Nothing to report

State of Jalisco—The enemy, under the command of Bañuelos, Caloca, Parra and others, was engaged while retreating by General Novoa, at Providencia, on November 5, obliging them to fall back to Magdalena, where the column of General Ramon F. Iturbe, pursuing them compelled the enemy on November 6 to evacuate Magdalena and retreat towards Hostotipoquillo. At this place the enemy was obliged to give combat, and was attacked by the combined forces of Generals Iturbe, Amaro, Espinosa, Cardona, Novoa and Colonel Fernandez de Lara; the losses of the enemy being 80 killed, and 12 prisoners, together with two cannon, arms and ammunition. The enemy was pursued at once through Ahualulco by the cavalry of General Amaro, with a loss of 22 killed, and losing the same number of arms, together with 25 horses that were captured from them. The reactionaries in complete rout, fled through the ravine of San Pedro Analco. Rebel leaders have surrendered at Ameca with armed and mounted troops, taking advantage of the amnesty. Communication between Jalisco and Tepic has been reestablished, the line being protected by forces of General Pedro Morales. The troops of this officer have been reinforced by state troops as well as those of Generals Amaro and Torres. An active campaign for the annihilation of the reactionary bands still operating in the states of Jalisco and Colima has begun. Part of the forces of General Amaro have left Guadalajara in order to accelerate the campaign against the bandits in Colima who left Jalisco on being pursued. Later a campaign will begin against the bandits that may be found in Guanajuato state. Forces of General H. Alvarez, under the command of Lieutenant Colonel Severo Rodríguez, have defeated some bandits near Santa Regina, chasing them towards the Hacienda Providencia in which vicinity they were again defeated, the forces of Colonel Tranquilino Mendoza defeating the leader Lorenzo Ortiz, near Salvatierra, capturing arms, horses and ammunition. The same band is being pursued towards Valle de Santiago. Colonel Tranquilino Mendoza also engaged the leader, Buenaventura Maldonado at Tarimoro, capturing ammunition, arms and important documents. The leader Maldonado, was killed; together with two so-called captains and five soldiers.

State of Michoacán—Nothing to report.

State of Aguascalientes—The governor, General Triana, has circulated the decree of amnesty issued by these general headquarters, throughout the state, and some rebels have come in at once to surrender, turning in their arms. The entire state is quiet, and under the control of our forces. Forces of Aguascalientes under the command of General Vidal Silva, have defeated the rebels encountered at Tlaltenán, Colotlán (Jalisco) and Villanueva (Zacatecas), dislodging them and obliging the surrender of the so-called Colonel C. Jaime, with his officers, Salvador P. Huisar, Simon Gonzalez and David Avila with three hundred men, who will come to Aguascalientes to be paroled in conformity with the amnesty decree. The surrendering leaders have stated that the rebels who are in the Juachila canyon, are not able to leave that region, on account of the very bad condition of their horses, which are used up and unshod. Many of the men of these leaders, Bañuelos and Caloca, have surrendered, and the rest are negotiating for their own surrender.

State of San Luis Potosi—General H. Alvarez concentrated all his forces in the city of San Luis in order to begin a campaign against the reactionaries commanded by Carrera Torres, the Cedillos, Argumedo, and Almazán, who are in various places in the state, and on the line of the railroad from San Luis Potosi to Tampico. Rebels under the command of Argumedo, Almazán, and Cedillo have been engaged at Tablas Station by the vanguard of these forces, and were defeated with heavy losses in killed and prisoners, some of the rebels flying towards the north and others towards the south. The column under the command of General Alvarez is continuing its advance in order to engage the reactionaries who are in possession of some points near and upon the line of the railway to Tampico. Forces of the Lárraga brigade under Colonel Olvera in combination with forces of General H. Alvarez, occupied Conoas Station Nov. 3 pursuing the enemy to the mountains. Forces from this same brigade (Lárraga) pursued the bandits to the Sierra of Tamasopo. They got as far as Verástegui Nov. 3, and afterward as far as Espinosa, working in combination with forces of Soberón. The line from San Luis Potosi to Tampico and from San Luis to Cárdenas again was open Nov. 9, the enemy, who occupied fine positions in the canyon of Guerrero, being dislodged from them by forces of the Lárraga brigade, the same who co-operated in the engagement at Tablas against the reactionaries of Argumedo and Almazán. General Rosales Cuellar in his advance upon Ciudad del Maíz, recovered 104 Mauser rifles from the enemy, together with ammunition.

State of Coahuila—Nothing to report. The entire state is dominated by our forces.

State of Tamaulipas—Nothing to report. The entire state is dominated by our forces.

State of Nuevo León—Nothing to report. The entire state is dominated by our forces.

State of communications—The lines cited in the former report are in working order.

Respectfully,

ALVARO OBREGON.

Commanding General

5761-1054

Brownsville, Texas, Dec. 12, 1915.

From absolutely reliable source following information learned:

Large force of "Conventionist and Carrerista" troops, estimated at from 3000 to 6000 attacked Tula, Tamaulipas, Nov.15, the fighting continuing until the 17th. Only the scarcity of ammunition and the bad condition of the men and horses prevented the defeat of General Manuel [Miguel] Zapata of the Carranza forces, who defended the place. The enemy at last fled to the northwest, crossing the Nuevo Leon border, and attacked the town of Doctor Arroyo. Repulsed there, he turned southwest and attacked Matehuala; repulsed there, he marched northwest and occupied Parras, Coahuila 150 miles distant, his forces there being driven out Nov. 27 by Gen. Luis Gutiérrez and other forces. The last report was that they were in the mountains making their way towards the United States, the "Generals" remaining together with a small escort and some wounded men. The main force seems to have split in every direction. One group reported to north of Aldamas, Nuevo Leon, is said by Ricaut's information office to be from this defeated force. Other advices via Reynosa where more of the so called "bandits" have friends and sympathizers than almost any other place, say that Luis de la Rosa and perhaps Aniceto Pizaña are with this Aldamas crowd. General Ricaut's man said that General Ricaut was looking this crowd up on his inspection trip.

My informant arrived at Matamoros yesterday by automobile from Ciudad Victoria. He says the road built by Governor Arguelles some years ago is still fine, although for six years it has had no attention. He said that although it is through the hills, that from Victoria to Soto la Marina on the river no trouble was experienced. At Soto the oil boats are running from the Barra, the river having 20 feet of water if the bar can be crossed. From Soto la Marina to Matamoros the road follows the coast, and is now good, at places he ran 30 to 40 kilometers without slacking speed. He crossed the Rio de San Fernando (Conchos) below its junction at Adjuntas with the Rio Presas on a ferry. He says business is good at Victoria. General Luis Caballero, who is governor of the state and also commander of the 5th Division of the Army Corps of the Northeast. (J. B. Treviño's), is at Victoria. General of Brigade Emiliano Nafarrate, with the 1st Brigade, and General Brigadier Eugenio López de Lara with the 2nd Brigade, both of 5th Division, are at Ciudad Victoria. They may go to Tula or to Tampico and thence to Cerritos, San Luis Potosí. Colonel Eutemio Cantú is with López. He was formerly commander at Reynosa, opposite Hidalgo, Tex. Colonel Alonso Velazco, commander at Reynosa at height of bandit troubles, is in San Luis Potosi. Colonel Procopio I. Elizondo was lately at Tula. Colonel Pedro A. Chapa is at Tampico. He has some knowledge of aviation. It is said Alberto Salinas, who claims to have flown the new Mexican made aeroplane "Anáhuac" over Ajusco mountain, 10,000 feet above sea level, let his imagination loose in making his report, and that he did not go into the Zapata region as he claimed. He is a brother of Gustavo Salinas, who went to Agua Prieta with Obregón as chief of artillery. Colonel Jesus Ramirez Quintanilla of Guerrero, Tamps. now colonel of the 19th Cavalry Regt. is commander at Matamoros. He is about 55 years old and appears to be a man of steady character and good common sense. One of his officers is Captain Carlos Belloc of French descent, who speaks some English.

W. L. Gibson,

Brownsville, Tex

Dec-11-15

[Penciled in: "My informant is a leading business man of Ciudad Victoria"]

[Penned in attached: "Recd by *La Republica* from Laredo, Tex (Villista source) says that "Carrera" forces still hold Doctor Arroyo, in southern Nuevo León, and yesterday or day before cut the National R. R. at Vanegas, S.L.P., and one attacking a small Carranza garrison at Matehuala. Will send more about this later. W.L.G."]

5761–1057

Brownsville, Texas, Dec. 15, 1915

Dear Friend:

I have seen a telegram purporting to be from Monterrey saying a new conspiracy is being formed with Hq. there at Monterrey and that 600 men on both sides of the border are implicated. I will send you some letters I expect in a day or two from there and Laredo. The Villa and Huerta crowds here on the border are naturally very "hot" over the recognition of Carranza and will do all they can to keep rumors alive of possible trouble. Now it must not be forgotten that practically everyone down here wants all the troops (with the pay day attachment and detachment of said pay) down here indefinitely. Also a lot of gentlemen who enjoyed themselves in Aug + Sept during the "open season" for Mexican speaking residents are now wondering what might happen if too many guards left, of course, as I heard two local gentlemen remark "all the soldiers are good for is to shoot *our* quail" but they still like the pay day of the "quail killers."

At last account *Luis de la Rosa* was at or near Jiménez, Tamaulipas, between San Fernando and the R.R. from Victoria to Tampico.

I don't think Luis Caballero, comdg 5 Div. nor Nafarrate nor López will try to get him. Maybe Ricaut can get him by some stratagem into his 'zone.'

Things are fairly quiet as far south as the S.L.P./Tampico line, which the Carreristas are trying to break as often as possible. They seem to be holding the town of Doctor Arroyo in Southern Nuevo León and raiding from there.

I am satisfied Ricaut is doing all he can to help, also Carranza has a very good I.O. [Intelligence Officer] in Daniel Gómez La Madrid, a man of 40 yrs old, tall, black moustache, white, formerly commissioner of 6th Police Dist. Mex. City. He was at N. Laredo + Laredo Monday. Left yesterday for Saltillo to join V.C. They go to S.L.P. thence perhaps Aguascalientes Guadalajara etc.

I think Jacinto B. Treviño has 8000 to 10,000 men from Escalón to Santa Rosalía, Chih, via Jiménez about now. Villa will try to hold him with 4000 to 5000 at Bachimba pass about 30 miles. S.E. of Chihuahua City.

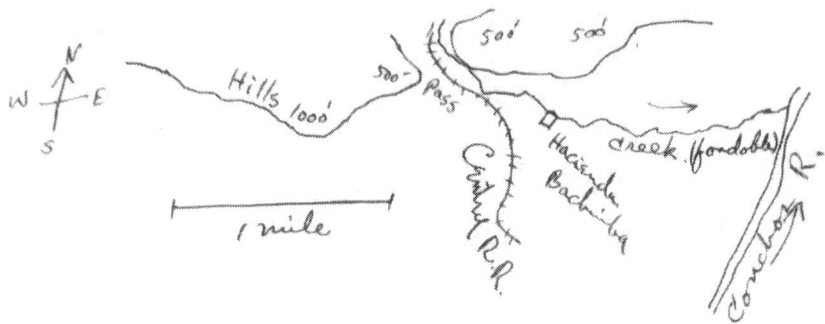

Map 3.1 Hand-drawn Map of Bachimba

This pass can easily be turned by following the Conchos for some 20 miles then turning to N.W. + going via Santa Eulalia to Chihuahua City. I was at the engagement there July 3, 12, so was Capt W. A. Burnside, 26th Inf, now at Kingsville, Tex.

Santa Rosalía is the name of the R.R. station. The official name of the town is Ciudad Camargo.

Will write again soon as I hear from some of my friends. Only small garrison. 19th Mex. Cav Col. J. R Quintanilla, across river. Almost no men in hospital there.

as ever yours truly, W.L.G.

5761-1058

The following is memorandum to Captain Robert from Mr. Gibson, received December 20, 1915:

(Carrancista source)

Saltillo, Coahuila, Mex.,

14 Dec., 1915

A few days ago reports were received here of a tremendous defeat suffered by Benjamín Argumedo at the hands of a Constitucionalista force, at the Hacienda de la Gruñidora, situated southwest of Mazapil, state of Zacatecas. Argumedo's losses were about 800 in killed, wounded and prisoners. He also lost 6 machine guns, 100 saddled horses, the great part of his impedimenta and about all his ammunition. The prisoners have just arrived at this place and are:

General Federico Tafolla (civil engineer),

General Manuel Pedere,

Colonel Louis Ocampo,

Colonel Pedro Villavicencio,

Colonel J. Antonio Herrera Falomir,

Major Manuel Domínguez,

Major Antonio Ruiz,

1st Capt. Pablo Mojarre,
1st Capt. Miguel Balderrama,
2nd Capt. Manuel Rodríguez Chao.,
2nd Capt. Isaac Segura,
Lieut. Jose Maria Monroy,
Lieut. Alberto Valencia,
Lieut. Anastasio López,
Lieut. Eduardo Meraz,
Sub-lieutenant Juan Davila,
Sub-lieutenant Antonio Davila,
1st Sergt. Federico Gomez,
   "   Miguel Romero,
Paymaster Vicente Macias,
   "   Herón Olvera,
Storekeeper Marcos Lezama,
Steward H.C. Emilio Valadez.
- - - - - - -

Civilian Prisoners:
Atty. Francisco Encinas,
   "   Wenceslao Urbina,
   "   Porfirio Flores,
C. Engr. Pedro J. Jauregui,
   "   Domingo López,
   "   Porfirio Parra.
Clerks Manuel S. Garcia,
   "   Leopoldo Reyes,
   "   Gonzalo Alverde,
   "   Manuel Jauregui,
   "   Leopoldo Freg,
   "   David Olives,
   "   José Quiroz,
   "   Jose Zamacona,
Chauffer Ramón Grifel.

From information given out at the headquarters here it is learned that both the military enlisted prisoners will be at the disposition of the military authorities of Parras, by whom they will be tried.

The officers and civilians will be tried by court martial at Saltillo and punished according to the rank and station of each prisoner.

- - - - - - - - - - - -

U.S. Army Intelligence in the Mexican Revolution

(Clerical source) City of Mexico, Dec. 10, 1915.
correspondence

It should be noted that in spite of the censorship existing in official circles as to news about the revolution that is devouring the country, that the newspapers of the capital are not denying the version, going about for some days past, about the advance of the troops of Argumedo towards the frontier of the Republic, but on the contrary they are commenting in diverse ways, but always give the matter importance and consider it of some gravity. "El Democrata" for example, says Argumedo knows the Laguna region (around Torreón) perfectly, and that it would be lamentable if, for a lack of care by the Carranzistas, that the brave ex-federal should attain that stronghold, as it is certain that he would there display an astonishing activity and it would be more difficult for Carranza to dominate him. "We must desire," says El Democrata, "That Argumedo shall be stopped in Zacatecas, where he is operating at present, because his defeat at Gruñidora (in the south part of the state) may lead to the breakup of his forces. And it goes on with more or less impassioned comments, affirming that the losses suffered by Argumedo in La Gruñidora amount to more than two thousand men, which is about admitting, tacitly, that the column commanded by Argumedo is more than ten thousand strong."

- - - - - - - - - - -

Benjamin Argumedo was a tailor in Torreón. He joined the Maderistas in 1910 and was a Colonel in March, 1912. He joined Orozco then and became a "general." Upon Orozco's joining Huerta in February 1913, Argumedo became a general of federal irregulars, went south and landed in Oaxaca, and about October started north through Hidalgo, as far as Tula, in Tamaulipas (Nov. 15-16-17). Then went west, Mazapil is in north east part of Zacatecas. La Gruñidora is S.W. of Mazapil, Lat. 24° 30' N, Long 102° W.

Confidential

B35-50 Dept. Engr.   1st Ind.   HHR-vve

Hdqrs. So. Dept., Ft. Sam Houston, Texas, Dec 20, 1915. – To The Chief, War College Division, General Staff, Washington, D.C.

Malvern Hill Barnum

Major, General Staff,

Chief of Staff

In the absence of the Department Commander.

5761-1059

Fort Sam Houston, Texas,

December 20, 1915.

Documents Cited

Memorandum to Captain Robert:

Herewith extracts from a letter to me from Mr. Gibson, which accompanied the memorandum handed you this morning. The following notes may be of interest as showing a few "side lights" on some of the personalities.

W.K. Adams

"Brownsville, Texas, December 16, 1915.

My dear Mr. Adams:

Well, I inclose a memo. It shows Argumedo is still alive. They must have had a good fight at La Gruñidora. He was evidently headed for the Laguna and will probably arrive. Treviño seems to have taken all he could of the Central to Santa Rosalía. He claims 12,000 there and 22,000 all told. Quien sabe?

Villa has Fidel Avila and Limon with him at Chihuahua. Whether they will make a stand at Bachimba is yet a problem. I see a column is going from Piedras Negras via Eagle Pass to Marfa, thence *overland* to Ojinaga and "on to Chihuahua." Another column has or is going in at Palomas across the border from Columbus, New Mexico. Máximo García surrendered to Treviño the other day at Corralitos, between Rellano and Jiménez, and has been added to the "happy family."

Gens. Urbalejo and Trujillo of Villa's forces have surrendered in Sinaloa but "Generals" José M. Acosta, Fructuoso Méndez, Ramón Sosa, Fortunato Zenorio, Juan A. García and some others are still "Luchando." It is also said Eduardo Ocaranza, Gen. Delgado and Ignacio Morelos Zaragoza, our "friend of Tampico," are also with Villa's forces yet, but Arnoldo Casso López has quit, he was a pretty good fellow in Morelos even if he did get "good and licked." I saw the entertainment "some of it I was and all of it I saw." José Rodríguez seems to be at Casas Grandes and Ochoa is still on deck at Juárez. Ricaut is Carranza's nephew. He is stout, dark, about 35, and 'hits the booze' fairly often. Colonel Jesús Ramírez Quintanilla, the Jefe de Armas at Matamoros is from Guerrero, Tamps., a man of 55 or 60, a good quiet big blonde man. Carlos Belloc, his secretary, is a Frenchman by descent and a Baptist. Major Flores at Reynosa I don't know well. Col. Marciano González of Mier is from Cerralvo, N.L. He talks English well. Quite a friend of Capt. McCoy at Roma, Texas. Everything is "mucho amigo" now except the rangers are pretty sullen over not being allowed on or near the river, ditto deputy sheriffs, Nafarrate ditto, everyone else pleased."

*   *   *   *   *   *   *

6269-7
Headquarters Second Brigade.

Laredo, Texas, Nov. 6, 1914

From: Commanding General.

To: The Commanding General, Southern Department, Fort Sam Houston, Texas.

Subject: Aviation equipment of General Carranza.

1. Mr. C. F. Niles, Aviator, is now in Laredo on his way to the City of Mexico. He is employed by General [sic] Carranza as his Chief Aviator and gives me the following information, to the effect that General Carranza has the following aeroplanes in the City of Mexico:

Two Deps. 80 – H.P.;

One 50 – H.P. Moisant-Bleriot;

Three 80 – H.P. Moran's Monoplanes; and

One 70 – H.P. Farman Bi-plane.

2. The 80 – H.P. Moran's cost $8200 each. He (Carranza) has ordered 1 Curtis Bi-plane and one Kirkham Tractor Bi-plane, to be delivered in six weeks. Carranza has spent in the last six months about $70,000 for aviation material. He has one carload of dynamite bombs, six for each machine in each flight, in Mexico City. His entire investment for aviation in the last year is about $300,000 for hangars, and all equipment.

3. Mr. Niles is taking 1 carload of spare parts, and 1,000 gals. of special gas, to the City of Mexico. Common gas cannot be used with the planes. Mr. Niles has all the machines in Mexico City, including four specialy equipped aviation cars.

4. Mr. Niles is one of the best known and daring aviators in America. I have known him for some time, and have seen him giving flying exhibitions. He tells me that he is employed by Gen. Carranza as his Chief Aviator. I know that he has made several trips to Mexico lately. I believe his statements to be reliable. While Mr. Niles is perfectly willing to give this information for the use of the War Department, yet, for obvious reasons, he requests that it be held as confidential. Mr. Niles is here for a few days getting through a carload of spare parts for aeroplanes and special gasoline. He intends to go to the City of Mexico via Monterrey and Tampico.

R. K. Evans,

Brig. Gen. U.S.A.

8529-12

REFER TO FILE
NO. 141-CF--DE

HEADQUARTERS SOUTHERN DEPARTMENT,
OFFICE OF DEPARTMENT ENGINEER,
FORT SAM HOUSTON. TEXAS.

March 25, 1915.

From: Department Engineer.
To: Chief, War College Division, General Staff, Washington, D.C.
Subject: Information on Mexico.

Documents Cited

1. The following is brief memorandum of information as given by Mr. A. J. Ruckman, Manager, Mexican Coal and Coke Company, who came from the mines at Las Esperanzas, March 22, 1915, after a stay of about one week there, having made the trip by automobile. Ruckman's connection with this affair should be kept especially confidential as he expects to return to Mexico from time to time.

2. From Eagle Pass he followed Rout 1A to Sabinas thence crossing the Sabinas River at the ford near the railroad bridge, thence he followed a road, not shown on the maps, alongside the railroad for about two kilometers, passing Hacienda Soledad. This road is on the west and runs almost directly to Las Esperanzas.

Returning, Mr. Ruckman traversed Route 1 to San Juan de Sabinas, thence via 1A to Sabinas and Eagle Pass.

The road over which he passed is in fine condition, the auto making 25 miles an hour throughout. Mr. Ruckman says that "if the road via Morelos is any better than this, it must be a dandy."

3. Alamos Arroyo, which crosses Route 1 between Rosita and San Juan de Sabinas and Route 1A a little west of Cloete, is crossed at Cloete by means of a ferry-boat. Mr. Ruckman understands that there is flowing water there at all times. The Arroyo was fordable for autos, the water being about two feet deep there he crossed. This arroyo is wild and treacherous during the rainy season, which usually starts in April. It drains a large area in the Burro Mts. to the northwest of Sabinas.

4. Villistas are now in full control of the district and the railroad is in operation from Asarco and Durango to Sabinas via Torreón. The track was badly damaged during the retreat of the Carrancistas, especially north of Sabinas but is being rapidly repaired. Telegraphic communication is now open from Eagle Pass southward.

5. On their retreat, Gutiérrez adherents dynamited the Cloete, Rosita and Agujita coal mines and burned all the bridges on the Múzquiz Branch of the Railroad. These bridges are now being rebuilt with material furnished by the Villa forces, and brought from the Durango district.

6. Villistas are threatening to confiscate the mines of the Mexican Coal & Coke Company at Las Esperanzas and the Coahuila Coal Company at Palau, which are the only ones now in condition to operate. Mr. Zamora, General Manager of National Railways under Villa control, tried to make contract with Mr. Ruckman for coal from these mines. Mr. Ruckman refused to contract, owing to doubt of the permanency of Zamora's jurisdiction but offered to operate the mines for the benefit of the Villa Rys. on a royalty of 12½ ¢ gold per ton. Later he made an offer at a fixed price. Both offers refused by Zamora who made an offer of $1.50 gold and $3.00 Mex., which Mr. Ruckman accepted verbally subject to confirmation by the New York office. This means a direct loss to the Company of about $1.00 gold per ton or perhaps $30,000 per month but seems preferable to confiscation which means the ruin of the mines on account of

improper methods of removing coal, etc. It is understood that confiscation is the only alternation offered although the Villistas are hesitating to go to such extremes just at present.

7. The Asarco smelter, located at the station of the same name on the line Torreón-Durango, and owned by the American Smelting & Refining Co., is in operation but badly hampered by inability to secure a steady supply of coke, which they ordinarily get from the mines at Las Esperanzas.

8. Mr. Ruckman confirms statements by other parties to the effect that the anti-American feeling among the people of northern Mexico is increasingly evident and considers their attitude as contemptuous, insulting and overbearing.

9. With reference to the foregoing information from Mr. Ruckman, Mr. Walter K. Adams, former bridge engineer, National Rys. of Mexico, who was present at interview, states as follows:

> (a) With reference to ford mentioned in line 2, 2nd paragraph, page 1, "A good ford, ordinarily about 18" deep, but with a very steep approach on the north side, bottom hard, gravelly, water about 150' wide."

> (b) With reference to "Hacienda Soledad" mentioned in line 6, 2nd paragraph, page 1 "The Hacienda Soledad is located at Mesquite station, not Soledad station, although the latter is on the same property, which belongs to the International R.R. of Mexico."

> (c) With reference to 3rd paragraph of paragraph No.2 "The auto travel appears to be almost entirely by Route 1A so far as Sabinas and the south are concerned. South of Allende, there are two roads, one about two miles west of the railroad, which is perhaps a little better than the one close to the track but a little longer."

Henry H. Robert
Captain, Corps of Engineers.

8529-13

Eagle Pass, Texas.
May 3, 1915.

From: Intelligence Officer, Eagle Pass, Texas.
To: The Department Engineer.
Subject: Rain fall.

1. Replying to 204-CF

2. I have seen a number of travelers from Mexico. They report that during the last month the streams near the railroad as far south as Monclova have been several feet high, some of them overflowing their banks. I am informed that just as soon as this

exceptionally rainy season has ceased the streams will rapidly go down. No important washouts on the railroad are reported.

3. The only change of permanent nature likely to be brought about by the recent rains will be the filling of the tanks to a higher level, thus increasing the water supply.

The actual amount of increase in tank water cannot be ascertained at present but will be reported at a later date.

John W. Wright

Captain, 17th Infantry

Acting Adjutant.

204-D-C.F.   1st Ind.   HHR-waf

Office Dept. Engr., So.Dept.,Ft.Sam Houston,Texas,May 4,1915.- To Chief of Staff, So. Dept., for forwarding to Chief, War College Division, General Staff, Washington, D.C.

Henry H. Robert

Captain, Corps of Engineers,

Department Engineer.

204 D, D.E.   2d End.

Hq. Southern Dept., Fort Sam Houston, Texas, May 5, 1915.

To Chief War College Division General Staff Army War College, Washington, D. C.

W. H. Hay

Major, General Staff,

Chief of Staff.

WHH-HS

8529-14

Maps forwarded as enclosures to this report filed in Map Section as Map No. 258=Mexico

223 C.F.

HEADQUARTERS SOUTHERN DEPARTMENT

DEPARTMENT ENGINEER,

FORT SAM HOUSTON TEXAS.

May 19, 1915.

From: Department Engineer.

To: Chief, War College Division, General Staff, Washington, D.C.

Subject: Information on Mexico.

U.S. Army Intelligence in the Mexican Revolution

1. Herewith certain water supply and road notes compiled for this office by Mr. Walter K. Adams, former bridge engineer, National Rys. of Mexico, and Doctor W.E. Quinn, former resident of Monclova. Doctor Quinn has just returned from trip to Monclova by rail. He reports road from Piedras Negras to Monterrey in operation by Villistas and reports further that arrangements are now being made for concentration for various Carrancistas commands, including that at Las Vacas, for the purpose of defeating in detail various Villistas commands between Piedras Negras and Monterrey. He reports the first movement from the troops at Las Vacas will be to proceed to Allende, and begin destruction of railroad in that vicinity.

2. Attention is invited to the fact that these notes are in themselves incomplete and must be considered in connection with other information on file at the War College pertaining to the same locality. Paragraphs followed by letter "Q" in parenthesis contain notes by Doctor Quinn. Information followed by letter "A" in parenthesis contain notes from Mr. Adams. These notes are to accompany map C.F. 1-B 3-2 – Portion of State of Coahuila, 1915, by T. S. Abbott - on which information shown in red has been added by Mr. Adams based on his own information and from notes furnished by Doctor Quinn. Attention is invited to the following note on map:

> "Information shown in red hereon was added by Mr. Walter K. Adams at office of Department Engineer, Southern Department, Fort Sam Houston, Texas, May 13, 1915, from his own observation and from information furnished by Doctor Quinn. This information is not positive in every particular and must be considered in connection with other corresponding information on file at War College."

> Henry H. Robert
> Captain, Corps of Engineers.

LAS VACAS TO CLOETE:

Road south from Las Vacas to Cloete. Advantage more water found than road from Piedras Negras. Road runs almost directly from Las Vacas to San Juan Sabinas. First water is encountered about 18 miles south of Las Vacas flowing from the foot hills and being the headwaters of San Diego River. From headwaters of San Diego River to Remolino is about 20 miles, water being found at Remolino.

Goodman's ranch is to west of main road and there is a road between headwaters of San Diego leading to westward to Goodman's ranch and from Goodman's ranch to Remolino in addition to main road to Remolino. The advantage of going to Goodman's ranch and not direct south is to escape large gravel bed in main road. (Note: This evidently refers to the Ranch Headquarters, the exact location of which is not indicated "X") Water at Remolino is from San Rodrigo River. Main road passes west of Remolino.

Proceeding south, are encountered the headwaters of the San Antonio River at Barrancas. Distance from Remolino to Barrancas is about 15 miles.

Proceed southward 18 miles from Barrancas to Macho where there are wells and tanks and a dry creek. From Macho a road branches to Zaragoza.

There is only a dry creek at Macho and probably no water is obtainable by digging in its bed. From Macho the road runs south to Cloete via San Jose Ranch about 18 miles when there is water (in tanks). The distance from Macho to Cloete is about 30 miles and pools of water are found in rainy season only. From Cloete across Arroyo Alamos to San Juan Sabinas, or to Sabinas station, thence to Barroterán where there is only tank of water. Then follow road (Route 1) to Monclova, Hermanas, Abasolo, Monclova. Alternate route to west of main road, route 1. The road from Hermanas to Abasolo is very bad at present. Route is from Hermanas to Rodríguez to Abasolo Viejo.

It is now necessary to go from Hermanas to Rodríguez to New Abasolo Viejo.

It is now necessary to go from Hermanas to Rodríguez to New Abasolo thence to Abasolo Viejo and Monclova.

"Q"

From San José's Ranch, notes now on hand, in connection with Lt. Seoane's notes of Route 1, will apply. An alternate route, however, follows:

Advancing eastward from Cloete; along the Sabinas River to Sabinas, about one Quarter mile west of Sabinas station, proceed southward thro. center of town to a good ford across the Sabinas River. The north bank is clay and gravel, and quite steep. The south bank is on a gradient of perhaps 5% and also clay and gravel. The crossing is about 350 ft. wide, hard gravel bottom, in ordinary water passable for automobiles. The shallow water is over a gravel ridge curving downstream somewhat from the entrance to the stream. Bearing rapidly upstream from about the middle of the channel to the road on the south bank the main channel is on the south side.

From the Sabinas River, proceed southward on the west side of the railroad about one mile, passing Mesquite Station and the Soledad Ranch houses and corrals which are on the east side of the track. The road diverges to the westward a little south of Mesquite, running thence almost directly to Las Esperanzas, a distance of approximately 18 miles. There is ordinarily no water between Sabinas and Las Esperanzas. The road is used to some extent by autos and although entirely unimproved, may be considered a good road. Two small arroyos are crossed, usually dry, and passable almost anywhere. The road crossing of the more southerly one is steep and narrow with a good deal of large gravel and boulders.

Water at Las Esperanza is pumped from the mines of the Coahuila Coal and Coke Co. The tank here has been described elsewhere.

Southward from Las Esperanzas a good road passes close to the foot hills thro. old Barroterán and Aura; joining Lieut. Seoane's road at Obayos. This road is good all the way, and is probably as good as the route over Boca del Aura and Puerto de Obayos besides being much shorter. The river crossing at Aura is direct and cannot be missed. Ordinarily there is no water on the surface except in pools here and there and in an irrigation ditch on the north side. This is understood to go completely dry at times, but it is though [thought] that water can always be obtained by digging shallow wells in the river channel,- probably near the south bank. It is doubtful if water will be found in the arroyo de Obayos. The road crossing is not known but the arroyo can be crossed almost anywhere in the vicinity, easily at the railroad track about one half mile east although there is no road there.

Arroyo Barroterán usually carries a small quantity of water, exact amount not known. About one mile east of the road on this arroyo is located the railroad reservoir described elsewhere. A passable road leads from the road crossing to the reservoir on the north side of the arroyo.

This road is practically parallel to the railroad, averaging about one and one half miles to the westward from Barroterán to Lampacitos.

Especial emphasis must be imposed upon the fact that water from Salado River at Hermanas must not be used by either men or animals as it will cause severe colic at least. The water from the hot springs here is usable.

"A"

Monclova to Paredón

Follow road shown to Reata from Reata to Sauceda along railroad at Sauceda turn eastward going around point of mountain, follow Mexican Central R.R. to Paredón. This is a plain road and the only one. Some water will be found in rainy season and some small streams and springs are permanent. It is a very bad road. From Paredón to Monterrey road follows Patos River. This road cannot be missed and crosses river four or five times.

Anhelo to Paredón

No possible wagon road Anhelo to Paredón on account of deep arroyos and Patos River with deep steep banks. Cavalry can make it. There are, on the railroad, four big iron bridges in six kilometers between Anhelo and Paredón and three wooden trestles.

Reata to Monterrey

(Reata- Anhelo – Ixtle – Mina – Mont-y.) Road follows railroad diverging to enter towns near but not on railroad. Road is very bad Anhelo to Ixtle.

"Q"

With regard to Dr. Quinn's road, Monclova to Paredón, and without personal knowledge of the roads themselves, it is thought that a preferable route should be as follows:

Leaving the route in question at or near La Perla station on the railroad pass behind the ridge east of the railroad thro. Pilillal to Anhelo, or follow the railroad Reata to Anhelo.

Reports are conflicting regarding the route from Anhelo to Paredón. It is undoubtedly bad by reason of the numerous arroyos. This stretch can be crossed however by making use of the railroad embankment and bridges if they are not destroyed.

If this route is found impassable troops of all arms can be moved from Anhelo via Delgado Canyon to Ixtle. Thence by either route to Monterrey and Saltillo.

Delgado Canyon is a narrow crooked pass with high steep walls, commanded on both sides. The railroad is now abandoned thro this canyon but it is thought that the railroad bridge has not yet been removed. In any event the stream is passable near the bridge site. The road bed is largely in side hill cut, close to the stream in which there is a small amount of water most if not all the time.

Judging entirely from a general knowledge of the railroad lines and having no specific knowledge of the roads, it is thought that the following consideration will apply as to choice of route in this locality.

It is understood that the immediate objective points to be reached from Reata are Monterrey, Torreón and Saltillo.

Excluding the direct railroad route, Reata – Anhelo – Paredón, Saltillo, the most direct and feasible cross country route is Reata – Sauceda – Patos River – General Cepeda or El Sauz – Saltillo. Torreón is not under discussion here but will be reached via Sauceda and Jaral.

From Reata to Monterrey, it may be said that:

The road Reata – Treviño – Sauceda is good with sufficient but not abundant water.

Sauceda or rather Arizpe, to Paredón, the route is through rough mountain country – bare and rocky and practically dry. From Paredón to Ixtle the road is good with a fair supply of water.

Distances are roughly as follows:

Reata to Sauceda.................... 25 miles good road,
Sauceda to Paredón................ 27 " bad   "
Paredón to Ixtle...................... 9 " good   "
Total... 61 "

On the other hand Reata to Anhelo, about 15 miles is through a mountain valley largely, but fairly level going.

Anhelo to Ixtle is about 12 miles, 6 miles of which are in Delgado Canyon.

The total distance of 27 miles Reata – Anhelo – Ixtle – is decidedly better road and better supplied with water than the 27 miles Sauceda to Paredón.

There is apparently no object in going to Paredón except to rebuild the railroad toward Saltillo. This can be done directly from Anhelo and the railroad route Reata Anhelo Paredón is beyond question far superior to the railroad route, Reata-Sauceda-Paredón, even without considering the difference in distance. These routes are described elsewhere.

Ixtle to Monterrey:- Regarding this route it would seem that the more southerly road is preferable, that is, via Fraile and García. The railroad can be rebuilt and maintained with comparative ease as there are no serious defiles or high and long bridges. The question of water is doubtful however as after leaving Icamole and the Patos River, there is none known definitely until García is reached, and from García to Monterrey there is none until in the immediate vicinity of Monterrey.

By the northerly road, that is, via Mina and Hidalgo the route follows the Patos and Salinas Rivers about to Chipinque. Near Arista the railroad crosses the Salinas River at a place about 30 feet deep. From this point the river drops down into a deep valley which is perhaps two kilometers wide, while the railroad follows along the side hill well above the bottom. Bridges at Hidalgo Chipinque and Topo Grande are long and over deep channels with steep rocky banks, difficult and slow to rebuild if destroyed. From Chipinque the roads are good and there is abundance of water all the way to Monterrey.

"A"

Route Lampazos to Monclova:
Road: Good wagon road all the way and not merely a trail out from Lampazos as shown on Abbot map.

Romero Rubio is but another name for Candela, they being the same place.

Permanent water is found at Lampazos and Candela from river Arroyo Candela which originate in Puerto de Carroza which is the pass out from Candela toward Monclova. Permanent water is obtainable from the river in Puerto Carroza, said Puerto being about 500 meters wide and the river flowing on one side at Saucillo permanent water consisting of a stream about 8 inches in diameter. At Las Armadas (probably Las Enramadas, W.K.A.) there is a well of water now being operated by hand wind lass and also a tank. The only permanent water is in well and is not in great quantity. From Saucillo to Salitrillos no water is found. At Salitrillos there is permanent water from a stream. The next water is at La Mota then El Oro San José and Monclova.

Documents Cited

"Q"

It is understood that, from Puerto Carroza there is a road through San Nicolás and Pánuco rejoining the above described road at Las Mota.

At Mineral de Pánuco there is a copper mine owned entirely by American (San Antonio) capital.

It was thought that the road, La Mota – Pánuco – San Nicolás – Puerto Carroza – was the main road rather than the La Mota – La Enranada – Saucillo – Puerto Carroza road. Both are through rugged mountains and further investigation will be made to determine, if possible, which is preferable.

"A"

8529-15

Eagle Pass, Texas.
May 14, 1915.

FROM: Intelligence Officer, Eagle Pass, Texas.
TO: Department Engineer, Southern Department.
SUBJECT: Rainfall

1. Referring to 204-CF and to my communication of May 1, 1915, I have interviewed a number of travelers, cattlemen, and fishermen and it appears that the water level in tanks in northern Coahuila has been raised about two feet by the recent rains.

John W. Wright
Captain, 17th Infantry
Acting Adjutant

Recd.O.D.E., S.D. May 15, 1915.

204-E-DE-CF   1st Ind.   HHR-waf
Office Dept,Engr., So.Dept.,Ft. Sam Houston,Texas,May 15, 1915.- To Chief of Staff, So. Dept. for forwarding to Chief, War College Division.

Henry H. Robert
Captain, Corps of Engineers,
Department Engineer.

204 E, D.E.   2d End.
Hq. Southern Dept., Fort Sam Houston, Texas, May 17, 1915.
To Chief, War College Division, General Staff, Army War College, Washington, D. C.

Major, General Staff,
Chief of Staff.

WHH-HS

Refer to file
246-C.F.

HEADQUARTERS SOUTHERN DEPARTMENT
DEPARTMENT ENGINEER,
FORT SAM HOUSTON, TEXAS.

June 9, 1915.

From: Department Engineer.
To: Chief of Staff, Southern Department.
Subject: Mexican information.

1. The following are extracts from letter received this date from the Intelligence Officer at Camp Eagle Pass, Texas, for forwarding to Chief, War College Division, General Staff:

"The Mexicans across the river, in and near Piedras Negras, are very much excited over the situation. They believe that an intervention is near and public opinion seems to be entirely opposed to intervention of any kind on our part.

"The Carrancista Consul here, Mr. Sequin, is opposed to our rendering any aid to the Mexicans in the form of Red Cross supplies. He said that such aid has always been closely followed by intervention and he points to Cuba as an example. He says that Mexico is not starving; that there is plenty of corn and beans and other food stuffs and that the reports of the people starving is manufactured by the foreigners and old Federalists for the purpose of forcing intervention. He says that any food introduced into Mexico will have to be followed by American agents to prevent it being seized by the soldiers or being reshipped to the United States, sold and the money used to purchase ammunition. He says that Mexico is still exporting food stuffs which shows that there is no starvation.

"While it is true that there is exportation of beans and corn, I believe that the Mexican people are actually starving. The cause of the exportation is the better prices obtained in the United States.

"Mr. Sequin told me yesterday that the force of Villistas at Piedras Negras will retire south towards Monclova next Saturday. I cannot verify this statement. He said further that General Villa is retiring all of his forces towards Chihuahua City as he has been defeated and will be safe in Chihuahua among his old sympathizers. This will put things back to where they were a year ago."

*　　*　　*　　*

"Four Englishmen, who have been doing business in Mexico near Piedras Negras, called on the Commanding Officer here yesterday and stated that the Mexican people were becoming very excited and expected an intervention. The Englishmen wished to know if the Commanding Officer thought it advisable for them to go back into Mexico. These Englishmen stated that if the lower classes of Mexicans believed that any intervention was coming there would be an uprising, resulting in the killing of foreigners, particularly Americans, in Mexico."

Henry H Robert

Captain, Corps of Engineers.

1st End.

Hq. Southern Dept., Fort Sam Houston, Texas, June 10, 1915.

To Chief, War College Division, General staff, Army War College, Washington, D.C.

W H Hay

Major, General Staff

Chief of Staff.

WHH-HS

8529-18

Camp Eagle Pass, Texas.

July 12, 1915

From: Intelligence Officer.

To: Engineer Officer, Southern Department.

Subject: Report of Barroterán Engagement

1. I enclose herewith a rough sketch, made by a civilian of the railroad station at Barroterán and vicinity.

The civilian was present at Barroterán when an engagement took place on July 8th between a force of 1200 Carrancistas and an equal number of Villistas.

This sketch is enclosed to illustrate the description of the engagement which follows. This description is forwarded as it serves to give an idea of Mexican tactics.

2. A Carrancista force of approximately 1200 men occupied the heights on the south end of the sketch. They had previously burned some railroad bridges farther to the south and were advancing northward towards Barroterán. As their advance troops reached the heights referred to, they were discovered by the Villistas, who decided to attack them immediately. Sixty Villista Cavalry, armed with rifles and carbines, rode against their center, deployed as foragers. When they had reached a point about 600 yards from the enemy's position, they opened fire and continued to fire as they advanced at a trot. In the meanwhile a force of two hundred and fifty cavalry, executed an enveloping movement to the left of the Carrancistas. This force also advanced as foragers with about five foot interval between foragers. There were no supports or reserves.

An Infantry force in the meanwhile, of about 200 men, was kept in the town of Barroterán for the purpose of protecting the town and a train of cars on the track.

The frontal attack, made by sixty cavalry, reached a point half way up the slope when they retired in confusion. The enveloping movement reached the top of the slope on the left of the Carrancistas and held the heights for possibly five minutes, when this force also retired in confusion in a north-westerly direction. It appears they ran out of ammunition upon reaching the enemy's position.

After the retirement of the Villista cavalry the Carrancistas advanced and took a part of the town of Barroterán but were driven out afterwards by Villista reinforcements arriving from Piedras Negras by rail.

3. The Carrancista force were practically all mounted and were reinforced by two machine guns.

4. I am told that the foregoing is practically a description of the usual order of attack. If so, it would appear from the above that mounted attacks are made in a single line without supports; that an enveloping movement usually accompanies the frontal attack and mounted men advance and open fire at 800 yards and continue to fire as they advance at a trot.

5. I am given to understand that the railway leading south from Piedras Negras will be closed, in all probability, for a considerable time on account of the numerous bridges which have been destroyed during this last raid of the Carrancistas.

John W. Wright

Captain, 17th Infantry

1 Incl.

286-Confidential   1st Ind.   HHR-waf

Office Dept.Engr.,So.Dept.,Ft.Sam Houston,Texas,July 14., 1915.-To Chief of Staff., So.Dept.,for forwarding to Chief, War College Division, General Staff, Washington, D.C.

Henry H Robert

Captain, Corps of Engineers,

Department Engineer.

1 inclo.

286- C.F.   2d End.

Hq. Southern Dept., July 15, 1915. - To Chief, War College Division, General Staff, Army War College, Washington, D. C.

W H Hay

Major, General Staff

Chief of Staff.

1 encl.

whh-kls

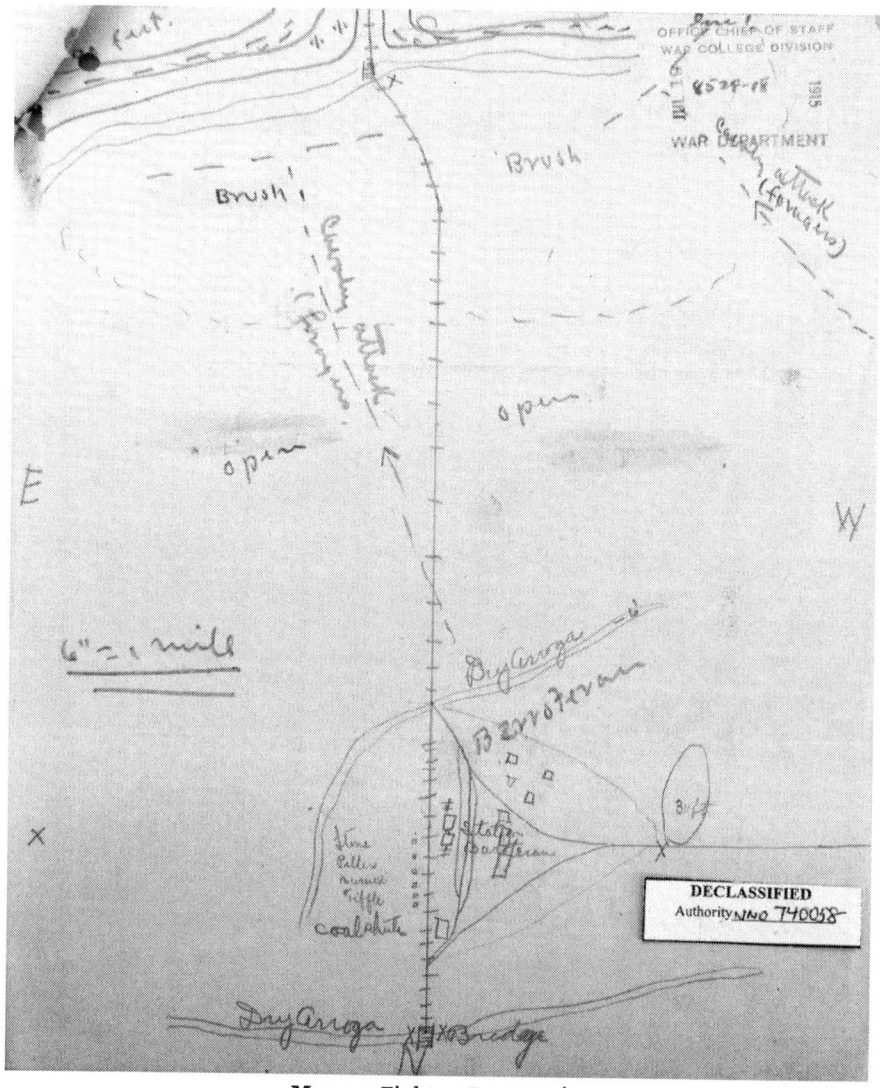

Map 3.2 Fight at Barroterán

8529-21

Camp Eagle Pass, Texas.

October 4, 1915.

From: Intelligence Officer, Camp Eagle Pass, Texas.

To: Department Engineer, Southern Department, Fort Sam Houston, Texas.

Subject: Plan de la Ciudad de Monclova Distrito de Monclova, Coah., Mex.

257

1. Dr. Quinn was interviewed and went over the map with the undersigned. His ideas were very general but I believe are correct as far as they go. I have separated the two plans and pinned them together as they should go, marking a north line on each part.

Dr. Quinn has indicated the location of the street car barn, electric light plant, continuation of the Tramvia, a wagon road, a hill, Loma de Bartola and an irrigation ditch. He has also located the railroad station.

2. I did not trace the two sections and combine them, due to the very general ideas on the subject of Dr. Quinn, so general in fact that I concluded that any drawing which indicated accuracy might be misleading.

3. Should you desire to connect the two plans the basis should be that the distance from where the Tramvia crosses the Arroyo de la Estación (at the electric light plant) to the Panteón de San Francisco is approximately two and one half (2½) miles, and the two sketches as pinned together are approximately correct relatively as to direction.

4. The Loma de Bartola is an important feature. The Doctor informs me that it is about one hundred and forty feet above the wagon road running at its base and it has always been fortified and is in fact the citadel of the town.

5. In view of this, I believe it would be advantageous to secure more information regarding the hill as it will be intimately connected with any defense or attack of the town.

John W. Wright
Captain 17th Infantry.

Confidential

B4-3 C.F.   1st Ind.   HHR-vve

Office Dept. Engr., So. Dept., Ft. Sam Houston, Texas, Oct. 11, 1915.- To the Chief, War College Division, Washington, D. C., for consideration in connection with War College Division 8529-20 of September 8, 1915, addressed to the Department Engineer.

Henry H. Robert
Captain, Corps of Engineers,
Department Engineer.

ADDITIONAL NOTES ON PLAN OF MONCLOVA
Furnished by Dr. Quinn.

(A) Wagon road crosses railroad. Road branches to right along railroad leading to Polka Ranch and to Estancia San Juan.

(B) Wagon road crosses the railroad at B. At about 100 yards turns to right and crosses irrigation ditch.

(C) Road to Romero Rubio.

(D) Wagon road leaves street car line, turns to right and goes over into main road leading to Castaños.

(E) Little to right of "C." Marks location of Hermita. A church on hill. Hill about 80 feet high and church is surround by wall. This church is fortified.

(F) Main wagon road between Monclova and Estancia San Juan.

($X^1 - X^2$) A line connecting these two points crosses at its center Gloria Hill. This hill is about 100 feet high and is fortified.

(M) Wagon road from Monclova station to Monclova.

(P) Road leading from San Buena Ventura.

[written in pencil: "as stated in my letter, all of this data is general in character and I do not think it is accurate enough to draw in plan. It might appear in 'notes.' JWW"]

8532-3

U.S. Camp at El Paso, Texas.
September 4, 1915.

From;   Major F. A. Wilcox, 6th Infantry, Intelligence Officer.

To;   Department Engineer, Southern Department, Fort Sam Houston, Texas.

Subject; Forwarding information data.

I submit herewith the following military notes and data prepared by me in pursuance with instructions received. All data is submitted in duplicate.

RIO GRANDE RIVER. River bottom, fords, etc.

WATER SUPPLY OF EL PASO, TEXAS.

WATER SUPPLY OF JUAREZ, MEXICO.

Form 1. MEXICAN CENTRAL R.R. BRIDGE AT EL PASO. With photo.

    "   MEXICO-NORTHWESTERN R.R. BRIDGE AT EL PASO. With photo.

    "   EL PASO & S. W. R. R. BRIDGE near EL PASO. With photo.

    "   S. P. R. R. BRIDGE near EL PASO. With 2 photos

Form 2. WAGON BRIDGE TO JUAREZ, MEXICO. (South). With photo.

    "   WAGON BRIDGE TO JUAREZ, MEXICO. (North.) With photo.

    "   SUSPENSION FOOT BRIDGE near EL PASO. With photo.

Blue print diagram of El Paso & S.W.R.R. Bridge.

Blue print diagram of S. P. R.R. Bridge.

I am also forwarding, under separate cover;

Blue print map, induplicate, showing Rio Grande River, Railroads and region from the western limits of El Paso to the crossing of the E.P & S. W. and S. P. railroads from Texas into New Mexico.

2. I am expecting to have other data ready to forward within a few days.

32 Inclosures.   F. A. Wilcox

Recd: O.D.E., S.D.
Confidential
B8-2   1st Ind.   HM-vve
Office Dept. Engr., So. Dept., Ft. Sam Houston, Texas, Sept. 7, 1915. – To the Chief of Staff, Southern Department, Ft. Sam Houston, Texas for forwarding to Chief, War College Division, Washington, D.C.

> Henry H Robert
> Captain, Corps of Engineers.

B8-2 D.E.   2 End.
Hq. Southern Dept. Fort Sam Houston, Texas, Sept. 10, 1915.
To Chief, War College Division, General Staff, Army War College, Washington, D.C.

> W H Hay
> Major, General Staff,
> Chief of Staff.

32 encls.
WHH-HS

U.S. Camp at El Paso, Texas.   CONFIDENTIAL.
September 4, 1915.
By Major F. A. Wilcox, 6th Infy.

RIO GRANDE RIVER.
RIVER BOTTOM, FORDS, etc.

The RIO GRANDE, as it flows by El Paso, is a muddy stream varying in width up to 250 yards and varying in depth from dry to 8 or 10 feet in deepest channel, according to conditions.

The VELOCITY of flow varies from 2 miles per hour to 5 or 6.

The BOTTOM of the river is variable, though generally sandy. Often there is a 6" or 8" strata of sand on a similar strata of silt, and vice versa.

The sand in places, though not the true quick-sand, exhibits the properties of the same in a more or less degree and is a source of danger.

Documents Cited

The BANKS of the river are retained and erosion of the stream restrained by protected embankments from 8' to 12' above the river bottom. On the Mexican side, timber revettment is used in the region of the Juarez bridges.

FERRIES. There are no ferries and boats or rafts are not to be found.

FORDS. The river at low water can be forded along most its length, but heavy heeled vehicles would have difficulty at any time and many wagons passing over the same trail would quickly wear down the bottom soon making crossing at that point impractical.

In the vicinity of the bridges and principal streets leading to the river, the embankments could be scarped without difficulty to permit the passing of wagons and guns.

Crossing by way of the river bottom however, if anticipated, can be made a very simple problem, for the gates of the upstream dams may be closed and the water diverted, leaving the river bottom practically dry.

The BEST METHOD of CROSSING, in case of serious resistance, is First; to communicate with the U. S. Reclamation Service, (1218 Mills Building) who will provide for cutting off the water at the upstream dams, leaving the river bed dry, over which foot troops could pass at will and where cavalry and light field artillery could easily find practical crossings. Second: to seize the bridges, if unimpaired, reserving them for use of heavy field artillery, vehicles and trains. Third: should the bridges be badly damaged, to prepare a roadway across the river bottom by laying sleepers and flooring for use of heavy field pieces and vehicles.

The EL PASO MILLING COMPANY, said to be the largest of its kind in the world, is by the river just east of the lower wagon bridge to Juarez. From it lumber of all descriptions can be obtained.

DAMS. The Dams are four in number. They are of permanent construction, of concrete and stone.

The first dam is at the western limits of El Paso proper. It connects U.S. and Mexico and impounds the river and diverts the water into irrigation canals.

The second dam is nearing completion at Mesilla Park, about 43 miles north and west of El Paso in New Mexico. It is also a diversion dam pertaining to the irrigation projects. It is due to be completed in January or February, 1916.

The third dam is at Leesburg, about 60 miles north and west of El Paso in New Mexico. This is a large diversion dam and part of the irrigation projects.

The fourth dam is the gigantic Elephant Butte Dam, about 125 miles north and west of El Paso in New Mexico. It is of solid concrete and rises about 200 feet above the river bottom. It is backing up the Rio Grande for miles and will form an artificial lake from 40 to 50 miles in length by 2 to 7 miles in width.

By turning off the water at Elephant Butte only, when the water is at flood, it would require from three to four days to expose the river bottom at El Paso. By closing all the dams and diverting the water into the irrigation ditches, the bottom of the river would be bared within 24 hours at almost any season of the year.

U.S. Camp at El Paso, Texas.	CONFIDENTIAL.

September 4, 1915.

By Major F. A. Wilcox, 6th Infy.

Class D.

Dept. Engr., So.Dept.

3-14-15

NOTES ON WATER SUPPLY OF EL PASO, TEXAS.

The Permanent Supply is about 10,000,000 gallons each 24 hours from what is known as the Mesa Supply. This supply is obtained from deep wells - 600 feet deep - with water at pumping level of 230 feet. These wells are pumped by the air system to the surface, then repumped by steam pumps to the City and the City Reservoir.

There is no Temporary Supply as the same facilities exist all the year round.

The City has no Artificial Supply.

The Quantity does not vary with the season, being the same in the dry as well as in the rainy season, i.e. 10,000,000 gallons per day.

There is also a supply that can be obtained from a large well about 250 feet from the bed of the Rio Grande from which can be pumped about 2,000,000 gallons per day. This supply is only to be used for fire protection should the Mesa Supply become for any reason inadequate (a remote contingency).

The water from the Mesa wells is rated as absolutely pure for drinking purposes. The following in the official chemical analysis.

Parts per 1,000,000

Total solids	338.0
Fixed solids	30.0
Chlorine	32.0
Hardness	163.0
Required oxygen	0.8
Free ammonia	0.04
Albuminoid ammonia	0.06
Nitrates	3.0
Nitrites	0.0

The water from the well near the river is reported by chemists to be pure but had the taint of alkali with which the water becomes in a degree impregnated while flowing in the river bed, either surface or underflow. It is healthful however and was used for years by the city before the Mesa water was supplied.

There is no liability of contamination of the Mesa water from natural sources and it is never exposed excepting in the enclosed reservoir within the City limits where it is well protected.

The water from the large well previously mentioned is practically the underflow of the Rio Grande filtered through 250 feet of sand.

Both these supplies are of clear water without sediment.

The Mesa Supply is situated just north of the Fort Bliss Military Reservation, about 6 miles distant from the center of the city. The water reaches the city through two main force lines.

The pumping plant if not guarded could be seriously injured by persons maliciously inclined. The well near the river has its separate pumping plant and is subject to the same dangers as the Mesa plant excepting that it is less isolated and connects direct with the city distribution system, not needing any main force lines.

The depth of the wells at the Mesa pumping plant is 600 feet, the wells being from 8 to 14 inches in diameter.

The water level, when pumping is in operation, is 230 feet below the surface.

The total capacity of the city's supply is then approximately as follows;

From Mesa plant -------- 10,000,000 gals. per day.
From city well --------- 2,000,000   "   "   "
                Total   ------ 12,000,000   "   "   "

This supply never varies with any season of the year, whether wet or dry.

The city and the additions thereto are very generally piped with large mains.

There are numerous wells privately owned or controlled within the city limits and pumped by wind or by steam or gas engines. There are numerous hydrants and public and private drinking places for man and animals.

The water supply has been developed for the needs of the present and near future. The supply can be increased at any time to meet the needs of a city of several thousand inhabitants, all sources being practically within the city limits.

U.S. Camp at El Paso, Texas.   CONFIDENTIAL.
September 4, 1915.

U.S. Army Intelligence in the Mexican Revolution

By Major F. A. Wilcox, 6th Infy.

Class D.
Dept. Engr., So.Dept.
3-14-15

NOTES ON WATER SUPPLY OF JUAREZ, MEXICO.

The City of Juárez has three sources of water supply;

The first and oldest is from a canal or ditch running through the city, being diverted from the Rio Grande by a dam about one mile to the north-west of the city.

The second is by the individual sinking of wells (mostly driven wells), the water being the filtration from the river flow and is obtained from a depth of from 20 to 30 feet.

The third is the city water works supply. The water is taken from a dug well 60 feet deep and about 20 feet in diameter and is the filtration from the river flow. The water is pumped directly to a high tank (a conspicuous landmark) of about 75,000 gallons capacity.

The canal supply may be considered as temporary. The balance of the supply is constant and independent of whether the season be rainy or dry.

The water from the main city well is reported as pure water though it has the taint of alkali which from a healthful standpoint is not considered injurious.

The water from the canal is most liable of contamination and should not be drunk by Americans without boiling. The water from the private wells is much less liable to contamination while the water of the main city well is considered quite safe and sanitary.

The canal supply to the city could be easily broken. As the water of the main system goes directly to the high tank, any damage to the same or to its connections would put the system out of commission. It would not be difficult to make arrangements to pump directly into the pipe system.

The water or pumping level is approximately 30 feet below the surface.

The water supply from the city pumping plant is about 400,000 gallons per day and does not vary with the season.

Though there is not the established water system that pertains in El Paso yet water is more easily obtainable naturally in Juárez than in the former city. As one travels to the south, up to forty or fifty miles, the water level is reported near the surface and easily reached with wells.

In case of occupation of Juárez by U.S. troops the water supply could be augmented, if necessary, by piping from El Paso.

U.S. Camp at El Paso, Texas.
Submitted by Major F. A. Wilcox, 6th Infy.
September 4, 1915.

Documents Cited

Form 1 - Dept. Engr, So. Dept.

RAILROAD LINE: "Mexican Central R. R."
Location of bridge: No miles (direction) El Paso, Texas.
Obstacle crossed: "Rio Grande River opposite Juárez, Mexico"
Character of bridge superstructure: "Timber supported on wooden trestles."
Character of piers and abutments: "Wood and earth abutments."
Total length of bridge: "180 yards."
Number of spans: "36 bays, fifteen feet each."
Character and length of each span (beginning on side toward United States)

    1.   See accompanying photos

    4.   C

Character of approaches (beginning on side toward United States)

    1.   Low earth embankment.

    2.   Low earth embankment.

Height of rail above bottom of obstacle: "About fifteen feet."
Depth of water in obstacle in various seasons

    (a)   During rains: 5 to 10 ft. (b) Ordinary: 2 to 5 ft. (c) Dry: Dry.
Current: 2 to 5 or 6 miles per hour.
Width of water in obstacle at various seasons.

    (a)   During rains 100-250 yds. (b) Ordinary: 75-150 yds. (c) Dry: Dry.
Height and character of banks (beginning on side toward United States)

    1. 6 to 12 feet; from steep to gradually sloping.

    2. 6 to 12 feet; steeper on Mexican side.

Character of bottom: "Sand, silt, and in places a species of quick sand. Bottom variable and condition uncertain."
Capacity of bridge in normal condition: "Standard loads."
Present condition of bridge: "Fair. Trains must run slowly and carefully."
Capacity of bridge in present condition: "Ordinary loads."
Nature of quick repairs or structure suggested in case of damage or destruction of present bridge: "Repairing and replacing of trestles or timber work. Adjacent railroad bridge could be used instead as there are the necessary rail connections on both sides of the river."
Repair or construction materials in vicinity: "Timber and rails can be found in nearby plants."
Remarks: "Accompanying photos and maps show character of bridge, location and surroundings."
Rough estimate of time required for emergency repairs or temporary construction to admit crossing of trains

(a)   In case of total wrecking or destruction of bridge: 2 to 3 weeks.

(b)   In case of damage to extent roughly estimated as most probable of encounter: from 24 hours up.

Location and description of nearest fords above and below: "Position is variable. Animals and men may cross when the river is sufficiently low. (See special report on river bottom and fords.)"

Location and description of nearest bridge above and below: "Mexican Northwestern R. R. bridge 150 yards above. (Separately described.) Wooden wagon bridge 150 yards below."

Defensibility: Report here under heading "Defensibility" estimate of

(a)   Measures likely to be taken by enemy to defend crossing: "Burning of wooden parts of bridge. A Mexican enemy is hardly liable to make an effective standing at the river."

(b)   Measures recommended to be taken by our forces to minimize damage to crossing and to seize and utilize same in best manner: "Constant guarding by troops and seizure of opposite shore and town of Juárez at the first outbreak of hostilities. The R.R. embankment parallel to and almost coincident with the left (U.S.) bank is a good natural parapet covering the bridges and opposite shore."

Photo 3.3 Mexican Central Railroad Bridge

U.S. Camp at El Paso, Texas.

Submitted by Major F. A. Wilcox, 6th Infy.

Documents Cited

September 4, 1915.

Form 1 - Dept. Engr, So. Dept.

RAILROAD LINE: "Mexico-Northwestern R. R."

Location of bridge: No miles (direction) El Paso, Texas.

Obstacle crossed: "Rio Grande River opposite Juárez, Mexico"

Character of bridge superstructure: "Steel Plate Girders; 100 ft. each."

Character of piers and abutments: "Piers of stone. Abutments of stone and earth."

Total length of bridge: "403 feet. (Approx.)"

Number of spans: "Four."

Character and length of each span (beginning on side toward United States)

1.	About 100 ft.
2.	Same.
3.	Same.
4.	Same.

Character of approaches (beginning on side toward United States)

1.	Low earth embankment.
2.	Low earth embankment.

Height of rail above bottom of obstacle: "About 15 feet."

Depth of water in obstacle in various seasons

    (a)   During rains: 5 to 10 feet. (b) Ordinary: 2 to 5 feet. (c) Dry: Dry.

Current: "2 to 5 or 6 miles per hour."

Width of water in obstacle at various seasons:

    (a)   During rains: 100-250 yds. (b) Ordinary: 75-150 yds. (c) Dry: Dry.

Height and character of banks (beginning on side toward United States)

    1. 6 to 12 feet; from steep to gradually sloping. (See photos)

    2. 6 to 12 feet; steeper on Mexican side.

Character of bottom: "Sand, silt, and in places a species of quick sand. Bottom variable and condition uncertain."

Capacity of bridge in normal condition: "Maximum loads."

Present condition of bridge: "Excellent."

Capacity of bridge in present condition: "Maximum loads."

Nature   of quick repairs or structure suggested in case of damage or destruction of present bridge: "First; use other bridge if it remains undamaged. Broken spans could be replaced by trestles of timber obtained at the large lumber plant near-by."

Repair or construction materials in vicinity: "Timber from lumber plant and material of the R. R. shops."

Remarks: "Accompany photos and maps show character of bridge, location and surroundings."

Rough estimate of time required for emergency repairs or temporary construction to admit crossing of trains

(a)   In case of total wrecking or destruction of bridge: "Time necessary to erect 400 feet of trestle."

(b)   In case of damage to extent roughly estimated as most probable of encounter: "From 24 hours up."

Location and description of nearest fords above and below: "Position is variable. Animals and men may cross when the river is sufficiently low. (See special report on river bottom and fords.)"

Location and description of nearest bridge above and below: "International wooden wagon bridge about 100 yards above. (Separately described.) Mexican Central R. R. bridge below about 100 yards."

Defensibility: Report here under heading "Defensibility" estimate of

(a)   Measures likely to be taken by enemy to defend crossing: "Dynamite an element, joint or pier of the structure. Mexican enemy would probably not make a stand at the river."

(b)   Measures recommended to be taken by our forces to minimize damage to crossing and seize and utilized same in best manner: "Constant guarding by troops and seizure of opposite shore and town of Juárez at the first outbreak of hostilities. The R.R. embankment parallel and almost coincident with the left (U.S.) bank is a good natural parapet covering the bridges and opposite shore."

Photo 3.4 Mexico-Northwestern Railroad Bridge

U.S. Camp at El Paso, Texas.
Submitted by Major F. A. Wilcox, 6th Infy.
    September 4, 1915.
                    Form 1 - Dept. Engr, So. Dept.

RAILROAD LINE: "EL PASO & SOUTHWESTERN R. R. CO."
Location of bridge: "2 miles N.W. (direction) from El Paso, Texas."
Obstacle crossed: "Rio Grande River."
Character of bridge superstructure: "Steel Truss, Girder, and Steel Trestle."
Character of piers and abutments: "Concrete Piers. Concrete and earth abutments."
Total length of bridge: "1692½ feet."
Number of spans: "5 large and 8 smaller."
Character and length of each span (beginning on side toward United States)
        1.   See accompanying blue print of the bridge.
Character of approaches (beginning on side toward United States)
        1.   High earth embankment in Texas.
        2.   Earth embankment and general contour level. In New Mexico.
Height of rail above bottom of obstacle: "85 feet. Close approx."
Depth of water in obstacle in various seasons
        (a)   During rains: 5' to 10' (b) Ordinary: 2' to 5' (c) Dry: Dry.
Current: "2 to 5 or 6 miles per hour."
Width of water in obstacle at various seasons:
        (a)   During rains: 100–250 yds. (b) Ordinary: 75–150 yds. (c) Dry: Dry.
Height and character of banks (beginning on side toward United States)
        1. Sandy 4' to 6' high with higher ground in rear. (See photos)
        2. Sandy 4' to 6' high with high hill rising in rear.
Character of bottom: "Sand, silt, and in places a species of quick sand. Bottom variable and uncertain."
Capacity of bridge in normal condition: "Heaviest loads."
Present condition of bridge: "Excellent."
Capacity of bridge in present condition: "Heaviest loads."
Nature of quick repairs or structure suggested in case of damage or destruction of present bridge: "The company has its specially trained engineers and workmen who are prepared to make all necessary repairs to the bridge in a minimum of time. By building a shunt at a practicable point about a mile to the West of the river, the trains could use either in case of damage or destruction of its neighbor."
Repair or construction materials in vicinity: "In R. R. shops, large supply houses and the large lumber works."

Remarks: "Accompanying maps, plan of bridge and photographs show location, character and surroundings of bridge. Though the bridge crosses into New Mexico, the western abutments are but about 500 yards from the Mexican line and this would be a route for getting into Mexico and flanking Juárez from the West."

Rough estimate of time required for emergency repairs or temporary construction to admit crossing of trains

> (a)   In case of total wrecking or destruction of bridge: "A matter of months according to conditions."
>
> (b)   In case of damage to extent roughly estimated as most probable of encounter: "4 days to 2 weeks and up."

Location and description of nearest fords above and below: "Position variable. Animals and men may cross nearby when the river is sufficiently low. (See special report on river bottom and fords.)"

Location and description of nearest bridge above and below: "Southern Pacific R. R. bridge 100 yards above. (Separately described.) Suspension foot bridge 300 yards below."

Defensibility: Report here under heading "Defensibility" estimate of

> (a)   Measures likely to be taken by enemy to defend crossing: "Dynamite an element joint of the structure. One of the most serious damages to be guarded against is the blowing up of a 965 foot tunnel about four-fifths of a mile to the West-North-West in New Mexico."
>
> (b)   Measures recommended to be taken by our forces to minimize damage to crossing and seize and utilized same in best manner: "A detachment of troops should be stationed at this bridge to guard it and the neighboring S. P. R. R. bridge, with posts both sides of the river and at the rail-road tunnel."

Photo 3.5 El Paso & Southwestern Railroad Bridge

U.S. Camp at El Paso, Texas.
Submitted by Major F. A. Wilcox, 6th Infy.
    September 4, 1915.
                Form 1 - Dept. Engr, So. Dept.

RAILROAD LINE: "SOUTHERN PACIFIC R. R. COMPANY."
Location of bridge: "2 miles N.W. (direction) from El Paso, Texas."
Obstacle crossed: "Rio Grande River."
Character of bridge superstructure: "Truss, Plate Girder & Wooden Trestle."
Character of piers and abutments: "Concrete Piers. Earth & stone abutments."
Total length of bridge: "1137 feet."
Number of spans: "6 spans not counting trestle."
Character and length of each span (beginning on side toward United States)
        1.   See accompanying blue print of the bridge.
Character of approaches (beginning on side toward United States)
        1.   High earth embankment.
        2.   Earth and stone embankment on general contour level.
Height of rail above bottom of obstacle: "66 feet. (Close approx.)"
Depth of water in obstacle in various seasons
        (a)   During rains: 5 to 10 feet. (b) Ordinary: 2 to 5 feet. (c) Dry: Dry.
Current: "2 to 5 or 6 miles per hour."
Width of water in obstacle at various seasons:

(a)   During rains: 100-250 yds. (b) Ordinary: 75-150 yds. (c) Dry: Dry.
Height and character of banks (beginning on side toward United States)
1. 4' and up. Banks sandy and higher ground to East.
2. 4' and up. banks steep and rough. See photos.
Character of bottom: "Sand, silt, and in places a species of quick sand. Bottom variable and condition variable."
Capacity of bridge in normal condition: "Heaviest loads."
Present condition of bridge: "Excellent."
Capacity of bridge in present condition: "Heaviest loads."
Nature   of quick repairs or structure suggested in case of damage or destruction of present bridge: "The company has its specially trained engineers and workmen who are prepared to make all necessary repairs to the bridge in a minimum of time. By building a shunt at a practical point about a mile to the West of the river, the trains could use either bridge in case of damage or destruction of its neighbor."
Repair or construction materials in vicinity: "In R. R. shops, large supply houses and at the large lumber works."
Remarks: "Accompanying maps, plan of bridge and photographs show location, character and surroundings of bridge. Though the bridge crosses into New Mexico, the western abutments are but about 600 yards from the Mexican line and this would be a route for getting into Mexico and flanking from the West."
Rough estimate of time required for emergency repairs or temporary construction to admit crossing of trains

(a)   In case of total wrecking or destruction of bridge: "A matter of months according to conditions."

(b)   In case of damage to extent roughly estimated as most probable of encounter: "From two days up."

Location and description of nearest fords above and below: "Position variable. Animals and men may cross nearby when river is sufficiently low. (See special report on river bottom and fords.)"
Location and description of nearest bridge above and below: "An old but practical wooden bridge about 1¼ miles upstream leading into New Mexico. El Paso & Southwestern R. R. bridge about 100 yards down-stream. (Separately described.)"
Defensibility: Report here under heading "Defensibility" estimate of

(a)   Measures likely to be taken by enemy to defend crossing: "Dynamite an element or joint of the structure or burn the wooden trestle at the east end of the bridge."

(b)   Measures recommended to be taken by our forces to minimize damage to crossing and seize and utilized same in best manner: "A detachment of troops should be stationed at this bridge to guard it and the neighboring E. P. & S. W. R. R. bridge."

Photo 3.6 Southern Pacific and E.P. & SW Railroad Bridges

Photo 3.7 Southern Pacific Railroad Bridge

U.S. Camp at El Paso, Texas.

Submitted by Major F. A. Wilcox, 6th Infy.

September 4, 1915.

Form 2 – Dept. Engr, So. Dept.

WAGON ROAD FROM: "El Paso, Texas." TO: "Juárez, Mexico"
Location of bridge: "Foot of Stanton St."

Obstacle crossed: "Rio Grande River."

Character of bridge superstructure: "King-post trusses, timber and iron rods."

Character of piers and abutments: "Piling, Wood and earth abutments."

Total length of bridge: "160 yards. (Close approximation.)"

Number of spans: "16 thirty foot spans."

Character and length of each span (beginning on side toward United States)

      1.   K. P. Truss 30' each.

Character of approaches (beginning on side toward United States)

      1.   Earth and wood on street level.

      2.   Earth and wood on street level.

Height of floor above bottom of obstacle: "12 to 15 feet. approx."

Width of roadway: "[drawn cutaway that totals 25 feet, see "Remarks" below]" Clear headroom: "No obstruction."

Depth of water in obstacle in various seasons

      (a)   During rains: 5 to 10 ft. (b) Ordinary: 2 to 5 ft. (c) Dry: Dry.

Current: "2 to 5 or 6 miles per hour."

Width of water in obstacle at various seasons:

      (a)   During rains: 100-250 yds. (b) Ordinary: 2 to 5 ft. [obviously an error] (c) Dry: Dry.

Height and character of banks (beginning on side toward United States)

1. 6 to 12 feet; from steep to gradually sloping.

2. 6 to 12 feet; steeper on Mexican side.

Character of bottom: "Sand, silt, and in places a species of quick sand. Bottom variable and condition uncertain."

Capacity of bridge in normal condition: "Weight of field guns and heavy wagons."

Present condition of bridge: "Very good. Heavy loaded trolley cars pass over this bridge many times daily."

Capacity of bridge in present condition: "Weight of field guns and heavy wagons."

Nature of quick repairs or structure suggested in case of damage or destruction of present bridge: "First step would be the use of neighboring bridges if undamaged. With suitable flooring the neighboring R. R. bridges could be used as wagon bridges. New trestles and flooring would probably be the repairing necessary."

Repair or construction materials in vicinity: "Timber from the adjacent lumber plant. (El Paso Milling Co.)"

Remarks: "This bridge has a 15' roadway for wagons and a raised space 10' wide for the electric cars."

Rough estimate of time required for emergency repairs or temporary construction to admit crossing of trains

      (a)   In case of total wrecking or destruction of bridge: "Five to ten days."

Documents Cited

(b)   In case of damage to extent roughly estimated as most probable of encounter: "One to four days."

Location and description of nearest fords above and below: "Position is variable. Animals and men may cross when the river is sufficiently low. (See special report on river bottom and fords.)"

Location and description of nearest bridge above and below: "Wooden R. R. bridge of Mex. Cent. 150 yards upstream. No bridge downstream for many miles."

Defensibility: Report here under heading "Defensibility" estimate of

(a)   Measures likely to be taken by enemy to defend crossing: "First: Destruction of bridge by burning. Second: Possible barricade and intrenchment on Mexican end. It is most probable a Mexican enemy would not make a stand at the river."

(b)   Measures recommended to be taken by our forces to minimize damage to crossing and to seize and utilize same in best manner: "Constant guarding by troops and seizure of opposite shore and town of Juárez at the first outbreak of hostilities. The R.R. embankment parallel to and almost coincident with the left (U.S.) bank is a good natural parapet covering the bridges and opposite shore and could be easily be prepared for cover for riflemen and machine guns. Note: The four Juárez bridges are within a quarter mile length of the river and the defense of the four results in one problem. If necessary the four dams above El Paso may have all their gates closed preliminary to crossing so that all foot troops at least could rush across the river bottom leaving the bridges for animals and wagons and field artillery."

Photo 3.8 Stanton Street Wagon Bridge

U.S. Camp at El Paso, Texas.   September 4, 1915.
Submitted by Major F. A. Wilcox, 6th Infy.
Form 2 - Dept. Engr, So. Dept.

WAGON ROAD FROM: "El Paso, Texas." TO: "Juárez, Mexico"
Location of bridge: "Foot of Santa Fe Street."
Obstacle crossed: "Rio Grande River."
Character of bridge superstructure: "Timber trusses with iron rods."
Character of piers and abutments: "Wooden trestles. Wood and earth abutments."
Total length of bridge: "150 yards. (Approx.)"
Number of spans: "8 King-post spans, 30' each; and 3 wooden & iron 22 yd. trusses."
Character and length of each span (beginning on side toward United States)
      1.   to 8. as described supported on wooden piling
      2.   9 to 11, Wood and iron rod trusses supported on piers of wood and
      stone.
Character of approaches (beginning on side toward United States)
      1.   Earth and timber.
      2.   Earth and timber.
Height of floor above bottom of obstacle: "10 to 15 feet."
Width of roadway: "[drawn cutaway that totals 28 feet, see "Remarks" below]" Clear
headroom: "Unobstructed."
Depth of water in obstacle in various seasons

Documents Cited

(a)   During rains: 5 to 10 ft. (b) Ordinary: 2 to 5 ft. (c) Dry: Dry.
Current: "2 to 5 or 6 miles per hour."
Width of water in obstacle at various seasons:

(a)   During rains: 100-250 yds. (b) Ordinary: 60-150 yds. (c) Dry: Dry.
Height and character of banks (beginning on side toward United States)

1. 6 to 12 feet; from steep to gradually sloping.

2. 6 to 12 feet; steeper on Mexican side. (See photos.)

Character of bottom: "Sand, silt, and in places a species of quick sand. Bottom variable and condition uncertain."

Capacity of bridge in normal condition: "Weight of field guns or heavy wagons."

Present condition of bridge: "Old and rather shaky but serviceable. Heavy loaded trolley cars pass over the bridge many times daily. Additional trestles could be placed as supports if necessary."

Capacity of bridge in present condition: "Weight of field guns and heavy wagons."

Nature of quick repairs or structure suggested in case of damage or destruction of present bridge: "First step would be to use neighboring bridges if undamaged. With suitable flooring the neighboring R. R. bridges could be used as wagon bridges. New trestles and flooring would probably be the repairs needed."

Repair or construction materials in vicinity: "Timber from the adjacent lumber plant. (El Paso Milling Co.)"

Remarks: "This bridge has a 15' roadway for wagons, 5' for pedestrians and 8' space with rails for street cars. (See photos.)"

Rough estimate of time required for emergency repairs or temporary construction to admit crossing of trains

(a)   In case of total wrecking or destruction of bridge: "Five to ten days."

(b)   In case of damage to extent roughly estimated as most probable of encounter: "One to four days."

Location and description of nearest fords above and below: "Position is variable. Animals and men may cross when the river is sufficiently low. (See special report on river bottom and fords.)"

Location and description of nearest bridge above and below: "Suspension foot bridge and 2 large R. R. bridges about 2½ miles up-stream. (Separately described.) Mexico N. W. R. R. bridge 100 yards down-stream. (Separately described.)"

Defensibility: Report here under heading "Defensibility" estimate of

(a)   Measures likely to be taken by enemy to defend crossing: "First: Destruction of the bridge by burning. Second: Possible barricade and intrenchment at Mexican end. It is most probable a Mexican enemy would not make a stand at the river."

(b)   Measures recommended to be taken by our forces to minimize damage to crossing and to seize and utilize same in best manner: "Constant guarding by troops and seizure of opposite shore and town of Juárez at the first outbreak of hostilities. The R.R. embankment parallel to and almost coincident with the left (U.S.) bank is a good natural parapet covering the bridges and opposite shore and could be easily prepared for cover for riflemen and machine guns. Note: The four Juárez bridges are within a quarter mile length of the river and the defense of the four results in one problem. If necessary the four dams above El Paso may have all their gates closed preliminary to crossing so that all foot troops at least could rush across the river bottom leaving the bridges for animals and wagons and field artillery."

Photo 3.9 Santa Fe Street Wagon Bridge

U.S. Camp at El Paso, Texas.   September 4, 1915.
Submitted by Major F. A. Wilcox, 6th Infantry.
Form 2 - Dept. Engr, So. Dept.

CROSSING FROM: "El Paso, Texas." TO: "New Mexico, near Int. Boundary."
Location of bridge: "2 miles North West (direction) from El Paso (proper.)"
Obstacle crossed: "Rio Grande River."
Character of bridge superstructure: "Suspension: 12" steel cable; iron rods and wood."
Character of piers and abutments: "Timber trestle. No abutments."
Total length of bridge: "About 350 feet."
Number of spans: "One long span."
Character and length of each span (beginning on side toward United States)

Documents Cited

1.   See photo.

Character of approaches (beginning on side toward United States)
    1.   Earth: practically ground level.
    2.   Earth: practically ground level.

Height of floor above bottom of obstacle: "6 to 15 feet."
Width of roadway: "3 feet" Clear headroom: "No obstruction for pedestrians."
Depth of water in obstacle in various seasons
    (a)   During rains: 5 to 10 ft. (b) Ordinary: 2 to 5 ft. (c) Dry: dry.
Current: "2 to 5 or 6 miles per hour."
Width of water in obstacle at various seasons:
    (a)   During rains: 100-250 yds. (b) Ordinary: 75-150 yds. (c) Dry: Dry.
Height and character of banks (beginning on side toward United States)
    1. Gently sloping bank of sand and silt.
    2. Gently sloping bank of sand and silt.

Character of bottom: "Sand and silt and in places a species of quick sand. Bottom variable and condition uncertain."
Capacity of bridge in normal condition: "Weight of men in single file."
Present condition of bridge: "Shaky but serviceable."
Capacity of bridge in present condition: "Men in single file."
Nature of quick repairs or structure suggested in case of damage or destruction of present bridge: "Re-erection of wooden trestles as piers; reanchorage, etc. In case of destruction the neighboring R. R. bridges would afford easy crossing which could be facilitated by laying heavy planks along the ties."
Repair or construction materials in vicinity: "Material easily obtained in the supply houses and shops of El Paso."
Remarks: "This bridge appears to have been constructed for use of the laborers who live on the Texas side but who work in the brick yards directly across on the right bank. Photograph accompanies."
Rough estimate of time required for emergency repairs or temporary construction to admit crossing of trains
    (a)   In case of total wrecking or destruction of bridge: "One to two days."
    (b)   In case of damage to extent roughly estimated as most probable of
        encounter: "One day or less."
Location and description of nearest fords above and below: "Position is variable. Animals and men may cross when the river is sufficiently low. (See special report on river bottom and fords.)"
Location and description of nearest bridge above and below: "El Paso and Southwestern R. R. bridge about 300 yards above. (Separately described.) International wooden wagon bridge about 2½ miles downstream. (Separately described.)"
Defensibility: Report here under heading "Defensibility" estimate of

(a)   Measures likely to be taken by enemy to defend crossing: "Burning or cutting down the wooden trestle piers. The region is rather desolate with poor communication with Juárez on the Mexican side. The position would hardly be occupied by a Mexican enemy but roving bands might hover in the vicinity with an object of damaging the splendid R. R. bridges above."

(b)   Measures recommended to be taken by our forces to minimize damage to crossing and to seize and utilize same in best manner: "Continuous guarding by a detail from the detachment that will be necessary to guard the two large R. R. bridges nearby. As only U.S. territory is concerned, both banks can be occupied before the outbreak of hostilities."

Photo 3.10 Suspension Bridge Northwest of El Paso

8532-4

U. S. CAMP AT EL PASO, TEXAS.

September 18, 1915.

From;   Major F. A. Wilcox, 6th Infantry, Intelligence Officer.

To;   Department Engineer, Southern Department, Fort Sam Houston, Texas.

Subject;   Forwarding information data.

1. I submit herewith the following military notes and data prepared by me in pursuance with instructions received. All data is submitted in duplicate;

MEXICO NORTHWESTERN R. R. FROM JUAREZ TO CHIHUAHUA.

MEXICAN CENTRAL (NATIONAL) R. R. FROM JUAREZ TO TORREON.

Documents Cited

MEXICAN TROOPS IN OR ABOUT JUAREZ, MEXICO.

2. Maps of El Paso and Juarez are being compiled and prepared, accompanied with panoramic photographs. I expect to forward these with notes on the two cities at an early date.

[signed] F. A. Wilcox

6 Inclosures.

Confidential

B8-10   1st Ind.   HHR-vve

Office Dept. Engr., So. Dept., Ft Sam Houston, Texas, Sept. 20, 1915. – To the Chief of Staff, Southern Department., Fort Sam Houston, Texas, for forwarding to Chief, War College Division, Washington, D. C.

Henry H Robert
Captain, Corps of Engineers.
Department Engineer.

3 Inclos.

2d End.

Hq. Southern Dept., Fort Sam Houston, Texas, Sept. 21, 1915.

To Chief, War College Division, General Staff, Army War College, Washington, D. C.

W H Hay
Major, General Staff,
Chief of Staff.

3 encls.

WHH-HS

U.S. Camp at El Paso, Texas.	CONFIDENTIAL.
September 17, 1915.	Class E.
By Major F. A. Wilcox, 6th Infy.	Dept. Engr. So. Dept.

MEXICAN TROOPS IN OR ABOUT JUAREZ, MEXICO.

1. Locations, numbers and composition of troops at present in town;
    a. Infantry- Estimated at 300, located as follows:-
        150 at Barracks, 100 yards west of City water tank.
        100 at Barracks, Eastern suburbs of City.
        50 at Barracks, about 100 yards west of Grand-stand at the Race Track.
    b. Cavalry-Estimated at 300, located as follows:-
        150 at Barracks, 100 yards west of City water tank.
        100 at Barracks, Eastern suburbs of City.
        50 at Barracks, about 100 yards west of Grand-stand at the Race Track.

    c. Artillery- None about Juárez.

    d. Machine Guns- 6 located as follows:-

        4 at Main Barracks at City Hall near City water tank.

        2 at Barracks in Eastern suburbs.

    e. Technical Troops- None excepting medical men, number unknown. These are constantly changing and their places of service are mostly on trains.

2. Names of principal leaders;

    General Tomás Ornelas, Commander-in-chief.

    General Ochoa, Second in command.

    Major De la Paz, Adjutant or Executive Officer.

3. Character of personnel and allegiance;

    Ornelas is pleasant but lacks strength of character.

    Ochoa and De La Paz are arbitrary and over-bearing and much stronger in character than Ornelas. Ochoa is considered by many as a "bad one."

    The subordinate officers are for the most part of mediocre type and are not of high principal or training.

    The enlisted personnel is of the usual peon class. Some are decrepit old men and there are boys not over 16 years of age.

    Allegiance;- These troops are at present Villaistas but a secret plot appears to be under way to turn over to Carranza. The power of Villa in this vicinity is waning and it is probable that by October 1st the garrison will all be Carranzistas.

4. Kind of arms and approximate amount;-

    Mauser, 7 m-m: Winchester 30-30 and 30-40.

    All officers and non-commissioned officers carry pistols of different makes in addition to their rifles or other arms. All appear to be fully equipped and armed. The arms and equipment on hand in excess of this appears to be small.

    At the Main Barracks, there is a quantity of home-made hand grenades made from 2" pipe, estimated at half a car load.

5. Kind and approximate amount of ammunition;-

    Figures are not at present obtainable but the amount appears to be moderate. During the past six months much ammunition is reported to have passed over the border but here such ammunition has been mostly shipped on further south.

6. Field telegraph and telephones.

    No such field instruments are known. Military operators are located at all important points on both the Mexican Central and Mexico North- western rail-roads. Both of these railroad telegraph lines are directly connected with military headquarters on Calle Lerdo, near cor. of Calle Comercio.

7. Aeroplanes;-

None in or about Juárez.

8. Wireless:-,

One tall mast has been completed and erected and a second one is under construction. The wireless is not believed to be in condition for any possible use at present.

9. General condition of troops;-

The men are fairly well clothed, fed and equipped. They appear generally to be uneasy and discontented.

10. Sources of supply of food and clothing;-

Practically all clothing is purchased in the United States.

Food is supplied from United States and from Mexico. The latter source of supply is becoming rapidly depleted.

11. Stores, factories or other sources of supply of arms, ammunition or explosives;-

None in or about Juárez. There is a dynamite factory which is located on a spur, about 26 kilometers in length, branching off the main line of the Mexican Central at about kilometer No. 1179.

12. Probable attitude of troops in event of invasion by the United States;

Judging from the character of the personnel of the forces at present at Juárez, a strong resistance could not be offered against an invasion from the United States and it is not probable that resistance could be very serious. It is an accepted fact that a great number of the private soldiers at present enrolled in the Northern Revolutionary movement would voluntarily retire from military service if permitted to do so.

13. Important military operations in progress or in prospect;-

Just at present there are no operations in progress in this vicinity. From authentic and confidential sources it appears Villa, with his remaining forces, is planning to withdraw from his present field of operations. He is trying to arrange for the Mexican Central rail-road to bring him north to Juárez, thence taking the Mexico Northwestern R.R. to the south-west with an apparent view of getting into Sonora Province, where he appears to have his greatest hold and where he will be in more friendly territory.

The indications are that should Villa not make this move before October 1st, conditions in Juárez may have changed so that he would find it very difficult to pass through Juárez.

U.S. Camp at El Paso, Texas.   CONFIDENTIAL.

September 7, 1915. [should be September 17?]   Class B.

U.S. Army Intelligence in the Mexican Revolution

By Major F. A. Wilcox, 6th Infy.   Dept. Engr. So. Dept.

MEXICO NORTHWESTERN R. R. FROM JUAREZ TO CHIHUAHUA

1. This road is at present operated by the Mexico-Northwestern Railway company with general offices in El Paso and Juárez.
   J. O. Crockett is the general head of the company with headquarters in El Paso.
   J. J. Pruett is the general manager with headquarters in El Paso.
2. The entire line is being operated at this time.
3. The duration of the present service depends largely upon the whims of the faction in control of the territory through which the rail-road runs.
4. The present traffic consists of about five trains daily into El Paso. About one-fourth of this is passenger traffic.
5. The rails and road-bed are in good condition. The ties are in bad condition. The rolling stock is in very bad condition and very scarce, most of it having been confiscated for military purposes. The motive power is in very bad shape.
6. All repairs are being made that are absolutely necessary and such repairing is generally made by the Company itself.
7. The present condition of the telegraph lines paralleling the tracks is very good.
8. These telegraph lines are operated by the rail-road company.

U.S. Camp at El Paso, Texas.   CONFIDENTIAL.
            September 17, 1915.   Class B.
By Major F. A. Wilcox, 6th Infy.   Dept. Engr., So. Dept.

MEXICAN CENTRAL (NATIONAL) R. R. FROM JUAREZ TO TORREON

1. This road is now being operated by that faction of the Revolutionists controlled by General Villa, as far as the City of Torreón. From there to the south it is operated at irregular intervals by forces under the leadership of Carranza. It is probable that in the near future the control of the road will be extended from Torreón to the north by Carranza.
2. The operation of this road between El Paso and the towns and cities to the south is very irregular and is limited mostly to passenger traffic.
3. About two passenger trains per day arrive in Juárez, and on an average of twice per week, a special freight train arrives, consisting principally of confiscated property being brought to El Paso for disposal.
4. The duration of the present service is dependent entirely upon the progress of the revolution. General Villa will probably cause the road to continue to be operated substantially as it is now, so long as he can maintain control of the territory through which the line runs.

5. The road-bed is in fair shape and the rails are in good condition. The ties are, however, in very bad shape. The motive power is very scarce and in very bad condition and greatly in need of repairs. The rolling stock is also very scarce and in bad condition.

6. Only such repairs are made as are absolutely necessary. The road-beds and bridges are continually being torn up by small bands of bandits or by parties opposed to the faction in control of the road.

7. The telegraph lines paralleling the tracks are in fairly good condition.

8. These telegraph lines are operated by the Military faction in control of the road. Their use is reserved almost exclusively for military purposes.

<u>8532-5</u>
U. S. CAMP AT EL PASO, TEXAS.

September 25, 1915.

From: Major F. A. Wilcox, 6th Infantry, Intelligence Officer.
To: Department Engineer, Southern Department, Fort Sam Houston, Texas.
Subject: Forwarding information Data.

1. I submit herewith the following military notes and data prepared by me in pursuance with instructions received;

    MEXICAN TROOPS IN OR ABOUT JUAREZ, MEXICO. (In duplicate.)

    6 Photographs of views in El Paso, Texas, and Juárez, Mexico, showing country along the Rio Grande and the high ground commanding the two cities. (In duplicate.) [not in file "Maps forwarded as enclosures to this report filed in Map Section as Map No. 618 Mexico 121 Texas, 619 Mexico 121 Texas"]

    GEOLOGY AND WATER RESOURCES of TULAROSA BASIN, NEW MEXICO. (With Maps.)

2. Under separate cover I am forwarding by registered mail the following;

    MAP OF JUAREZ, MEXICO. (In duplicate.)

    MAP OF EL PASO, TEXAS. (In duplicate.)

    MAP OF MEXICO-NORTHWESTERN RAILWAY.

    Panoramic View of El Paso and Juárez; Taken from Mount Franklin.

15 Inclosures.   F.A. Wilcox

Recd. O.D.E., S.D. SEP 27 1915
Confidential
B8-12   1st Ind.   WKA-vve

U.S. Army Intelligence in the Mexican Revolution

Office Dept. Engr.,So.Dept.,Ft. Sam Houston, Texas, Sept. 28, 1915. – To Chief of Staff, Southern Department, for forwarding to Chief, War College Division, Washington, D. C., with the following inclosures:

    1. Report; "Mexican Troops in or about Juárez."

Under separate cover are being forwarded by registered mail the following:

    1. Six photographs of views in El Paso, Texas and Juárez, Mexico.

    2. Map of Juárez, Mexico.

    3. Map of El Paso, Texas.

    4. Map of Mexico Northwestern Ry.

                    Henry H Robert
                    Captain, Corps of Engineers,
                    Department Engineer.

Confidential.

B8-12   2d End.

Hq. Southern Dept., Sept. 28, 1915. - To Chief, War College Division, General Staff, Army War College, Washington, D.C.

                    W H Hay
                    Major, General Staff,
                    Chief of Staff.

whh-kls

U.S. Camp at El Paso, Texas.	CONFIDENTIAL.
September 25, 1915.	Class E.
By Major F. A. Wilcox, 6th Infy.	Dept. Engr., So. Dept.

MEXICAN TROOPS IN OR ABOUT JUAREZ, MEXICO.

Military operations in progress.

1. During the week beginning Monday, September 20th, an extensive movement of the troops under General Villa has been in progress. Up to to-day (the 25th instant), 15 train loads of troops, arms, ammunition, horses, livestock, supplies and camp followers have arrived from the vicinity of Torreón on the Mexican Central Railroad and have departed from Juárez on the Mexico Northwestern Railroad with Casas Grandes as the undoubted objective or point of disembarkation.

From authoritative sources, 7 trains have arrived at Casas Grandes. The eighth train was badly wrecked on September 23-24, at Kilometer 175, with a considerable loss of life. This wreck occurred on a "Shoo-fly" that had been built around a bad break in the road.

It is reported the remaining trains will be blocked until another "Shoo-fly" is built around this wreck.

286

Another train wreck of lesser severity occurred on September 24th at Kilometer 122.

The total number of troops being moved is estimated at about 12,000. [this may be where estimate of Villista invasion force came from.]

From 10 to 15 trains are yet due to arrive and pass through Juárez.

2. The first refugee train arrived this morning (September 25) from Chihuahua. Refugees report General Villa has entirely abandoned the country about Torreón, which is being occupied by Carranza forces.

General Villa is reported still at Chihuahua.

They report that at Chihuahua, every bit of rolling stock obtainable is being utilized to bring the remainder of Villa's army and several additional train loads of refugees to the north. The whole trackage thereabouts is congested and blocked with these trains.

The rolling stock is in poor condition and the locomotives in wretched shape and having continual break-downs.

3. Most of the cars used are box-cars with a lesser number of flat-cars and occasional passenger coaches.

The first trains out of Juarez bore field artillery - amount not ascertainable.

The fifteenth (and last) train out had nine cars of ammunition. This ammunition was brought up with the troops from the south and was probably the source of the report appearing in the daily papers, that a large amount of ammunition had been sent over the border and received at Juárez.

This report was erroneous.

4. Great quantities of horses, cattle, goats and other live-stock are being transported, the box-cars being practically filled with it. This is further evidence that the plan of Villa is to break away from his trains at Casas Grandes and march westward into Sonora Province where the people are better off and the country less devastated by the war, and where he may expect to count on the support of the Yaqui Indians.

5. It is reported Villa plans burning all his rolling stock before leaving Casas Grandes. Reports also continue to be received that many of Villa's subordinates are becoming more and more dissatisfied, and may soon turn over to Carranza. It is further reported this element is plotting to save the trains at Casas Grandes so as to be of use in turning over to Carranza.

6. With field glasses at short range and by other confidential means, I have been able to observe several of the Villa trains. The box-cars are mostly filled with horses and livestock. The roofs of the cars, and even the trusses underneath, are thickly occupied by soldiers and a large number of women and children. Beds are improvised beneath the cars and cooking is done on the tops. Others find places in the flat cars.

7. The condition of all the personnel is dirty, and for a greater part, filthy. The soldiers in general appeared listless and like so much cattle. Boys not over 13 or 14 years of age were observed in uniform.

There was much drunkenness among the soldiers in Juárez and it is believed there were many desertions from the ranks.

The women, many of whom had with them young children, seem to be the inseparable companions of the men.

Many of the officers were clean in appearance and well dressed.

All arms were taken away from the soldiers and locked up in the cars.

Much new clothing and other supplies were issued in Juárez.

8. No changes appear to have been made in the permanent garrison of Juárez (previously reported on).

The rumor grows that this garrison will turn over to Carranza if he continues to gain in power.

F. A. Wilcox

8532-6

U. S. CAMP AT EL PASO? TEXAS.

September 28, 1915.

From: Major F. A. Wilcox, 6th Infantry, Intelligence Officer.
To: Department Engineer, Southern Department, Fort Sam Houston, Texas.
Subject: Forwarding Information Data.

I submit herewith the following military notes and data prepared by me in pursuance with instructions received;

EL PASO, TEXAS (Class C.)   In duplicate.
JUAREZ, MEXICO. (Class C.)   " " "

F.A. Wilcox

4 Inclosures.

Recd. O.D.E., S.D. SEP 30 1915
Confidential
B8–13   1st Ind.   HHR–vve

Office Dept. Engr.,So. Dept., Ft. Sam Houston, Texas, Sept. 30, 1915. – To Chief of Staff, So. Dept., for forwarding to Chief, War College Division, Washington, D. C.

Henry H Robert
Captain, Corps of Engineers,
Department Engineer.

Confidential   2nd End.

B8-13

Hq. Southern Dept., Sept. 30, 1915. – To Chief, War College Division, General Staff, War College, Washington, D. C.

W H Hay

Major, General Staff,

Chief of Staff.

whh-kls

U.S. Camp at El Paso, Texas.	CONFIDENTIAL.
September 28, 1915.	Class C.
By Major F. A. Wilcox, 6th Infy.	Dept.Engr., So.Dept.

EL PASO, TEXAS.

– – – – – – – –

1 & 2. EL PASO and its location are generally known.

3. The population is estimated by those in authority as about 65,000, including the Smelter District which is really outside the city limits to the west.

Of these there are at least 35,000 of the Mexican race, about 4,000 of which live at the Smelter. It is estimated there are 5,000 Mexican men in El Paso, mostly unemployed and not permanent residents, who have either been in the Mexican Army or who are here to escape service in the same. The bulk of these men are reported to be of bad character.

4. Names of all streets are shown in the City map submitted.

5. There are practically no frame buildings in El Paso. The better buildings are of brick. In the Mexican section of the town, along the river, the greater part of the buildings are of one story adobe. A few of the largest buildings are of concrete and steel, such as the 12 story Mills Building. (See Panoramic Photograph.)

6, 7 & 8. The location of depots, stations, offices, shops, and warehouses are shown on the City map or given in the notes on or attached to the map. They are shown also, for the most part, in the Panoramic Photograph.

9. Large open spaces suitable for camping are shown in yellow on the City map. The space by the river, now occupied by the 6th and 16th Infantry, has about 250 acres available for camp sites. The space east of Golden Hill contains about 50 acres suitable for camping.

At Fort Bliss and on the land adjoining, two or more divisions could be encamped. The City, Fort Bliss and Railroad pumping plants are all nearby.

The Panoramic Photograph shows the large buildings that stand out by themselves.

10. The water supply has been separately reported on. See also the Geology and Water Resources of Tularosa Basin, New Mexico.

11. The mesa and hills to the immediate north of El Paso command the entire City. This is shown on the maps and photographs. There are no artificially constructed fortifications.

Juarez is within long artillery range from these heights.

12. Maps and photographs have been submitted.

F. A. Wilcox

U.S. Camp at El Paso, Texas.	CONFIDENTIAL.
September 28, 1915.	Class C.
By Major F. A. Wilcox, 6th Infy.	Dept.Engr., So.Dep.

JUAREZ, MEXICO.

1. JUAREZ. The name is pronounced as though spelled with an "H" instead of a "J." The pronunciation is generally known.

2. Its location is on the Rio Grande River opposite El Paso, Texas. Its position is shown on all maps of Mexico and on the maps accompanying the reports from this station and vicinity.

3. The population of Juarez and immediate vicinity in Mexico is about 8,000. These are practically all Mexicans. Americans or other foreigners having dealings or occupation in Juarez live, for the most part, in El Paso or are but temporarily or transiently in Juarez.

Formerly the bulk of the people were small farmers and laborers. Many are now supported by the wastes of lawless occupations in Juarez and many have little apparent means of livelihood.

4. Names of principal streets are shown on the map of Juarez submitted.

5. Nearly all the buildings are one-story adobes. Some of the principal buildings in and near the business center of the town are of brick.

A few of these are two stories in height. In the more settled parts of the town the buildings are built close against each other.

6. The location of the principal buildings and offices is indicated on the map submitted.

7 & 8. There are no large store houses or elevators. Most of the food supplies now come principally from United States though cattle are occasionally brought up through Mexico. There seem to be no important blacksmith shops or wagon shops. Outside of the railroad yards there appear to be no machine shops.

9. From Calle de la Constitution to the east and south-east toward, and including, the Race Track, the country is quite open and level with but few buildings. Two or three brigades could easily camp here.

The city car line passes through this area. Water is available from the railroad tanks and the very large tank supplying the Race Track buildings and stables. There are about 2/3 of a mile of stables around the race track furnishing unusual accommodations for mounted troops.

The most conspicuous outstanding buildings are the school house near the Juárez Monument and the large stands and buildings of the Race Track.

10. The water supply of Juarez has been separately reported upon.

11. The mesa to the south and west touches the town limits and commands the town. This is shown on maps and photographs submitted.

There are no artificial fortifications. The dykes of the existing and abandoned canals and ditches afford very good cover for riflemen.

The large number of brick and adobe walls and buildings are easily adapted for defense against rifle fire.

12. Maps and photographs of the region have already been submitted.

A large panoramic photograph, about five feet in length, is being prepared of Juárez, being taken from a very high water tank near the international bridges. This photograph and the large one already submitted of El Paso will show practically all the ground and structures commanding the city of Juarez.

F.A. Wilcox

8532-7

Refer to File
No. B35-12 DE.

War Department
Headquarters Southern Department
Fort Sam Houston, Texas.
Sept. 25, 1915.

From: Chief of Staff
To: Commanding General, Eighth Brigade, El Paso, Texas.
Subject: Translation of certain extracts from Mexican newspapers.

1. Herewith translations of certain extracts from local Mexican papers furnished this office by Mr. W. L. Gibson.

2. Please have your Intelligence officer verify the information herein contained so far as practicable and return it to these Headquarters with any additional information of value that may be available, at the earliest practicable date.

By command of Major General Funston:

W. H. Hay
Major, General Staff.

1st Ind.

Hq. 8th Brigade, Fort Bliss, Texas, October 1, 1915. To the Comdg. Gen., Southern Dept., Fort Sam Houston, Texas.

1. Returned· Attention invited to confidential memorandum herewith.

2. There seems to be little doubt that General Ángeles and his staff have determined to leave Villa's service.

John J. Pershing
Brig. Genl., Comdg.

Rec'd Back A.O.S.D. OCT 5 1915 To Dept. Eng.

Recd.bk. O.D.E., S.D. OCT 5 1915

B35–12 D.E.   2d Ind.   HHR–vve

Confidential

Office Dept. Engr., So. Dept., Ft. Sam Houston, Texas, Oct. 5, 1915.- To Chief of Staff, for forwarding to Chief, War College Division, Washington, D. C.

Henry H. Robert
Captain, Corps of Engineers,
Department Engineer.

B35–12 D.E.   3d End.

Hq. Southern Dept., Fort Sam Houston, Texas, Oct. 5, 1915.

To Chief, War College Division, General Staff, Army War College, Washington, D. C.

W H Hay
Major, General Staff,
Chief of Staff.

3 encls.

WHH–HS

Fort Sam Houston, Texas,
September 25, 1915.

MEMORANDUM FOR CAPTAIN ROBERT:

A special telegram to *La Prensa*, a Mexican daily of San Antonio, dated El Paso, Texas, September 24th, reads as follows:

a. "Another brigade of Villista troops arrived today at the neighboring city of Ciudad Juárez, leaving at once for Casas Grandes over the Mexico Northwestern Railway.

b. Villa is trying to show the Americans who live in the neighboring city that he has an army of more than twenty five thousand men and the movement of troops which has been in the course of carrying out for some days is due to this idea.

It is said that in order to produce this effect the Chief of Operations (Villa) has availed himself of a stratagem which is giving him good results, because he has caused it to be believed that he has this number of troops. The stratagem referred to is as follows: To order troops along the line of the Central Railroad of Mexico, which links the neighboring city of Ciudad Juárez with the capital of Chihuahua, from Chihuahua City to Juárez and ordering them to leave at once over the Mexico Northwestern Railway for Casas Grandes; but upon arriving at Mesa (twelve and one half miles from Juárez) these forces leave their train and cross to the Central line, which is crossed by the Mexico Northwestern near that station, and there they occupy waiting empty trains, being at once again brought into Ciudad Juárez. In this way the same troops are arriving and leaving the nearby city (Juárez) and making it appear that the Chief of Operations (Villa) has at hand a large number of troops.

c. It has become known in this city that General Felipe Ángeles, who with five of his officers were stopped here when they came over to this side a few days ago, declared under oath to the American authorities that they had left the Villa cause and also that the ex-Federal artillerist has also declared that Villa was not the man called upon to pacify the Republic. The former chief of the Villa artillery also said that since the defeats of León and Celaya, he did not receive any further orders from Villa, neither for himself nor the officers of his staff, and that taking this fact into consideration, he had resolved to separate himself completely from the Villa cause, although in a friendly manner. The fact that Ángeles has received instructions from Villa to go to Washington on a special mission shows the foregoing declarations, which have been attributed to Ángeles, when the officers of his staff were detained in this city, to be without weight."

A telegram from El Paso to *El Presente*, a daily of San Antonio, claiming to be independent of all Mexican factions, is as follows:

d. "El Paso, Sept. 24. Yesterday the division of General José D. Rodríguez, composed of more than two thousand five hundred men, arrived at Ciudad Juárez and left immediately for the District of Casas Grandes.

Within the past few days between eight and ten thousand men have passed through Ciudad Juárez for Sonora, all well-armed and munitioned.

Considerable interest has been awakened in this great mobilization that the Commander of the Division of the North has been making, and it is believed that it is sure that he will abandon the Laguna region (around Torreón) and perhaps part of the State of Chihuahua.

e. It is also believed that in the vicinity of Torreón, as well as between Torreón and Jiménez (Chihuahua), Villa has left numerous detachments that will begin to fight in guerrilla bands as the Carrancistas advance, thus greatly complicating the situation."

A telegram from El Paso to *La Raza*, a Carrancista daily of San Antonio, is as follows:

"El Paso, Texas, Sept. 24. Last night it was reported in this city on good authority that the Villista leader José Rodríguez, who was reported to have been killed in a fight near Aguascalientes, arrived at Ciudad Juárez with some of the men of those who compose his Division.

f. The commentaries that have been made in this city upon the newest phase of the campaign in the State of Sonora, have been most varied but the general opinion is that the campaign will be another failure for the reactionaries, because apart from the fact that they are going to operate in an unknown territory, they will encounter a great number of troops of the Constitutionalist (Carrancista) forces, who have invaded that state in order to fight Maytorena as soon as possible, and not give Villa time to arrive with his hordes.

At present no one doubts that if at the end Villa should carry out these movements, the evacuation of Torreón and Chihuahua would of necessity follow.

g. Today a message of Miguel Díaz Lombardo was published in which the news that Villa was about to evacuate the cities of Torreón and Chihuahua was denied, but this news does not agree with everything else that has been noted on seeing the flight of the hostile forces, from the places where they have been dominant for so long a time.

It is also confirmed that as sent in my last message last night, that Villa is staying in the neighborhood of Ciudad Juárez, but has not made his entrance into that city.

A telegram from El Paso to *El Presente*, dated September 24, further says:

"The Minister of Foreign Affairs in the Villa Cabinet, Miguel Díaz Lombardo, telegraphed from Chihuahua denying in a positive manner that Durango and Torreón had been abandoned by the Villistas who yet remain in those places.

h. Villa himself is south of Juárez and according to passengers who have talked with him and who came from Chihuahua, he denies that his forces have left Torreón. It is almost sure that those places have been left with very small garrisons, just enough to preserve order, and with instructions that as soon as the Carrancistas approach to evacuate the places and retire to the north.

An El Paso telegram to *La Prensa*, dated September 24, says:

i. "The military authorities have suspended passenger train service for one week between Ciudad Juárez and Chihuahua. Yesterday the authorities of our neighboring city received an order from Villa that passenger traffic would remain suspended because it was necessary to use all the rolling stock on hand for military operations and the transportation of troops. The train which should have left last night for the City of Chihuahua, was annulled. This fact gave rise to rumors that the suspension of traffic was due to the fact that the line had been torn up by the enemy, but these rumors were dissipated when other trains came into Juárez bringing more troops from Chihuahua. Villista forces have continued to leave for the district of Casas Grandes. It is calculated that in the last four days something of more than five thousand men have left for that district. All will go to Sonora in compliance with a new plan that Villa is carrying out or is trying to carry out.

j. A Villista captain named Holguin was deported yesterday by the American military authorities, who turned him over to the authorities of Ciudad Juárez with the notice that on three occasions this captain had been arrested on American territory and that he must abstain from crossing the frontier in future, without getting previous permission from the Government of the United States. This captain belongs to the forces now garrisoning Ojinaga and each time he has had to come to Juárez on military business, he has made the trip across American territory because it is easier. Holguin was brought to this city under guard of an escort of the 13th Cavalry and taken to the International bridge by the same patrol.

k. Due to the great number of troops which have been coming to Ciudad Juárez from the south and in obedience to orders of Villa, all the places where alcoholic liquors are sold in Juárez have been closed and will remain closed until further orders. The authorities are exercising a strict watch in order to stop the sale of a drop of alcohol to the soldiers and even to civilians, because it is desired to avoid any disorder. The soldiers of Villa are becoming converted into teetotalers, because in every town that they come into they find the drinking places closed by orders of Villa."

W. L. Gibson

Confidential

B35-13 D.E.   1st Ind.   HHR-vve

Confidential

Office Dept. Engr., So. Dept., Ft. Sam Houston, Texas, Sept. 25, 1915.– To Chief of Staff, for his information and for forwarding to the Chief, War College Division, General Staff, Washington, D. C.

Captain, Corps of Engineers.

U.S. CAMP AT EL PASO, TEXAS.

September 29, 1915.

From: Major F. A. Wilcox, 6th Infantry, Intelligence Officer.
To: Chief of Staff, Southern Department, Fort Sam Houston, Texas.
        (Through Commanding General, 8th Brigade.)
Subject: Verification of extracts from Mexican newspapers.

I am forwarding herewith, in duplicate, a report on the probable truth of extracts appearing in Mexican newspapers and referring principally to the movements of troops in and about Juarez, Mexico, and to the south of that city.

F. A. Wilcox

3 Inclosures.

U.S. Camp at El Paso, Texas.   CONFIDENTIAL.

U.S. Army Intelligence in the Mexican Revolution

September 29, 1915.

by Major F. A. Wilcox, 6th Infy.

Class E.

Dept.Engr., So. Dept.

MEXICAN TROOPS IN OR ABOUT JUAREZ, MEXICO.

1. MEMORANDUM FOR CAPTAIN ROBERT: dated Fort Sam Houston, Texas, September 25, 1915, having been referred to me for verification, I submit the following as the best information now obtainable.

For convenience of reference I have designated with blue pencil subparagraphs of the MEMORANDUM.

a. On September 17, I reported that plans were being made for the movement of Villa's troops into Sonora on the Mexican Central and Mexico-Northwestern Railroads through Juárez. On September 25, I reported with some detail the movements that had taken place up to that date. 15 train loads of troops, arms, ammunition, horses, live-stock, supplies and camp followers had then arrived en route to Casas Grandes. Up to date 19 such trains have passed and more are en route from Chihuahua.

b. The story that Villa is making this movement to show off to the Americans is not believed by anyone conversant with facts. As stated in previous reports, he appears to be moving fully 12,000 troops; this is not counting the camp-followers.

The story that he is taking troops down on one railroad only to be brought back on another is too absurd to be considered. The trains have all been accounted for and their location vouched for by a high and confidential official of the Mexico-Northwestern Railroad, which is furnishing the transportation for the movement.

The troops were undoubtedly brought through Juárez on account of better trackage conditions and for re-outfitting at that city. It is known that a considerable amount of uniforms and other supplies were issued in Juárez.

c. There seems to be no authority for the story that General Felipe Ángeles and his staff had left the Villa cause. In a dispatch to the New York Times made on arrival of General Ángeles in Washington, September 25, General Ángeles issued a statement "professing his continued loyalty to the Villa Government and denying that he had deserted the Conventionist Army."

d. The report of the movement of between 8,000 and 10,000 troops through Juárez appears to be substantially correct.

e. The report that Villa has left numerous detachments about Torreón and Jiménez can not be substantiated at this time. It does not seem probable.

f. Discussion of the possible campaign in Sonora is little more than commentary. Villa is undoubtedly getting out of Chihuahua because his credit and power are weakening, because the country is so depleted by the war and because he will find a country richer in harvests and people more friendly in Sonora. He is undoubtedly counting on the support of the Yaqui Indians.

g. Reports from refugees and railroad officials with inside information seems to leave but little doubt but that Villa is abandoning the region from Torreón to Chihuahua.

From the last authentic reports Villa was still at Chihuahua. He has not yet been near Juárez. This morning's papers reported he had gone by a direct route from Chihuahua to Casas Grandes. This has not yet been verified by the officials of the Mexico-Northwestern Railroad.

h. My remarks under "e." and "g." appear to cover this.

i. This report seems to be fully in accordance with facts. The congestion of tracks due to the bringing in of troop trains and trains of refugees, together with the bad condition of rolling stock and the breaking down of engines, has made the running of scheduled passenger trains quite impossible.

j. The taking and deporting of the Villista captain named Holguin has been verified.

k. Orders may have been issued closing saloons in Juárez but of my own knowledge much liquor has been drunk in Juárez since the arrival of the troop trains and drunken men have been much in evidence. Immigration and customs officials have reported to me that many drunken men have come down to the bridges but have been sent back into the town of Juárez.

Saloons *have* been open during the occupation of the town by troops.

2. From the latest authentic reports, 19 troop trains have passed through Juárez up to date en route to Casas Grandes. Three more are reported on the way from Chihuahua, one of which however has now arrived in Juarez. 8 locomotives, with all cabooses, water tank cars and all foreign cars have been ordered back from Casas Grandes, presumably to go back to bring up more trains from Chihuahua.

F. A. Wilcox

AGOVR Box 7646, October 1, 1915
WESTERN UNION SPECIAL

Copies furnished to
>Telegraph Division to Secretary of War;
>Ass't. Secretary of War;
>Chief of Staff;
>Ass't. Chief of Staff.

Number 21 W WM Sheet 135 Govt. Night

Dated   Fort Sam Houston, Texas, Oct. 1, 1915.

To Adjutant General Army,
>Washington, D.C.

Number seven twenty three period.

Following received ten fifty P.M., this date from Colonel Frier Nogales, Quote:

Governor Maytorena and Staff officers released six thirty this evening in conformity your telegram date period. Maytorena accompanied by Alberto Morales legal adviser Castillo Brito ex-Governor State of Campeche and Roberto Almeda leave here tomorrow for Washington period.

General Leyva and number other officers endeavoring to quit indicate disintegration period.

Acting Governor Randall informed American Consul tonight Maytorena had left only three hundred and fifty odd dollars in State Treasury. If the desertion of cause by Maytorena's staff officers had crippled the office he could not do the work nor control the Army left with him in Nogales and would quit himself if conditions did not improve in next few days, unquote.

Funston.

[penciled in: "Copy furnished State Dept. 10/2/15."]

9:27 A.M., October 2nd.

WESTERN UNION SPECIAL

Number 1 A WM Sheet 168 Govt.

Dated   P Fort Sam Houston, Texas, Oct. 1, 1915.

To Adjutant General Army,   Copies furnished to

Washington, D.C.   Telegraph Division to Secretary of War;

Number seven twenty three period.   Ass't. Secretary of War;

Chief of Staff;

Ass't. Chief of Staff.

Reference my seven nineteen Commanding Officer Nogales reports Governor Maytorena crossed International boundary nine fifteen last night was detained and placed under guard of American soldiers in his house Nogales Arizona period.

Colonels De Lavega and Flores were arrested at ten twenty last night period. Each had paper purporting to be a discharge from service in Villista army period. They were released on parole to report to Commanding Officer Nogales ten o'clock this morning period. Both state they desire remain in United States on same footing as other Mexican citizens with authority to cross to and from Mexican territory as they see fit period.

Maytorena stated when arrested that he was en route to Washington to attend Pan American conference period.

Carlos Randall former state treasurer State of Sonora and General Urbalejo were announced as acting Governor and Commander in Chief Villa forces State of Sonora respectively period.

Instructions relative disposition General Maytorena and Colonels De Lavega and Flores requested.

Funston.

[Penciled in: "C of S says Sec War is taking this up personally with the State Dept + [illegible] will be sent this office later."]

Time 1:20 P.M.

11. The Bandit War

<u>5671-1020</u>

John J. Hainsworth

Chief Engineer

PORT BROWNSVILLE SUGAR LANDS COMPANY

    Located in Cameron County, Texas.   Rec'd W.C.D., G.S., DEC 3 1914

Brownsville, Texas., August 31, 1914.

Brig. Gen. D. C. Kingman,

        Washington, D. C.

Dear Sir:

I have been detained in getting out my new map of Mexico Coast.

This is my latest Mexican "War Dope" received direct today from Saltillo and Monterey through a personal friend of mine who makes my office here his headquarters. He is a French Engineer and Geologist employed by the "Madero" Group. This is the News:

Villa is with the "Madero" Group.

Villa will take such action as will force the U.S. to intervene in December next if possible.

Carranza will be forced out and Ernesto Madero former Minister of Finance takes his place as President of Mexico.

These matters were discussed by the "Maderos" as late as Saturday last.

This letter is of a confidential nature.

The Maderos know well that "Carranza" was in open revolt against "Francisco" before he was killed.

Yours truly,

John J. Hainsworth.

P.S. I am in close touch with all the prominent Engineers in Mexico.

John J. Hainsworth

Chief Engineer

    PORT BROWNSVILLE SUGAR LANDS COMPANY

U.S. Army Intelligence in the Mexican Revolution

Located in Cameron County, Texas.

Brownsville, Texas., September 7, 1914.

Brig. Gen. D. C. Kingman,
> Chief of Engineers, U. S. Army,
>> Washington, D. C.

Dear Sir:

Gen'l R. Cuellar formerly in command of the 6th Military Division (under Huerta) with Headquarters at Celaya, Mexico, and who retreated into Mexico City and came out of Mexico City with President Carbajal arrived here yesterday and is now at his son's home next door to me where he has been for past 3 years off and on.

He is a power among the Federals and a former advisor of Huerta. He says *neither* "Carranza or Villa" can bring peace to Mexico.

He can not state "openly" that intervention by U.S. is the *coming event* but he practically says so and that within 90 days.

Gen'l Cuellar was one of General Diaz fighters against Maximillian and is a man of wealth in State of Tamaulipas.

Gen'l Cuellar states in June the Rebels pressed them hard in the mountains and he never left the saddle for 6 days and nights in retreat towards Mexico City. I gave Captain Foy also Major Rice some "Dope" yesterday P.M.

Yours sincerely,

John J. Hainsworth.

P.S. U.S. better get on a good ready for intervention at once, before a lot of Americans get killed in Mexico.

5761-1030

Fort Sam Houston, Texas, October 5, 1915.

Memorandum for Captain Robert:

A special telegram to La Raza, of San Antonio, says:

Brownsville, Texas, Oct. 4, 1915.

General Eugenio López de Lara ordered to relieve Emiliano P. Nafarrate, as commander of the Constitutionalist garrison of Matamoros, had arrived at that port bringing a column of three hundred men with him, and it is thought that when General Nafarrate leaves Matamoros that he will take his own brigade with him to the place to which he may be ordered. Among the soldiers arriving with General López, there are about one hundred Tehuantepec Indians, who are considered to be warriors of a first class quality, especially for brush fighting and broken ground, and it is said that they were sent at once to patrol the river for the purpose of stopping the Mexican-Texan rebels from taking refuge on Mexican soil. (La Raza is the Carrancista organ in this part of Texas).

A special telegram to El Presente of San Antonio, says:

Brownsville, Texas, Oct. 4.

Last night unusual movements of the Carrancista troops in Matamoros were noted, without precedent since that place was attacked by the Villistas last March.

It is reported that a column under the command of General Eugenio Lopez has arrived and that he will relieve Nafarrate, who will have to leave later, as the American authorities and public sentiment in general accuse him of being the real author of the disorders that have taken place along the banks of the Rio Grande.

It is also said that these forces come with the plan of attacking the bandits who have concentrated on the Mexican side, opposite Progreso. Among the soldiers who arrived at Matamoros are more than one hundred Tehuantepec Indians, who are famed as fine guerrillas.

It is not known with certainty when Nafarrate will leave Matamoros, and meanwhile every day that passes (with him in command) holds a serious danger for the frontier.

(El Presente is said to be the Clerical paper of this region).

A special telegram to La Prensa, of San Antonio, says:

Brownsville, Tex. Oct. 3.

Nearly one thousand five hundred Carrancista soldiers have arrived at the port of Matamoros, coming from Monterrey and points between Monterrey and Matamoros. These fifteen hundred men will be stationed along the Rio Grande for a distance of seventy miles.

Fifty boxes of dynamite were exported two days ago to Matamoros. The consignment was invoiced to the commanding officer of that place.

Today it was told here that a train with cannon and artillerymen arrived yesterday at Matamoros. It is calculated that about two hundred soldiers came by this train arriving last night, and that several pieces of artillery came on the same convoy.

(La Prensa is the best newspaper in Spanish in Texas. It is the most impartial of all).

W. L. Gibson

5761–1032

Fort Sam Houston, Texas, Oct. 7, 1915.

Memorandum for Captain Robert: Rumors about Gen. Obregón.

U.S. Army Intelligence in the Mexican Revolution

Rumors from apparently reliable sources seem to indicate that General Obregón, commander of the Carrancista forces at Torreón, is keeping all his troops in that vicinity, waiting for the decision of the Pan-American Conference, and it is said, willing to be named as Provisional President of Mexico, in place of General Carranza.

It is said that Gen. Jacinto B. Treviño, commanding the Army Corps of the Northeast, who is under the command of Gen. Obregon, whose immediate command is styled the Army Corps of the Northwest, is in accord with Gen. Obregón, and has rather broken with Gen. Carranza with whom he took the field at Saltillo in Feb. 1913.

It is further rumored Treviño was very angry with Nafarrate, and that his promotion to general of brigade before being relieved was not in accordance with Treviño's ideas, but was a sop given to Nafarrate to keep him from breaking out at once. Nafarrate, it is now announced will go to Ciudad Victoria, Tamaulipas, and fit out his new command for a campaign against the hostiles under Alberto Carrera Torres, who are menacing the Tula district, in southwest Tamaulipas. Nafarrate was a member of the Rurales in the time of Diaz, and served in the Tula district for a long time.

The present governor of Nuevo León, Gen. Pablo A. de la Garza, is to be relieved by Niceforo Zambrano, and the governor of San Luis Potosí, Gabriel Gavira, is to be replaced by Gen. Vicente Dávila.

It is further announced that General Carranza is to leave Vera Cruz very soon for Tampico, thence to Monterrey and Saltillo. It is said he is going to make some reforms in the state of Coahuila, also, and that he will look after the disposal of the cotton crop of the Laguna, estimated to be worth eleven million dollars.

The idea that Gen. Carranza was about to proceed to the City of Mexico in person seems to have been abandoned of late.

W. L. Gibson

Confidential
B35-24 C.F.   1st Ind.   HHR-vve

Office Dept. Engr., So. Dept., Ft. Sam Houston, Texas, Oct. 7, 1915. – To Chief of Staff, for forwarding to Chief, War College Division, Washington, D. C.
                        Henry H. Robert
                        Captain, Corps of Engineers,
                        Department Engineer.

5761-1034
Fort Sam Houston, Texas, Oct. 7, 1915.

Memorandum for Captain Robert: Nafarrate ordered to Ciudad
                                                  Victoria, Tamaulipas.

Special correspondence to La Prensa of San Antonio dated Brownsville, Tex. Oct. 5, 1915 says:

The new military chief of the line of the frontier, General Eugenio Lopez, has taken over the command of all the troops that have been under the command of Nafarrate. The latter began to turn over yesterday, and finished up today. Nafarrate will go to Ciudad Victoria, where he has been ordered. But first, his friends and admirers will give him a banquet, where he will say farewell.

The former commander of the frontier line, who has given the press so much to talk about on account of his attitude in the Texan conflict [Plan de San Diego], will take charge of the campaign against the forces of General Carrera Torres, which have begun to display much activity in the Southwest part of the State of Tamaulipas. Nafarrate will take with him the men in his service who were with him in his former campaigns, as with them he can succeed in his task. In Victoria he will organize the troops that he takes with him to fight the Carreristas, and which will be turned over to him by the Governor, General Luis Caballero, who is interested in finishing up the revolutionists, that have invaded a large district of the state that is under his charge.

The Carreristas, as is known here, through private advices, have defeated various forces of Carrancistas, that were garrisoning the towns along the border of the state of San Luis Potosí, and have captured arms and ammunition, with which they are ready to even make more of a campaign against the troops of Caballero, who up to now have not succeeded in putting the revolutionists down.

Nafarrate was picked out to take charge of the campaign against the Carreristas because he knows the country well where he is to operate, having been stationed there some years, when he was in the Rurales in the administration of President Díaz.

These troops of General Alberto Carrera Torres, known in Mexico as the "Carreristas" have had their headquarters in Peotillas, State of San Luis Potosí. They claim to be an independent faction, but are hostile to the Constitutionalists or "Carrancistas". Very little seems to be known of their aims, plans, numbers or organization. W. L. Gibson

Confidential
B35-25 C.F.   1st Ind.   HHR-vve
Office Dept. Engr., So. Dept., Ft. Sam Houston, Texas, Oct. 7, 1915. – To Chief of Staff, for forwarding to Chief, War College Division, Washington, D.C.

Henry H. Robert
Captain, Corps of Engineers,
Department Engineer.

5761-1035

John J. Hainsworth

Chief Engineer

Brownsville, Texas, 10/4 1915

H H Robert Capt Corps Engr. U.S.A.

Ft Sam Houston

Dear Sir.

With the arrival at Matamoros of Gen'l Lopez from Torreón last nite there arrived with him a staff whose name I will get tomorrow also about 400 soldiers I should judge several of whom have worked around San Benito in Texas. etc in past. Gen'l Lopez is a Matamoros man where Father is now a milkman there.

The rifles of the men were in bad condition and about 2 wagon loads were hauled to the Carranza Armory and exchanged for repaired rifles.

Please refer to the International Rdy map Nov 1912. Sheet 24 = S. E. Co 8th and Manuel García "Calle" is located their repair armory, their aero-plane also there. It's a 1 story brick bldg. and was formerly the Post Office.

Many of the new Mex troops are Huasteca Indians about ½. I gave all this data to Col Blocksom at noon. I think Carranza Army rifles etc. in North Mex are in bad shape generally is my observation. I am told Gen'l Nafarrate's time departure is not set.

Respy, John J. Hainsworth.

5761-1036

John J. Hainsworth

Chief Engineer

Brownsville, Texas, Oct 3. 1915

H H Robert Capt C. Engr. U.S.A.

Ft S- Houston, Tex.

Dear Sir.

I gave enclosed news to paper. If someone could follow up this man Cabrera I feel he would gain much information relative to Texas side also situation – He speaks good English – I could not do so myself. A Carranza secret service man in Matamoros yesterday again informed me that many of the Bandits on Mex side are from around Cerralvo Mex. I am confident some Spanish mine owners have caused them to start out – anything to retard Carranza and perhaps cause intervention by U.S. is their hope. 2 troop trains left Matamoros this A.M. for points along river probably total 250 to 300 men.

I presume Gen'l Nafarrate will remain here 2 weeks longer. I was paid $75.00 yesterday out of expenses fund by Col Blocksom for past services. Americans in Matamoros territory look for trouble unless Carranza is recognized Oct 9/15. Probably some ugly showing against Americans. Cabrera has been with Carranza army around Reynosa-Cerralvo etc.

Yours Respy- John J. Hainsworth.

5761-1045

John J. Hainsworth
Chief Engineer

Brownsville, Texas, 10/24 1915

H. H. Robert Capt Corps Engr U.S.A.

Ft Sam Houston, Tx.

Dear Sir.

The Chief of Genl Lopez army secret service took breakfast with me in Matamoros this a.m. He has just rtd from a trip up as far as Camargo + states "La Rosa" was in Reynosa in disguise a few days ago. He states Genl Lopez has about 800 men between Matamoros + Camargo of which 600 are in citizens clothes + has issued orders to kill on sight any bandits. None of those bandits stop in the Mexico towns. I have given above to Col Blocksom.

On Oct 21 – A Chicago banking firm wired me if I could take charge of $100,000.00 Construction Investment near Victoria Tex for them. I have done work for the firm for past 2 3 years.

So I will try to see you at San Antonio some. I do not know how much longer I will be able to do this secret work for your Dept. Pres. V. Carranza is expected in Matamoros about Tuesday from today's reports.

Yours John J. Hainsworth.

5761-1046

John J. Hainsworth
Chief Engineer

Brownsville, Texas, 10/25 1915

Capt. H. H. Robert Corps Engr. U.S.A.

Ft Sam Houston

Dear Sir.

"U Blanco" Del Rio Texas registered at the Miller Hotel noon today. He is reputed to be a brother of "Gen'l Lucio Blanco" whom I stated a few days ago wanted to start another revolution. I am almost positive Mr. "U Blanco" was talking with Mr. F. Rabb Collector Customs here but am not certain.

I intended in closing this telegram to you – please read and return to me. They are Chicago Bankers + I have done work for them for years past. Gen'l Lopez I notice has dis-armed many of his men who were at time living in Texas but are now in his army. He thinks it will aid him to capture bandits better.

Respy John J. Hainsworth.

5671-1053

U.S. Army Intelligence in the Mexican Revolution

Extracts from communication from Mr. W. L. Gibson to Mr. W. K. Adams, employee at Office of Department Engineer, Southern Department.

Brownsville, Texas, December 1, 1915.

My dear Mr. Adams:

I just came back at noon from Reynosa, Tamaulipas (opposite Hidalgo, Texas). I was on the Carranza special train from Reynosa to Matamoros, Saturday and went back with them yesterday. The trains ran very slowly, about 4½ hours from Reynosa to Matamoros, or about 80 kilometers. There were four sections, that of Ricaut as escort, then the scout train, then the "presidential section" and the last another guard train. The second escort was 1200 of the Brigada "Supremos Poderes," a fair lot, mostly Veracruzanos, with some Yaquis. There was also a battalion of "Telegraphers and Signalmen" together with some "engineer" and "pioneer" troops, of the latter there were 600 to 800 – also the usual per cent of "Soldaderas," kids, etc. I have covered this stuff in daily telegrams to the *Light* and just thought I would tell you about the little details, etc.

They have a wireless outfit on the presidential section – in a box car, the chief operator is a Mexican – There are no Americans with Carranza as officers to my knowledge.

The local forces along the border from Matamoros to Reynosa, belonging to the 2nd Brigade (Eugenio Lopez) of the 5th Division (Gen. Luis Caballero) of Gen. Jacinto B. Trevino's Army Corps of the Northeast, are soon to be replaced by the troops of General Alfredo Ricaut's brigade, which went as escort for Gen. Carranza to Ramones.

In talking with Joseph De Courcy of the New York Times, who is travelling with the "Royal Party" he said that Carranza is "stiffening up" wonderfully since he arrived at Tampico and that a great improvement is manifesting itself in a lot of the leading men.

Cándido Aguilar, general of division and governor of Vera Cruz, is absolutely a man of his word, and General brigadier Alberto J. Machuca, a Oaxaca man; now governor of Hidalgo (but soon to be made governor of Oaxaca) is another good man. Machuca is a civil engineer, a graduate of the Polytecnic school of Lyons, France, and a lawyer also. Colonel Aguiles Juarez, also a Oaxaca man and governor of Tabasco, is a full blood Zapotec (like Benito Juarez), and speaks English well. He has none of the "up in the air" Spanish ideas at all, a very humble quiet man, who knows just what he is doing all the time. Luis Caballero is a big, fine looking man, and knows all gambling games as well as any man in Mexico. He is governor of Tamaulipas.

I am going to start on a sketch of the Railroad from Matamoros to Reynosa tomorrow.

"Ojo de Agua" where the fight took place October 21 is Abram P.O. There is a small lake of 15 to 20 acres called Ojo de Agua about 300 yards s.w. of the P.O. The fight took place on the north bank of this lakelet. There is a Japanese buried together with two

Mexicans about 150 feet n.w. of the place of the fight. I had a long talk yesterday with Gen. La Madrid, chief of military police and also with Major Almarante of Ricaut's staff, who is local "bandit chaser." They both said practically that they knew who was to blame and seemed to think Nafarrate had not tired himself out doing anything and Lopez not much better. Well we can only hope and wait. They all went away leaving Lopez in command and Nafarrate on leave in Matamoros. * * * * * * * * * * *

Yours sincerely,
W. L. GIBSON.

Recd. O.D.E., S.D. DEC 4 1915
Copy-vve

U.S. Army Intelligence in the Mexican Revolution

Document Index

RG 59

812-2311/195: 208

RG 92

OQG-September 18, 1914: 204

OQG-September 26, 1914, #232: 205

OQG-September 26, 1914, #233: 206

OQG-September 27, 1914: 206

OQG-September 28, 1914, #234: 206

OQG-September 28, 1914, #237: 207

OQG-September 29, 1914, #237: 207

RG 94

AGOMI-Box 7473, April 20, 1914: 187

AGOMI-Box 7473, April 23, 1914: 187

AGOMI-Box 7473, April 25, 1914#1: 188

AGOMI-Box 7473, April 25, 1914#2: 190

AGOMI-Box 7473, April 29, 1914: 191

AGOMI-Box 7473, May 15, 1914: 192

AGOMI-Box 7473, May 19, 1914: 192

AGOMI-Box 7473, May 21, 1914: 193

AGOMI-Box 7473, May 30, 1914: 193

AGOMI-Box 7473, June 15, 1914: 194

AGOMI-Box 7473, June 17, 1914: 195

AGOMI-Box 7473, August 4, 1914: 196

AGOMI-Box 7473, August 10, 1914: 197

AGOMI-Box 7473, September 8, 1914#1: 197

AGOMI-Box 7473, September 8, 1914#2: 198

AGOMI-Box 7473, September 14, 1914: 198

AGOMI-Box 7478, September 17, 1914: 199

AGOMI-Box 7478, September 22, 1914: 200

AGOMI-Box 7478, November 7, 1914: 201

AGOVR-Box 7646, October 1, 1915: 197

RG 141

MGV-Entry 12, File 277, May 1, 1914: 201

MGV-Entry 12, File 277, May 29, 1915: 203

MGV-Entry 12, File 277, November 17, 1914: 203

MGV-Entry 12, File 236, November 23, 1914: 204

RG 165

5761-362: 65

5761-363: 65

5761-941: 66

5761-942: 67

5761-943: 67

5761-947: 68

5761-948: 72

5761-949: 72

5761-951: 73

5761-952: 73

5761-953: 74

5761-955: 76

5761-957: 77

5761-958: 81

5761-960: 82

5761-968 Cover: 83

5761-968: 84

5761-974: 89

5761-975: 93

5761-996: 121

5761-999: 208

5761-1010: 210

5761-1020: 299

5761-1023: 214

5761-1025: 218

5761-1027: 221

5761-1028: 222

5761-1029: 225

5761-1030: 300

5761-1032: 301

5761-1033: 226

5761-1034: 302

5761-1035: 303

5761-1036: 304

5761-1039: 227

5761-1040: 228

5761-1043: 230

5761-1045: 305

5761-1046: 305

5761-1048: 232

5761-1052: 234

5761-1053: 305

5761-1054: 237

5761-1057: 239

5761-1058: 240

5761-1059: 242

5761-1091/8: 123

5761-1091/19: 126

5761-1091/28: 127

5761-1091/31: 129

5761-1091/32: 141

5761-1091/49: 141

6269-7: 243

6931-68: 144

6931-79: 145

8529-1: 146

8529-2: 149

8529-3: 152

8529-4: 154

8529-6: 155

8529-7: 156

8529-10: 160

Document Index

8529-11: 165

8529-12: 244

8529-13: 246

8529-14: 247

8529-15: 253

8529-17: 254

8529-18: 255

8529-21: 257

8532-1: 168

8532-2: 178

8532-3: 259

8532-4: 280

8532-5: 285

8532-6: 288

8532-7: 291

U.S. Army Intelligence in the Mexican Revolution

Index

Adams, Walter K., (Engineer) 38, 246, 248, 306

Agua Prieta, Son., 6, 43, 48, 52, 187, 208, 221-22, 224, 230, 234-35, 238

Aguascalientes, (State) 224, 236; (City) 18, 36, 42, 195, 239, 294

Aguilar, Cándido, (Constitutionalist) 29-31, 35, 58, 126-27, 200-1, 209, 306

Aguilar, Higinio, (Conventionist) 30, 126

Aguirre Benavides, Eugenio, (Constitutionalist) 57, 87

Aldamas, N.L., 50, 238

Alessio Robles, José, (Federal) 29, 129

Allende, Coa., 8, 39, 147, 149, 152, 246, 248

Almazán, Juan Andrew, (Conventionist) 236-37

Alvírez, Felipe, (Federal) 104

American Embassy, 2, 3, 34, 63, 68, 78, 90-91, 145-46

American Red Cross, 39-40, 52, 210-13, 254

Ángeles, Felipe, (Federal) 35-37, 135-36, 208, 214, 292-93, 296

Argumedo, Benjamín, (Orozquista) 49-50, 57, 134, 236-37, 240, 242-43

Army Corps of the Northeast, (Constitutionalist) 6, 16, 23, 32, 38, 41, 43, 54, 223, 233, 235, 238, 302, 306

Army Corps of the Northwest, (Constitutionalist) 21, 23, 32, 63, 116, 121, 302

Arrieta, Domingo, 42, 87, 235

Arrieta, Mariano, 42, 87

Associated Press, 53, 230-32

Aviation, (aeroplanes) 120, 139, 238, 244, 283

Bachimba, Chih., 51, 103, 109, 113, 131, 171-73, 177, 239-40, 243

Banderas, Juan, (Conventionist) 46-47, 224, 227

Bandit War, 54, 57, 59, 64

Barroterán, Coa., 41, 147, 149, 153, 162, 211, 249-50, 255-57

Battle of Aguascalientes, 116

Battle of Celaya, 38, 109, 115, 293

Battle of Cuesta Sayula, 36

Battle of El Ébano, 38, 41

Battle of Hacienda Temascatío, 54

Battle of Icamole, 41-43

Battle of La Gruñidora, 50, 240, 242-43

Battle of Matamoros, (1915) 37-38, 52, 54, 232; (1913) 57

Battle of Nogales, 26, 207

Battle of Paredón, 17, 194

Battle of Ramos Arizpe, 36

Battle of Santa Rosa, 23

Battle of Trinidad Station, 39, 42

Battle of Zacatecas, 12-13, 17-19, 24, 87, 156, 194-95

Beach, Harrison L., (San Antonio Light) 53, 232

Biddle, John, 19-20

Blanquet, Aurelio, (Federal) 131-32

Bliss, Tasker H., (U.S. Army) 6-8, 16, 23-26

Boer(s), 96, 111

Braniff, Alberto, (Federal) 2, 90

Bravo, Ignacio A., (Federal) 29, 129, 133

Brownsville, Tx., 7, 24, 53-55, 165, 188, 197, 206, 210, 225, 233

Burnside, William A., (U.S. Army) 1-5, 9-14, 16, 21-22, 27, 33-34, 51, 61-62, 79-80, 130, 144-45, 240

Caballero, Luis, (Constitutionalist) 56, 58, 193, 238-39, 303, 306

Calles, Plutarco Elías, (Constitutionalist) 43, 48, 221-22, 224, 229-30, 234

Canada, William W., (U.S. State) 28-29, 127

Capitalino, 61-62

Caraveo, Marcelo, (Colorado) 29, 102, 129

Carbajal, Francisco S., 25, 55, 142, 300

Cárdenas, S.L.P., 237

Carothers, George, (U.S. State) 23, 51, 140

Carrancista(s), 29, 31, 44, 47, 54, 56, 78, 128, 214, 219-22, 227-31, 233, 240, 245, 248, 254-56, 293-94, 301-3

Carranza Doctrine, 53, 57, 59, 64

Carranza, Jesús, (Constitutionalist) 17, 24, 29, 142, 201

Carranza, Sebastián, (Constitutionalist) 35, 210

Carranza, Venustiano, 16-18, 24-28, 30-33, 49, 54-55, 57-59, 64, 70, 88, 135, 141-42, 155-56, 182, 192-93, 197, 200, 208, 210, 212-13, 216, 224-25, 228, 230-31, 233-34, 237-39, 242-44, 282, 284, 287-88, 299-300, 302-6; (First Chief) 26-28, 30, 42, 49, 57, 59, 64, 126, 225, 228, 234

Carrera Torres, Alberto, (Conventionist) 49, 56-57, 216, 236, 302-3

Carrerista(s), 49, 56, 237, 239, 303

Casas Grandes, Chih., 6, 10, 43, 45-49, 85, 181, 187, 218-19, 223-24, 227, 231, 243, 286-87, 292-94, 296-97

Casso López, Arnoldo, (Federal) 50, 243

Catholic, 29, 79

Cavazos, (Constitutionalist) 55-56, 225

Celaya, Gto., 300

Center, 38, 104

Central Railroad, 14, 44-45, 147, 168, 226, 243, 250, 265-66, 268, 275, 282-83, 286, 293, 296

Chihuahua, (State) 17, 19, 21, 23, 25, 29, 35, 42, 46-48, 50, 85-89, 99, 137, 194, 196-97, 209, 212, 215-16, 219-20, 223-24, 226-27, 293, 296; (City) 6, 11, 15, 44-46, 48, 51, 85-87, 103-4, 124, 133, 137-39, 141, 169-70, 172-74,

Index

177-78, 183-87, 212, 217-21,
224, 226-27, 231, 239-40, 243,
254, 287, 293-94, 296-97

Chilango, 62

Citadel, (Ciudadela) 13, 25, 73, 75

Ciudad Guerrero, Chih., 47, 227

Ciudad Juárez, Chih., 6-7, 11, 14-15,
17, 26, 43-48, 85, 87, 135, 139,
155, 168-70, 175, 178-79, 181,
183-88, 190, 194, 205, 215-20,
223-24, 226-27, 231, 243, 261,
264-68, 270, 273, 275-76, 278,
280-88, 290-97, 306

Ciudad Victoria, Tamps., 38, 56, 58,
156-60, 193, 233, 238-39, 302-3

Coahuila, 7, 15, 21, 32, 35-39, 41, 43,
50, 55, 88, 137, 154, 215-16, 237,
248

Coahuila Coal Company, 37, 151, 245,
249, 253, 302

Coahuila y Pacífico Railroad, 164

Colegio Militar (de Chapultepec), 136

Colorado(s), 29, 108, 116

Colt Firearms Company, 137

Columbus, N.M., 19, 48, 51, 86, 208,
243

Constitutionalist Army,
(Constitucionalista) 3, 6-8, 12,
14, 16-18, 21-33, 35-44, 46-58,
61, 63, 78-79, 84, 88, 93-95,
102-5, 108, 115-16, 120-23, 128,
135, 149, 151, 155, 169, 187-88,
191, 194-96, 200, 205, 215, 220,
228, 240, 294, 300, 303

Convention of Aguascalientes, 29

Conventionist(s), 35-36, 38, 41, 46,
49-50, 56, 237, 296

Córdoba, Ver., 14, 30, 80, 82, 130

Correa, Francisco, (Federal) 5, 77

Crawford, C., (U.S. Army) 16, 232

Creel Terrazas, Enrique, 29, 129

Cuba, 16, 39-40, 53, 61, 65, 212, 232,
254

Cuéllar, Rómulo, (Federal) 54-55, 300

Cuernavaca, Mor., 13, 73, 79, 82

D'Antin, Louis, (U.S. State) 2-3, 78

De la Llave, Gaudencio, (Federal) 29,
129

De la Rosa, Luis, (Magonista) 57-58,
233, 238-39, 305

Decena Trágica, 62

Del Rio, Tx., 213, 305

Del Toro, Francisco, (Colorado) 29,
129

Delgado, José, (Federal) 50, 243

Devol, C. A., (U.S. Army) 39-41, 52

Díaz, Porfirio, 56, 85, 131, 134, 300,
302-3

Diéguez, Manuel M.,
(Constitutionalist) 36, 42, 47-
48, 224, 228-30, 234

Diehl, Charles S., (San Antonio Light)
53, 232

Division of the Bravo, (Federal) 36

Division of the Center, (Federal) 54

Division of the Center (1st),
(Constitutionalist) 17

Division of the Center (2nd),
(Constitutionalist) 17

Division of the East,
(Constitutionalist) 35, 209

Division of the North,
(Constitutionalist) 14, 17, 23, 32,
35, 63, 215, 293

Division of the Northeast, 35

Division of the Northwest, (Constitutionalist) 35

Doctor Arroyo, N.L., 49, 237-39

Douglas, Az., 26, 48, 205, 221-22, 230

Durango, (State) 42, 85, 87-88, 112, 133, 235; (City) 235, 245-46, 294

Eagle Pass, Tx., 7, 15, 39, 51, 148, 150, 155, 166, 188, 192, 207, 210-13, 243, 245, 254

El Democrata, 35, 208-9, 242

El Imparcial, 77-78

El Paso Mill Co., 261, 274, 277

El Paso, Tx., 7, 14-16, 43-44, 48-49, 86, 169, 175, 183-4, 188, 194, 196-97, 205, 207, 210, 212, 217, 219, 223-24, 228, 231, 259-62, 264-65, 267, 269, 271, 275-76, 278-79, 281, 284-86, 288-93

Emerson, Edwin, 14, 23, 63, 93

Empalme, Son., 47, 230

Escalón, Chih., 50, 239

Federal Army, (Mexican Army) 2-8, 10-15, 18-27, 29-32, 35, 50, 54-55, 61-63, 72-77, 79-80, 82, 84-86, 90, 92-93, 95, 97-100, 102-8, 112-17, 121-22, 126, 128-32, 134, 136, 138, 140-44, 146-47, 149, 152, 155, 188, 192, 194-95, 207-8, 242, 289, 293, 300

Federal District, 29, 129

Federal Navy, 16

Felicista(s), 70

Fierro, Rodolfo, 49, 88, 224, 227

Fletcher, Frank Friday, (U.S. Navy) 2-3, 68, 70, 80

Fort Wingate, 17, 25-26, 141, 194, 206

Funston, Frederick, (U.S. Army) 10, 12, 16, 24-25, 27-31, 37, 44-46, 48, 52, 73, 84, 126-27, 130, 141, 202, 210, 212, 291

Galveston, Tx., 8, 79, 86, 133, 191

García Peña, Ángel, (Federal) 14, 80, 130

García, Máximo, (Villista) 50, 235, 243

Garza, José Z., (Consul) 37, 54-55, 211, 225, 233

Gibson, W. L., See MacKinlay, W. E. W.

Gómez La Madrid, Daniel, (Constitutionalist) 58, 239

Gómez Palacio, Dgo., 63, 133, 136, 139, 215, 235

González Salas, José (as Salas), (Federal) 21, 86, 88

González, Pablo, 6, 17, 21, 23, 32, 35, 40, 42, 52, 88, 112, 192-93, 234

Goroztieta, Enrique, (Federal) 29, 129

Griffith, T. W., (U.S. Army) 17-18, 192

Guadalajara, Jal., 18, 35-36, 156, 236, 239

Guajardo, Luis Alberto, (Federal) 191

Guaymas, Son., 6, 47-48, 112-13, 120-21, 187, 224, 228-30

Guerra, Manuel, 16, 196

Guerrero, 36, 114, 131-32

Guilfoyle, John, (U.S. Army) 26, 205

Gunboat Bravo, 228

Gunboat Guerrero, 47, 121-22, 230

Gunboat Tampico, 84, 132

Gunboat Zaragoza, 228

Gutiérrez, Eulalio, (Conventionist) 36

Gutiérrez, Luis, (Constitutionalist) 37, 43, 50, 220, 238, 245

Index

Hainsworth, John J., 54–57

Hart, John Mason, (Academic) 31

Hay, W. H., (U.S. Army) 53, 232

Hensley, W.N., (U.S. Army) 14

Hermosillo, Son., 48, 124, 229, 234

Hernández, Rosalío, (Conventionist) 37, 41, 43, 47, 210–11, 213, 226–27

Herrera, Luis, (Constitutionalist) 47, 227

Hill, Benjamín, (Constitutionalist) 26, 205, 217

Hodges, Henry Clay jr., (U.S. Army) 35, 144

Huerta, Victoriano, (Federal) 2–3, 5–6, 9, 11–14, 16, 18, 21–23, 25, 27, 30, 32–33, 52, 54, 58, 62–63, 66–70, 72, 76, 79–80, 82–84, 86, 88, 91–92, 103, 109, 113, 126, 130–31, 133–35, 141–44, 146, 187, 195, 209, 215, 239, 242, 300; (President)

Huertista(s), 2–3, 6, 12, 17, 22, 24, 30–34, 74, 146, 194

Icamole, N.L., 163–64, 167, 252

International Railroad, 235, 246, 304

Interoceanic Railroad, 14, 75–76, 130, 143

Jalapa, Ver., 80, 83, 130

Jalisco, 17, 35–36, 235–36

Japan, 60, 84, 306

Jiménez, Chih., 14, 50, 86, 168–69, 171–73, 175, 177, 219, 227, 239, 243, 293, 296

Jiménez, Tamps., 58, 239

Juárez, Benito, 306

Kingsville, Tx., 51, 240

Kosterlitzky, Emilio, (Federal) 26, 207

La Laguna, 14, 42, 46, 49–50, 63, 139, 209, 214, 216, 219–20, 223, 242–43, 293, 302

La Raza, 56, 214–15, 220, 293, 300, 301

Laredo, Tx., 7–8, 188–91, 206, 210, 212–13, 238–39, 244

Laubach, Howard L., (U.S. Army) 21, 144

Leyva, Francisco, (Maytorenista) 46, 224, 298

Lind, John, 3, 33, 51–52, 62–63, 70

Livestock, (Horses) 97, 194

López de Lara, Eugenio, (Constitutionalist) 54, 56–58, 233, 238–39, 300–1, 303–7

Los Aldamas, N.L., 50, 238

Maass, Gustavo Adolfo, (Federal) 82

Maass, Joaquín, (Federal) 20, 87, 130

Machuca, Alberto J., (Constitutionalist) 58, 306

MacKinlay, William E. W., (Gibson, W. L.) 44–51, 53–59, 64, 232, 234, 240, 243, 291, 306

Maderista(s), 55, 80, 242

Madero, Ernesto, 55, 299

Madero, Francisco I., 2, 55, 62, 72, 80, 85–86, 91–92, 146, 209

Madero, Raúl, (Conventionist) 41, 217

Madinaveytia, Manuel, (Villista) 46, 48, 223, 231

Magonista(s), 53–54, 232

Manzanillo, Col., 11, 75, 121, 132

Marfa, Tx., 48, 51, 243

Martin, Walter F., (U.S. Army) 34

Matamoros, Tamps., 7, 16, 53-54-59, 64, 159-60, 188, 190, 192, 215, 225, 233, 238, 243, 300-1, 304-7

Matehuala, S.L.P., 49, 237, 239

Maytorena, José María Jr., 26, 35, 42-43, 46, 218, 222, 224, 234, 294, 298

Maytorenista(s), 27, 36, 46, 112

Mazatlán, Sin., 47, 82, 113, 120-22, 132, 224, 234

Medina, Juan N., (Villista) 46, 223, 227

Mercado, Salvador, (Federal) 21, 87-88

Mexican Coal & Coke Company, 37, 245

Mexican Railway, (Railroad) 4, 13-14, 67, 75-76, 80-81

Mexican War Department, (Guerra) 5, 76, 80, 131-32

Mexico City, City of, 2-3, 5-6, 10-14, 18-19, 22-24, 32, 34-36, 40, 42-43, 50, 52, 54, 61-62, 66-70, 72-84, 86, 89-91, 114-15, 117, 123-25, 129-31, 133, 142, 145, 149, 156, 169, 173-75, 187, 195, 203, 208-9, 212, 225, 244, 300, 302

Millán, Agustín, (Constitutionalist) 29, 201

Ministry of the Interior, (Gobernación) 132

Mitchell, William, (U.S. Army) 22, 32, 93-94, 97, 114

Monclova, Coa., 6, 15, 35, 39, 147-53, 155, 163, 166, 187, 193, 210-13, 246, 248-52, 254, 257, 259

Monterrey, N.L., (Monterey) 6-7, 13, 16, 37-39, 55, 58, 124, 147, 149, 151, 154-55, 161, 164, 166-67, 187-88, 192-93, 210-11, 215, 220, 225, 228, 239, 244, 248, 250-52, 299, 301-2

Morelos, 50, 131, 243

Morelos Zaragoza, Ignacio, (Federal) 50, 243

Murguía, Francisco, (Constitutionalist) 8, 36, 42, 191-93, 220, 235

Múzquiz, Ramón, (Constitutionalist) 39

Naco Accord, 36

Naco, Son., 26-27, 36, 93, 109, 112, 205, 222, 234

Nafarrate, Emiliano P., (Constitutionalist) 37, 53-59, 215, 225-26, 233, 238-39, 243, 300-4, 307

Natera, Pánfilo, (Conventionist) 17-18, 87, 155-56

National Cartridge Factory, 12, 75

National Railways, 5-6, 38, 74, 76-77, 147, 168, 239, 245-46, 248

New Mexico, 19, 47-48, 51, 86, 124, 208, 227, 243, 260-61, 269-70, 272, 278, 285, 289

New York Times, 58, 296, 306

Nogales, Az., 7, 46, 188, 210, 228, 297-98

Nogales, Son., 6-7, 26, 43, 46-47, 187-88, 207, 218, 222-23, 228-30, 234, 298

Norteño(s), 12, 38, 49, 62, 111

North, 6, 13, 36, 49, 62, 64, 67, 69, 94-97, 101, 104-5, 109, 111-14, 116, 120-21, 133, 135, 220, 231, 304

Index

Northwestern Railroad, 14, 44-46, 178, 219, 224, 227, 266-68, 282-84, 286, 292-93, 296-97

Nuevo Laredo, Tamps., 6-7, 23, 37-38, 93, 187-88, 190, 192, 216, 239

Nuevo León, 35, 37, 49-50, 228, 237-39, 302

O'Shaughnessy, Edith, 3, 72

O'Shaughnessy, Nelson, (Charge) 3, 9, 62, 71, 90

Obregón, Álvaro, (Constitutionalist) 6, 17, 21, 23-24, 32, 35-36, 38-39, 42-44, 48-49, 53-54, 63, 88, 108-9, 112, 116, 121, 142, 187, 209, 214-17, 220-21, 223, 228, 234, 238, 301-2

Ocaranza, Eduardo, (Conventionist) 48, 50, 231, 243

Ojinaga, Chih., 25, 36, 51, 87, 210, 227, 243, 295

Orozco, Pascual jr., (Colorado) 21, 23, 51, 85-86, 88, 102-3, 109, 113, 141, 209, 242

Paredón, Coa., 43, 147, 149, 151, 155, 163-65, 167-68, 212, 214-15, 250-52

Parras, Coa., 50, 237, 241

Peraldí, Fernando, (Constitutionalist) 39

Pershing, John J., (U.S. Army) 45

Piedras Negras, Coa., 6-8, 16, 25-27, 32, 35-36, 39, 41, 51, 141, 147-55, 157, 161, 166, 168, 187-88, 190-93, 210, 214-15, 243, 248, 254, 256

Plan de San Diego, See also Bandit War

PLM, 54

Puebla, (State) 35, 130-31, 209; (City) 14, 24, 30, 35-36, 80, 117, 126, 131, 142

Querétaro, (City) 24, 142

Quevedo, Rodrigo, (Colorado) 16, 21, 88, 194

Quinn, W. E., 38, 248, 251, 258

Ramos Arizpe, Coa., 168

Randall, Carlos, (Maytorenista) 46-47, 228-29, 298

Rellano, Chih., 50, 116, 243

Reyes, Canuto, 49, 224, 227

Ricaut, Alfredo, (Constitutionalist) 50, 58-59, 238-39, 243, 306-7

Rio Grande, (River) 16, 44, 56, 81, 193, 225-27, 232-33, 259-60, 262-65, 267, 269, 271, 274, 276, 278, 285, 290, 301

Robert, Henry H., (U.S. Army) 37, 39-40, 49, 55

Robles, Juvencio, (Federal) 29, 129

Rodríguez, José, (Villista) 37, 45, 47, 54, 219, 227, 232, 243, 293-94

Rojas, Antonio, (Vazquista) 21, 88, 102

Ruckman, A. J., 37-38, 245-46

Rural Police Corps (Rurales), 56, 85, 94, 132, 134, 209, 233, 302-3

Ryan, Edward W., 93, 131

Salas, Gustavo, (Federal) 29, 129

Salazar, José Inés, (Colorado) 21, 86, 88, 108, 207

Salina Cruz, Oax., 5, 34, 75-76, 132

Salinas, Margarito, (Conventionist) 49, 224, 227

Saltillo, Coa., 6-7, 17, 35, 38, 42-43, 50, 87, 124, 149-50, 152, 155,

164, 166-68, 187-88, 192, 194, 210, 212-13, 215-16, 239, 241, 251-52, 299, 302

Salvador, Alvarado, 35

San Antonio Light, 44, 49, 53, 55, 57, 64, 225, 232, 306

San Luis Potosí, 17, 20, 35, 56, 152, 155, 192, 212, 236-38, 302-3,

San Pedro de las Colonias, Coa., 11, 20, 87, 133, 136, 138-40, 166, 214, 220, 223

Santa Rosalía, Chih., 47, 50, 104, 171-73, 177, 209, 226-227, 239-40, 243

Sinaloa, 42-43, 46, 224, 235, 243

Soldadera(s), 24, 86, 99-100, 103-4, 142, 306

Sonora, 6, 26, 42-49, 182, 206, 217-19, 222-24, 227-31, 283, 287, 293-96, 298

South, 35-36, 49, 62-63, 79, 101, 104, 113, 121

Southern Pacific Railroad, 47, 228, 270-73, 279

Sturtevant, Girard, (U.S. Army) 62

Suriano(s), 63

Tamaulipas, 35, 43, 49, 55, 233, 237, 242, 300, 302-3, 306

Tampico Affair, 89

Tampico, Tamps., 16-17, 23-24, 37-38, 50, 58, 79, 93, 155-56, 158-59, 192-93, 197, 228, 237-39, 243-44, 302, 306

Tehuantepec, (Oaxaca) 5, 29, 76, 201; (Indians) 56, 300-1

Tehuantepec National Railway, 5, 74, 76-77

Tejería, Ver., 3, 70, 90, 130

Tepic, (Territory) 6, 42, 187, 234, 235-36; (City) 234

Terrazas, Félix, (Colorado) 29, 129

Texas, 6, 52, 54, 56, 59, 79, 98, 124, 196, 210, 212, 215, 232, 260, 269, 279, 301, 304-5

Thorne, George E., (U.S. Army) 10, 130, 202

Tierra Blanca, Ver., 11, 80-82

Toluca, Mex., 49, 69

Torreón Accords, 22, 123

Torreón, Coa., 6, 14, 17-22, 35, 42-44, 46, 48-49, 53, 56, 63, 86-88, 103, 124, 133, 136, 139, 147, 154-55, 164, 166, 168-69, 187, 192, 194, 214-17, 219-21, 223-225, 227, 231, 235, 242, 245-46, 251, 284, 286-87, 293-94, 296-97, 302; (Battle) 93, 97, 100, 104, 107, 115, 133, 138, 140-41

Tramways Company, 12, 69, 76

Treviño, Jacinto B., (Constitutionalist) 38, 41-43, 50, 53-56, 215, 220, 223, 225, 228, 233-35, 238-39, 243, 302

U.S. Army, 6, 8, 11, 14-16, 18-19, 22-23, 29, 32, 35-36, 38-43, 45, 48, 51-55, 57-64, 88, 114, 212

U.S. Civil War, 51-52, 62, 96

U.S. Expeditionary Forces, 10, 12, 202

U.S. Marine Corps, 4, 10, 27, 71, 90, 93, 125, 144, 199

U.S. Navy, 1, 16, 46, 60-61, 70-71, 74, 80-81, 84, 89, 131, 199, 202, 229

U.S. Secretary of War, 2, 7-8, 12, 14, 22, 25-27, 29, 33, 40, 60, 72, 89, 188-91, 193, 196-98, 201, 205-6, 210, 244, 248, 299

Index

U.S. State Department, 2, 9, 21, 23, 26, 34, 40, 51, 60-61, 71, 78, 89, 145, 206, 208, 229, 299

Urbalejo, Francisco, (Maytorenista) 46-47, 228-29, 243, 298

Velasco, José Refugio, (Federal) 20, 82, 87, 133

Veracruz, (State) 14, 29-30, 35, 129; (City) 1-6, 8-14, 22, 24, 27-35, 62-63, 67-72, 75-80, 83, 89-91, 93, 114, 127-28, 130

Veracruz al Istmo Railway, 11, 75, 81-82

Villa, Francisco "Pancho", (Conventionist) 5-7, 14, 17-21, 23-27, 29-32, 35-40, 42-53, 55, 58, 63, 67, 69-70, 83-89, 91, 96-102, 104-5, 107-9, 112, 133-41, 155-56, 188, 192-95, 197, 201, 205, 208-210, 213-21, 223-24, 226-27, 230-31, 234-35, 239, 243, 245, 254, 282-84, 286-87, 292-300

Villarreal, Antonio I., (Constitutionalist) 36

Villista(s), 30, 33, 35-48, 51, 54, 115, 208, 210, 214-20, 222-224, 227, 229, 231-235, 238, 245-46, 248, 254-56, 287, 292, 294-95, 297-98, 301

War College Division, (WCD) 19, 21, 42, 143-45, 165-66, 225-28, 230, 232, 242, 247, 248, 253-56, 258, 260, 281, 286, 288-89, 292, 295, 302-3

West, Duval, 51

Wilcox, F. A., (U.S. Army) 43-46

Wilson, Henry Lane, (State) 62

Wilson, Woodrow, 2-3, 33, 39, 48, 52, 62-63

Winchester Repeating Firearms Co., 25

Wotherspoon, W.W., (U.S. Army) 22, 24

Wright, John W., (U.S. Army) 15, 41, 65-66

Yaqui(s), 23, 32, 46, 287, 296, 306

Ypiranga, 11, 74, 89

Zacatecas, (State) 49, 135, 209, 224, 236, 240, 242; (City) 20, 42, 155, 220

Zapata, Emiliano, 31, 43, 62, 69, 76, 91, 208

Zapata, Miguel, (Constitutionalist) 49, 56, 216, 237

Zapatista(s), 3, 6, 12-13, 17, 22, 32-33, 35, 42, 62-63, 66, 69, 73, 76, 79, 82, 104, 224, 234

Zuazua, Fortunato, (Constitutionalist) 35, 41, 210

Printed in Great Britain
by Amazon